Coatings Technology Handbook

Coatings Technology Handbook

Shirley Barnes
Ethan Walker

Kruger Brentt
Publishers
2024

Kruger Brentt Publishers UK. LTD.
Company Number 9728962

Regd. Office: 68 St Margarets Road, Edgware, Middlesex HA8 9UU

ISBN: 978-1-78715-141-3

For information on all our publications visit our website at http://krugerbrentt.com/

Preface

Coating technology and coating formulation play a crucial role in many industries, including automotive, aerospace, construction, and consumer goods. Coatings are applied to a substrate to protect it from corrosion, abrasion, and other forms of damage, as well as to improve its aesthetic appeal. In this essay, we will explore the basics of coating technology and coating formulation.

Coating technology refers to the methods and processes used to apply a coating to a substrate. These methods include spray coating, dip coating, roll coating, and electrostatic coating, among others. Spray coating is the most common method and involves the use of a spray gun to atomize the coating material and apply it to the substrate. Dip coating involves immersing the substrate in a tank of coating material, while roll coating involves passing the substrate through a series of rollers coated with the coating material. Electrostatic coating involves the use of an electric charge to attract the coating material to the substrate. Coating formulation refers to the development of a coating material with specific properties to meet the requirements of the intended application. A coating material typically consists of a binder, pigment, and solvent. The binder is the main component that provides adhesion and durability to the coating. Pigments provide color and opacity, while solvents are used to dissolve the binder and make the coating material easier to apply.

Coating formulation can be tailored to meet specific requirements, such as resistance to corrosion, UV radiation, or heat. For example, coatings used in the aerospace industry must be able to withstand extreme temperatures and environmental conditions. Coatings used in the automotive industry must provide protection against corrosion and abrasion, as well as improve the aesthetic appeal of the vehicle. In recent years, there has been a growing interest in the development of sustainable coating technologies and formulations. This includes the use of bio-based materials, such as natural resins and vegetable oils, as well as the development of coatings with improved environmental performance, such as low-VOC (volatile organic compounds) coatings. Coating technology and coating formulation are crucial components of many industries. Coating technology involves the methods and processes used to apply coatings to substrates, while coating formulation involves the development of coating materials with specific properties to meet the requirements of the intended application. The development of sustainable coating technologies and formulations is becoming increasingly important in the face of growing environmental concerns.

The present book focuses entirely on testing, experimental design and strategies for selecting processing techniques in the coatings, It covers the entire spectrum from coatings formulation and relevant production processes through to practical application aspects. It takes a journey through the industry's various sectors, placing special emphasis on automotive coating and industrial coating in general. This is an excellent book for scientists, engineers, graduate students and industrial researchers in the field of coating technology.

We are grateful to all those persons as well as various books, manuals, periodicals, magazines, journals etc. that helped in the preparation of this book. In spite of the best efforts, it is possible that some errors may have occurred into the compilation and editing of the book. Further queries, constructive suggestions and criticisms for the improvement of the book are always welcomed and shall be thankfully acknowledged.

Shirley Barnes
Ethan Walker

Contents

01 Introduction

Primitive men are credited with making the first paints about 25 000 years ago. They were hunters and cave dwellers and were probably inspired by the rock formations of their cave walls to outline and colour the shapes of the animals they hunted. It is possible that by creating these images they thought their power over their prey would be increased.

Chemical analysis of cave paintings discovered at Altamira (Spain) and Lascaux (France) show that the main pigments used by Paleolithic artists were based upon iron and manganese oxides. These provide the three fundamental colours found in most cave paintings, namely black, red, and yellow, together with intermediate tints. Carbon from burnt wood, yellow iron carbonate, and chalk may also have been used. Surprisingly, there is no trace of a white pigment (the commonest pigment in use today) at Lascaux, where the natural colour of the rock was used as a pale background. However, white pigments do occur in some prehistoric paintings in Africa.

The Egyptians developed the art of paint-making considerably during the period circa 3000–600 bc. They developed a wider colour range of pigments which included the blues, lapis lazuli (a sodium silicate–sodium sulphide mixed crystal), and azurite (chemically similar to malachite). Red and yellow ochres (iron oxide), yellow orpiment (arsenic trisulphide), malachite green (basic copper carbonate), lamp-black, and white pigment gypsum (calcium sulphate) all came into use during this period.

The first synthetic pigment, known today as Egyptian Blue, was produced almost 5000 years ago. It was obtained by calcining lime, sodium carbonate malachite, and silica at a temperature above 830°C. The Egyptians also developed the first lake pigments. These were prepared by precipitating soluble organic dyes onto an inorganic (mineral) base and 'fixing' them chemically to form an insoluble compound.

A red dye obtained from the roots of the madder plant was used in the first instance. This is no longer used other than in artists' colours ('rose madder') because it fades rapidly on exposure to sunlight, and it has been replaced by alizarin. Lake pigments still, however, represent an important group of pigments today. Red lead was used in preservative paints for timber at this time, but was more extensively used by the Romans. The resins used were almost all naturally occurring gums; waxes which were applied molten as suitable solvents were unknown. Linseed and other drying oils were known, but there is no evidence that they were used in paints.

By the late eighteenth century, demands for paints of all types had increased to such an extent that it became worthwhile for people to go into business to make paint and varnishes for others to use. The industrial revolution had a major effect on the development of the paint industry. The increasing use of iron and steel for construction and engineering purposes resulted in the need for anti-corrosive primers which would delay or prevent rusting and corrosion. Lead- and zinc-based paints were developed to fulfill these needs. It is interesting to note that one of the simplest paints based upon red lead dispersed in linseed oil is still probably one of the best anti-corrosive primers for structural steel. Lead-based paints are being superseded not because better products have been produced, but because of the recognition of their toxicity and the hazards attendant upon their use.

The most recent influences on coating developments are related to environmental considerations, and the need to conform to health and safety legislation. Cost/benefit relationships have also become more important in an increasingly competitive world market and have influenced formulation practice markedly. The terms 'paint' and 'surface coating' are often used interchangeably. Surface coating is the more general description of any material that may be applied as a thin continuous layer to a surface. Purists regard the term 'surface coating' as tautological.

It has long been recognized that it is difficult, if not impossible, to meet the requirements of many painting processes by the use of a single coat of paint. If one lists the requirements of a typical paint system it is easy to see why. Many, if not all, of the following are likely to be required: opacity (obliteration); colour; sheen (gloss); smoothness (or texture); adhesion to substrate; specific mechanical or physical properties; chemical resistance; corrosion protection; and the all-embracing term 'durability'. Durability is an important area that we shall return to in many contexts. The number of different layers that comprise the paint system will depend on the type of substrate and in what context the coated object is used. A typical architectural (gloss) paint system might consist of a 'primer', an 'undercoat', and a 'topcoat'. All three are likely to be pigmented compositions, and it is probable that there will be more than one coat (or layer) of each of these paints.

The primer will frequently be required to contribute to corrosion protection. Those formulated for use on steel are likely therefore to incorporate a chemically active anti-corrosive pigment. Corrosion protection may be achieved by yet another means, the chemical treatment of the substrate. Thus many industrial coating processes involve a chemical pretreatment of metal, mainly aluminum or ferrous substrates. The latter is most frequently treated with a phosphate solution that produces a crystalline phosphate layer. Subsequent paint application, i.e. priming, is therefore to a crystalline inorganic layer and not directly to an (uncoated) pure metal surface.

Surfaces are seldom what they seem. With the exception of the noble metals almost all surfaces that will be commonly regarded as 'metal' surfaces will present to the paint a surface that is not a metal but an oxide layer. Even so the purity or cleanliness of the surface may well be an unknown quantity. Since this surface will have an important effect on such properties as the adhesive performance of the paint system it is important to appreciate this point. Just as most surfaces will be 'dirty' and thus be ill-defined, it is necessary to produce paint systems that can accommodate the contamination and general variability of surfaces. These types of system must be 'tolerant' to all but excessive contamination and are often described as 'robust' if the required degree of tolerance can be achieved.

The deterioration of paints which occurs in many situations is largely due to changes in the chemical nature of the film former with consequent changes in its mechanical properties, and research continues unabated to formulate polymers and resins to improve the performance of paints in use. The development of new improved pigments may contribute to improvements in durability, but in most cases the weakest link in the system is the film former. One consequence of this is the development of systems for specific end uses. Such approaches are adopted when it is practicable to avoid the compromises that are otherwise likely to be required for a general-purpose product. For economic and marketing reasons the best product may not be available for a specific end use, and a compromise of cost and performance may be required. Indeed the cost-effectiveness of a particular coating composition will usually dominate other considerations particularly in the industrial paint markets.

02 Binders

OILS AND FATTY ACIDS

Vegetable oils and their derived fatty acids still play an important a role in surface coatings today because of their availability as a renewable resource, their variety and their versatility.

Oils are mixed glycerol esters of the long chain (generally C18) monocarboxylic acids known as fatty acids (Fig. 2.1); unrefined oils also contain free fatty acids, lecithin's, and other constituents, the first two finding their own application in coatings.

Oils useful in coatings (Table 2.1) include linseed oil, soya bean oil, coconut oil, and 'tall oil'. Many countries will find some application for their own indigenous oils as well as those detailed here. When chemically combined into resins, oils contribute flexibility and, with many oils, oxidative cross linking potential, one of the properties most exploited in paint.

It should be added that in searches for a synthetic alternative to drying oils, only allyl ethers have been found to be both accessible and possess similar and useful autoxidative abilities. These have found limited application in high solids finishes and unsaturated polyesters.

Oils are classified as drying, semi-drying, or non-drying, and this is related to the behaviour of the unmodified oil, depending on whether it is, on its own, able to oxidize and crosslink to a dry film. This behaviour is directly related to the concentrations of the various fatty acids (Table 2.2) contained in the structure. These fatty acids contain, with few exceptions, 18 carbon atoms including the terminal carboxyl.

Fatty acids may be saturated (no double bonds), mono-unsaturated (one double bond) or polyunsaturated (two or more double bonds); to be considered 'drying' an oil must contain at least 50% of polyunsaturated acids. The fully saturated C18 fatty acid is stearic acid, of which some is found in all oils. The C12 saturated fatty acid, lauric acid, is the predominant component of coconut oil, where it constitutes 90% of its composition.

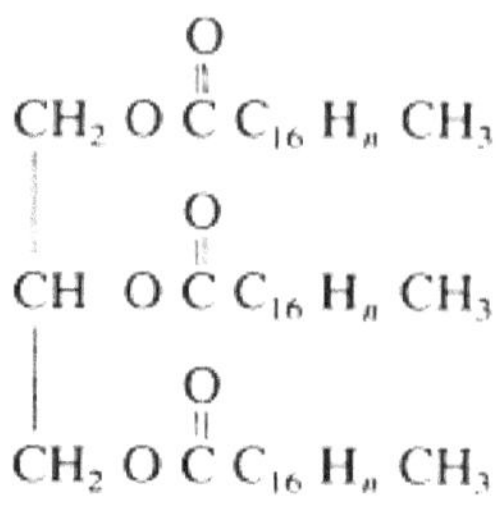

n may be 32, 30, 28, or 26

Fig. 2.1 — Representative structure of naturally occurring oil.

Thus linseed oil, a good drying oil contains over 60% of the polyunsaturated linoleic and linolenic acids, while soya bean oil, usually classified as semi-drying, contains just over 50% of linoleic acid. By contrast, coconut oil, a non-drying oil contains 90% saturated lauric acid and less than 10% unsaturated fatty acid.

Table 2.1 — Composition of major oils used in surface coatings

	Saturated acids	Oleic acid	9,12- Linoleic acied	9,12,15- Linolenic acid	Conjugated acid
Tung	6	7	4	3	80[d]
Linseed	10	20–24	14–19	48-54	0
Soya Bean	14	22-28	52-55	5-9	0
Castor oil	2-4	90-92[a]	3-6	0	0
Dehydrated castor oil	2-4	6-8	48-50	0	40–42[b]
Tall,	3	30-35	35-40	2-5	10-15[c]
Coconut	89-94	6-8	0-2	0	0

[a]Principally ricinoleic acid, not oleic.

[b]conjugated 9,11 linoleic acid.

[c]conjugated linoleic acid and octadeca 5,9,12- trienoic (pinolenic) acid proportions dependent on source and refinement.

[d]α- Eleostearic acid.

Polyunsaturated fatty acids may be conjugated or non-conjugated, and tung oil, now in short supply and little used, contains nearly 80% of the conjugated aeleostearic acid. By contrast 9, 12, 15-linolenic acid has three non-conjugated double bonds, while 9, 12-linoleic acid has two non-conjugated double bonds. High conjugation is reflected in higher drying ability and greater tendency to heat body over that expected from the double bond content; thus, tung oil can in fact readily be gelled by heating, and for this reason — if incorporated into an alkyd resin — would not be processed by the monoglyceride process, this being considered too risky. The drying oils, tung and linseed, may find application either unmodified or only heat bodied in coating compositions in simple blends with other resins. In contrast, most other oils find little application alone but only after chemical modification; soya bean oil, for example, finds its major application in chemically combined form in long oil alkyds. Oils such as coconut oil do not dry or heat body and have no application alone, but only find use when reacted into alkyds. In this case the oil component provides plasticization. Coconut alkyds have good colour and are used for stoving finishes and appliance enamels.

Table 2.2 — Structures of the commonly occurring unsaturated fatty acids

Octadeca-9-enroic (oleic)CH_3•(CH2)$_7$•CH=CH•(CH_2)7•COOH Octadeca-12-hydroxy-9-mole (ricinoleic)
Octadeca-12-hydroxy-9-enoic (ricinoleic) CH_3•$(CH_2)_4$–CH=CH=CH•CH_2•CH=CH•$(CH_2)_7$•COOH
Octadeca–9,11dienoic (conjugated linoleic) CH3•$(CH_2)_5$—CH=CH•CH=CH•$(CH_2)_7$•COOH
Octadeca-9,12,15-trinoic (linolenic) CH_3CH_2•CH=CH=CH_2CH=CH•$(CH_2)_7$•COOH
Octadeca-5,9,12-trinoic (linolenic) CH3• $(CH_2)_4$•CH=CH_2•CH=CH•$(CH_2)_2$• CH=CH•$(CH_2)_3$•COOH
Octadeca-9,11,13-trinoic (α-eleostearic) CH3• $(CH_2)_3$•CH=CH•CH•CH=CH• $(CH_2)_7$COOH

The two most important oils to the industry nowadays are probably soya bean oil and 'tall oil', the latter misnamed, as it is a fatty acid mixture containing over 50% polyunsaturated fatty acid, principally 9, 11-linoleic acid. Tall oil is a by-product of the paper industry, and to be useful in coatings must be refined to reduce its rosin content to 4% or less from the higher concentration in the crude product. However, it is the fractionally distilled form, which can now be interchanged with soya fatty acids (and in alkyds where soya bean oil is replaced with tall oil and polyol) depending on seasonal fluctuations of availability and price. Tall oil of Scandinavian origin may be more favoured than that from America owing to its higher unsaturated fatty acid content, which includes the triply unsaturated pinolenic acid. Tall oil is probably more variable in quality, dependent on source and refinement, than any other oil or fatty acid, and so needs careful evaluation and specification for any critical use.

Oil with a more distinctive fatty acid is castor oil which contains nearly 90% of ricinoleic acid, or 12-hydroxy oleic acid. It has some use as a polyol for polyurethane preparation; a major use exploiting the hydroxyl content is in the preparation of higher hydroxyl containing alkyds for cross linking with nitrogen resins or isocyanates, and for use as plasticizing alkyds for nitrocellulose where high polarity is necessary for compatibility.

Oils are generally purified by acid or alkali refining for the coatings industry in order to remove materials such as free fatty acids and lecithin's, to give a neutral clear product and to improve colour. When further reaction is required, oils may be separated into constituent fatty acids and glycerol, and though more expensive than oil, the free fatty acids of all the common oils are currently available to the coatings industry.

MODIFIED OILS

Drying oils may be heat-bodied at temperatures in the region of 290°C to produce 'stand oils', where the viscosity increase is due principally to dimerization reactions through double bonds of unsaturated fatty acid moieties. This reaction may be followed through decrease in the iodine value. 'Stand oils' find application in oleo resinous vehicles and in alkyd resins. The separated dimmer fatty acids also find specialized uses, for example in polyamide resin manufacture.

So-called boiled oils were once prepared by heating metal oxides in oil until fatty acids released during bodying solubilized the oxides as soaps. Nowadays a 'boiled oil' may be prepared more conveniently by blending commercial driers with'stand oil'.

Dehydration of castor oil is important, since the removal of the hydroxy group and an adjacent hydrogen atom creates an additional double bond, so increasing the residual unsaturation. Dehydrated castor oil (DCO) has distinctive characteristics since the second double bond generated is generally positioned relative to the original one to give both conjugated as well as non-conjugated polyunsaturation. Some bodying may occur on dehydration, although this can be minimized by use of a good vacuum and suitable catalysts, thus reducing the time necessary for the reaction.

By virtue of its unsaturated fatty acid content DCO is classed as a drying oil. DCO finds major application in stoving alkyds and in alkyds required for further modification by vinylation; it is little used in air-drying finishes because of the incidence of surface 'nip' (tackiness) in dried films. DCO fatty acids are also available to resin formulators.

Castor oil may also be hydrogenated, and hydrogenated castor oil (HCO) has some application in alkyd resins for stoving application with MF resins; HCO fatty acid, more usually named 12-hydroxy stearic acid, finds application particularly for graft copolymers for use in dispersants and surfactants. Isomerization of oils and fatty acids is used to improve the drying properties particularly of medium oil length alkyds. These processes include transformation of the double bonds of non-conjugated fatty acids (principally linoleic) to conjugated form, and in certain cases from the cis to the trans isomer; a number of processes exist for carrying this out. Improvements from using these oils/fatty acids include better colour, initial dry, and resistance to water, acid, and alkali. Weathering and gloss retention are also said to be improved.

Oils and fatty acids containing unsaturation can be reacted with maleic anhydride. Where conjugated double bonds are present a Diels–Alder reaction (Fig. 2.2) is possible, and this proceeds exothermically. Where the bonds are non-conjugated (linolenic or linoleic) the first addition is by the Ene reaction; this has the consequence of moving the double bonds from a non-conjugated to a conjugated configuration. There is then the possible side effect that a sluggish first Ene reaction (Fig. 2.3) may be followed by a faster exothermic Diels–Alder second reaction.

Maleinization is carried out to increase acid functionality of fatty acids, for example for subsequent water solubilization. It also occurs in situ in alkyd preparations where maleic anhydride is included in the for-mulation, and its effect is to increase viscosity by increasing acid functionality.

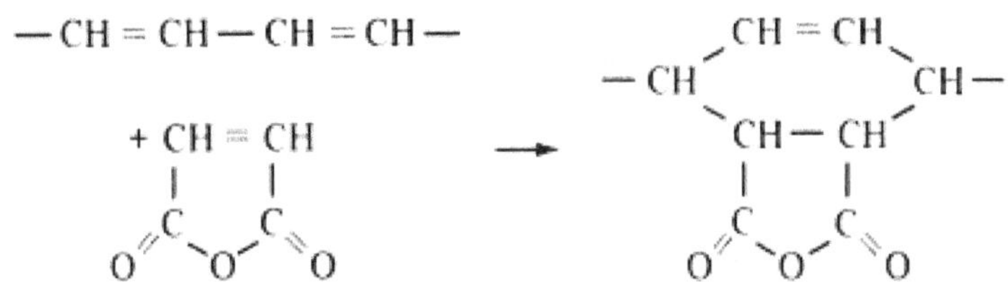

Fig. 2.2 — Diels–Alder reaction.

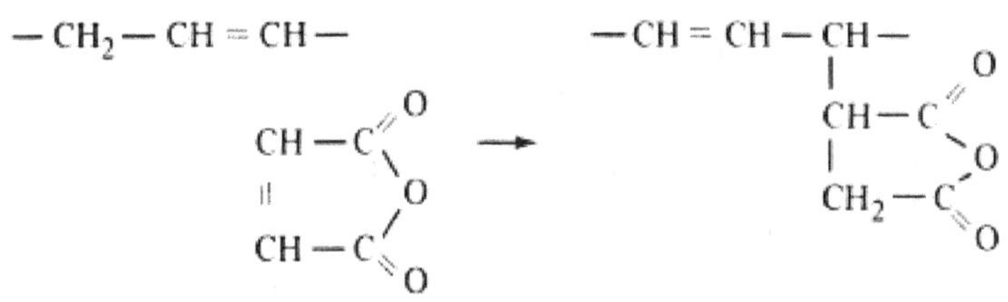

Fig. 2.3 — Ene reaction.

Care is needed in formulation since a small maleic addition can for this reason have an unexpectedly large effect. True 'urethane oils' are obtained where a monoglyceride produced by reaction of an oil with glycerol is reacted with a diisocyanate; most 'polyurethanes' of this type are, however, urethane alkyds where some condensation of monoglyceride with dibasic acid is first carried out prior to diisocyanate reaction.

Oils and fatty acids may be vinylated, but this alone has little application, though it may be a first step in the preparation of a water soluble or a vinylated alkyd.

DRYING OIL POLYMERIZATION

As previously stated, drying oil fatty acids may dimerize under heat treatment by Diels–Alder or Ene reaction of unsaturated groups. However, the most exploited reaction of these unsaturated groups is autoxidative cross linking, accelerated by the presence of heavy metal driers and also with the aid of heat in stoving compositions.

The theories of autoxidative dry have been well reviewed, and the mechanisms recently further investigated; only an outline is given here. It is now accepted that the first steps in autoxidative drying are free radical in nature and involve hydro peroxide formation. It is believed that in the process double bonds may shift from the cis to the trans structure; for linoleates hydro peroxidation can lead to a movement in the double bond adjacent to the position of attachment of the hydro peroxide leading to a conjugated structure that stabilizes the product. While Khan explained this by a non-radical reaction transition state, Fig. 2.4 shows a more likely mechanism with a delocalized radical transition state.

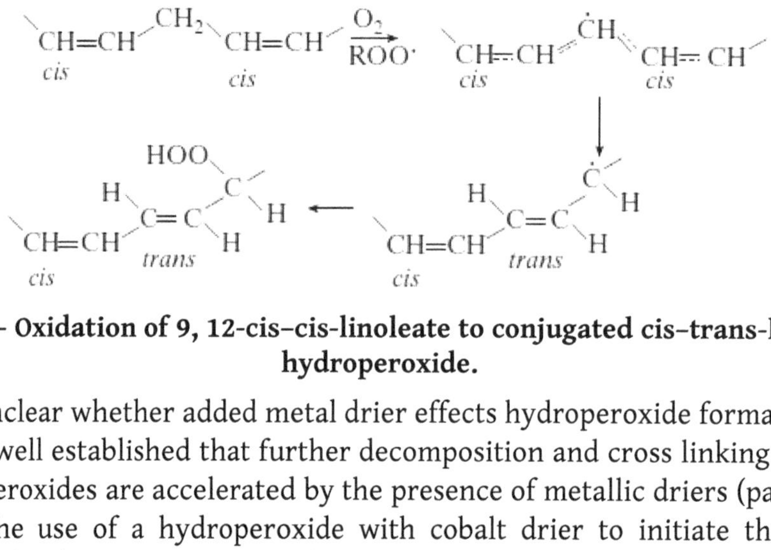

Fig. 2.4 — Oxidation of 9, 12-cis–cis-linoleate to conjugated cis–trans-linoleate hydroperoxide.

It is unclear whether added metal drier effects hydroperoxide formation. It is, however, well established that further decomposition and cross linking reactions of hydroperoxides are accelerated by the presence of metallic driers (particularly cobalt). The use of a hydroperoxide with cobalt drier to initiate the cure of unsaturated polyesters is an example of this reaction being used elsewhere. Just as this latter reaction is identifiable as free radical in nature, so the autoxidative cross linking reaction is also largely free radical in mechanism. This may be exploited by the copolymerization of high boiling or involatile vinyl monomers into an autoxidatively drying system to modify final properties.

ACRYLIC POLYMERS

Introduction

Since their introduction decades ago, acrylic polymers have gained a strong foothold in the coatings and allied industries as a result of their improved flexibility and adhesion compared to polyvinyl acetate emulsions, phenolics, and styrene-butadiene latex combined with their moderate cost. In addition, their significantly improved outdoor durability, including resistance to ultraviolet degradation, has mandated their use in several applications. In many respects, the name "acrylic" has become synonymous with a high performance level in a polymer system.

Presently, acrylics are available in three physical forms: solid beads, solution polymers, and emulsions. The emulsion form is by far the dominant form in use today. This is due generally to the ease of tailoring properties, and the lower hazards and manufacturing costs compared to the solid and solution polymers.

CHEMISTRY AND MANUFACTURE

Monomers

Acrylic monomers are esters of acrylic and methacrylic acid. Some common esters are methyl, ethyl, isobutyl, n -butyl, 2-ethylhexyl, octyl, lauryl, and stearyl. The esters can contain functional groups such as hydroxyl groups (e.g., hydroxyethyl methacrylate), amino groups (e.g., dimethylaminoethyl methacrylate), amide groups (acrylamide), and so on, in addition to the carboxylic acid functionality of the unesterified monomer. Acrylic monomers can be multifunctional (e.g., trimethylolpropane triacrylate, or butylenes glycol diacrylate, to mention two). The polymer chemist has a wide range of monomers to select from when designing a specialty polymer system.

Typically, mixtures of comonomers are chosen for the properties they impart to the polymer. Adhesive strength, for example, is increased by using monomers with low glass transition temperatures such as butyl acrylate or 2-ethyl hexylacrylate. Members of the carboxylic acid group of acrylic and methacrylic acids also tend to increase the adhesive properties of polymers. Cohesive strength is usually imparted by the harder acrylic monomers such as methyl methacrylate and methyl acrylate. Molecular weight is also a significant contributing factor, and these two parameters must be carefully balanced by the polymer chemist.

When functional groups are needed for postreaction, monomers such as hydroxyethyl methacrylate or N -methylolacrylamide are incorporated. Hydroxyl groups can be used in combination with melamine and epoxy curing agents to achieve cross-linking. Similarly, other functional groups (acid groups, amines, amides, etc.) can be incorporated.

POLYMERIZATION METHODS

Bulk Polymerization

As the name implies, bulk polymerization is accomplished by initiating acrylic monomers in the absence of solvents other than the monomers. Typically, peroxide or azo initiators are used. The major problem, of course, is the viscosity increase realized after approximately 30% conversion. Generally, heavy-duty mixers and temperatures exceeding 150°C are used to control viscosity. The well-known Tromsdorf gel effect is often observed at high conversion, leading to rapid exotherms, high molecular weight fractions, and increased polydispersity. Processes for the bulk polymerization of acrylic copolymers are described in the patent literature.

Solution Polymerization

Adding a solvent to a bulk polymerization recipe allows much easier control of viscosity at high conversions.

Again, peroxide (e.g., benzoyl peroxide, lauryl peroxide) or azo (azobisisobutyronitrile) initiators are used. Mercaptans and halogenated hydrocarbons are used to regulate molecular weight. Most solvents also act as chain transfer agents to some extent. Generally, acrylic polymers prepared by solution polymerization are less than 100,000 molecular weight units.

Suspension Polymerization

The process of suspension polymerization, as the name implies, involves the use of a dispersing agent to stabilize monomer droplets in a continuous phase, usually water. With continued agitation, the suspended droplets are polymerized using oil-soluble initiators, again, peroxides and azo compounds. In fact, the polymerizing droplets exhibit bulk polymerization kinetics, as each droplet is actually a small bulk reactor with the continuous aqueous phase acting as a heat sink. Typical suspending agents include polyvinyl alcohol, polyacrylic acid, and hydroxyethyl cellulose. As might be expected, suspension polymerization is usually limited to monomer compositions with glass transition temperature T g that is near to or greater than ambient temperature; otherwise, gelation or blocking easily results in the dry bead. Following polymerization, the suspension polymer particles, or beads, are dewatered, washed to remove impurities such as suspending agents and electrolytes, and dried.

Emulsion Polymerization

Perhaps the most complex of the polymerization processes, emulsion polymerization has been well studied and described. As the name implies, the acrylic monomers are emulsified using a surfactant and are suspended in a continuous phase, water. Water-soluble initiators are used to initiate polymerization.

Some common examples are ammonium or potassium persulfate, hydrogen peroxide, and redox pairs such as t -butyl hydroperoxide-sodium formaldehyde sulfoxylate. Polymerization is generally thought to initiate in the water phase. A growing radical precipitates to a micelle containing monomers and continues to polymerize. The growing micelle is resupplied with fresh monomer by diffusion from monomer droplets.

VERSATILITY OF ACRYLICS

Glass Transition Temperature

By selecting proper monomers, the glass transition temperature of the polymer and, therefore, the likely application area, can be varied. The glass transition temperature of a polymer is the simple average value in degrees Celsius representing a range of temperatures through which the polymer changes from a hard and often brittle material into one with soft, rubberlike properties.

Table 1.1— Glass Transition Temperature

Homopolymer	T8, (°C)
Acrylic acid	112
Methyl methacrylate	106
Methyl acrylate	8
Isopropyl acrylate	-8
Ethyl acrylate	-24
N-Butyl acrylate	-56
2-Ethyl hexylacrylate	-65

Although these average T g values sometimes vary with the test method used, they are reproducible within certain limits and represent specific polymer characteristics. The glass transition temperature is useful as a guideline for softness of hand, low temperature flexibility, and room temperature hardness and softening point.

Table 1.2 — Glass Transition Temperature versus Application Area

T8 (°C)	Suggested Application Area
80-100	High heat-resistant coatings
50-65	Floor care coatings
35-50	General industrial Coatings
10- 40	Decoratiw paints
25-35	Binders for inks
-60.25	Adhesives

The glass transition temperature should be used to compare hardness and softness of latex only within a simple polymer group. Table 1.1 illustrates the wide range in T g resulting from different monomer compositions.

Emulsion Acrylics

Emulsions have become the dominant technology in acrylic polymers, and we therefore focus much of our discussion on this category. Table 1.2 correlates T g ranges of emulsion acrylics with specific application areas. There is obviously overlap to be expected among the ranges. By varying the monomer compositions, we have a clear indication of the versatility of acrylics.

The wide range for adhesives encompasses pressure-sensitive polymers and heat-activating polymers, which dry to a tack-free state.

Physical Properties

On a physical basis, emulsions can be characterized in the following fundamental areas: solids content, viscosity, pH, particle size, minimum film forming temperature (MFT), and particle charge.

Solids content is determined by drying latex to constant weight, viscosity is determined by use of the Brookfield Viscometer, and pH is determined by pH meter. Particle size usually ranges from about 0.05 to 0.5 μm, depending on the type and amount of surfactant employed; measurement is usually made by laser light-scattering technique, electron microscopy, or ultracentrifuge.

As mentioned previously, surfactants stabilize liquid monomer droplets formed by agitation. Polymerization takes place in monomer micelles to form the solid dispersed particles of polymer in water.

Surfactants are classified into three types: anionic, nonionic, and cationic. The most common type, anionics, ionizes in water to leave a negative charge on the latex particle. Nonionics do not ionize but stabilize by a combination of hydrophobic and hydrophilic regions on the molecule. Cationics, which are not commonly used, ionize to give the particle a positive charge.

Surfactants help to improve mechanical and chemical stability of the emulsions but also tend to increase the water sensitivity of the dried film.

Minimum Film Forming Temperature

Unlike solution polymers, which have one homogeneous phase, emulsions are polymer particles dispersed in water, the continuous phase. Upon drying, the particles must combine or coalesce to form a continuous film. If the process is carried out at room temperature (25°C) with a polymer with a T g value above 25° C, film formation will not occur.

In this case, the polymer must be heated to above its T g or a coalescing agent must be added to the latex to soften or plasticize the particles so that they will combine to form a continuous film. Coalescents often used are high boiling liquids that have a solvating effect on the polymer but later evaporate so that the full physical properties of the polymer alone are achieved. Generically, many of these coalescents are classified as glycol ethers. Their relative toxicology has recently been studied.

APPLICATION AREAS

Coatings

The use of acrylic polymers for both emulsion and solution coatings is very diverse. Before discussing the areas in which acrylics might be used in coatings, let us first identify the fundamental reason for considering the use of a coating, namely, to protect the substrate being coated from the environment.

There are additional reasons, depending on the end-use area. Once the need for a coating has been decided, a further requirement is apparent; namely, the coating must adhere to the surface that has been coated because to decorate or protect any surface, the coating must remain in position. This phenomenon is known as adhesion. The reader is directed to an excellent review of the methods to determine satisfactory adhesion of a coating.

An additional point to consider is the air dry nature of properly formulated waterborne acrylics. Their full properties are developed not through extensive heat cure cycles but by room temperature drying through the use of coalescing agents, as discussed earlier. Where heat is available, increased solvent and heat resistance can be attained by the addition of cross-linking agents.

Each end-use product requires certain specific property characteristics to achieve the feature needed for the application; the type of substrate to be coated, as well as the manner in which the part is used, determine the property requirements. It is clear that plastics do not require a corrosion-resistant coating; therefore, a coating designed for plastics has property requirements different from a coating designed for metals. At the same time, adhesion of the coating, which is important for any substrate, becomes a key characteristic.

However, achieving adhesion to any specific substrate may require a totally different product requirement. Materials that adhere to metal may fail to bond to plastic. Therefore, coatings for metals may have to differ significantly from coatings for plastics or wood to achieve the desired performance. Each substrate area, therefore, requires the design of a specific coating to meet its individual end-use requirement for the intended application.

Returning to application of acrylics in coatings, we will center our discussion on five major areas:

- Automotive
- General metal
- Maintenance
- Wood
- Business machines

Automotive

With the increased use of plastics in automotive applications has come a surge in the use of coatings. Plastics are coated to improve resistance to chemicals, solvents, ultraviolet light, and abrasion, as well as exterior durability. Other applications for coatings include the engine enamels area, underbodies, and auto refinishing work. Performance properties required include grease resistance, durability, and adhesion to oily metals.

Coatings for Metal

Coatings are needed for metal casings and transformers. The requirements for such coatings include chemical resistance, hardness, and corrosion and humidity resistance. Because steel is still commonly used, the ability to protect this metal from corrosion remains an important requirement.

Maintenance Coatings

Coatings are required for bridges and storage tanks, where again, properties of corrosion and humidity resistance are required.

Wood Coatings

Coatings for boards destined to be used in furniture and kitchen cabinets require blocking and detergent resistance, sandability, and resistance to grain rising.

Machines

Coatings for calculators, typewriters, copy machines, and analytical instrumentation are common examples in the business machine category. Performance required includes chemical and solvent resistance and adhesion to plastics [e.g., polycarbonate, polyphenylene oxide, acrylonitrile–butadiene–styrene (ABS)].

ADHESIVES

Both solvent and emulsion acrylic adhesives are extensively used in the industry, but before discussing adhesives, we need to address the fundamental difference between a coating and an adhesive. A coating must adhere to only one substrate; an adhesive must adhere to one substrate and then to a second substrate. A coating, once applied, is exposed to the elements and must withstand abrasion, marring, solvents, water, and heat. It may require high gloss and other special properties, as well.

An adhesive is protected to a certain degree by being sandwiched between two substrates. It, therefore, does not have to have some of the performance properties that must be built into a coating. It must ideally have a bond strength high enough to fracture or tear at least one of the substrates. In many cases, the bond strength should not be materially affected by heat, solvents, or water. Therefore, an adhesive must not only have good anchorage to both substrates (adhesive strength), it must also have high enough cohesive strength to fracture or tear one of the substrates upon delamination. Thus, an adhesive must balance adhesive strength with cohesive strength. Another basic difference between emulsions (coatings and adhesives) is in their film formation properties. To have hard, tack-free, and heat-resistant coatings, the glass transition temperature of the polymer is intentionally designed to be higher than room temperature. The coating then requires a coalescing agent to form a clear continuous film. Adhesives form films at room temperature without the need for coalescing aids. A soft flexible polymer film is desired for an adhesive, and this film should be thermoplastic (i.e., able to soften and flow repeatedly upon the application of heat). The film can subsequently be cross-linked through functional groups if heat and solvent resistance are desired.

Acrylic-based adhesives are normally employed where improved specific adhesion and/or resistance to yellowing from exposure to ultraviolet rays is required. Acrylics are used in three main areas: heat sealable adhesives, laminating adhesives, and pressure-sensitive adhesives. These are discussed separately.

Heat-Sealable Adhesives

Heat sealing is used for bonding two substrates where one or both are impervious to water. Typically, the more heat- and solvent-resistant substrate is coated first, and the solvent or water is driven off in an oven. At this point, the web can be wound up on itself (if the dried adhesive is tack free or non blocking) and heat sealed to the secondary substrate at a later date. Alternatively, the adhesive-coated substrate can be laminated simultaneously to the second substrate to form the finished product. The latter case is the laminating adhesive technique, discussed later. When heat is applied to activate or soften the adhesive, two conditions must be met for adequate bonding:

- The adhesive must have sufficient flow at the activation temperature to properly wet out the secondary substrate.
- The adhesive must have a chemical affinity or specific adhesion to the particular substrate. This comprises the secondary chemical bonding forces, which give rise to what we call adhesive bonding.

Many applications involving food packaging fall into this category. Examples include lidding-type adhesives for coffee creamers and jams and the blister packaging of pharmaceuticals.

Laminating Adhesives

Laminating adhesives function much the same as heat-seal adhesives except that the temperature necessary to activate the adhesive is much lower. Where heat-seal adhesives may require an activation temperature of 120°C (250°F), laminating adhesives can be designed to function anywhere between room temperature and 90°C (200°F). Recently, new low temperature curing types of adhesive have, to some degree, replaced two-component solvent-polyurethane adhesives in the flexible packaging area.

Designing a room temperature curing mechanism into an acrylate system further enhances opportunities to replace solvent systems. Considerable progress has also been made using low temperature curing acrylics in the industrial area. Typical applications include vinyl-to-wood laminating and bonding vinyl to ABS for automotive interiors.

Pressure-Sensitive Adhesives

The pressure-sensitive method of bonding is often called the "one-way" bonding method. The adhesive is coated onto the substrate either directly or by transfer coating. The adhesive is protected by a release liner until it is ready to be used. When application is desired, the liner is removed and the adhesive coated substrate is bonded to the other substrate using pressure alone. Pressure actually activates the adhesive, hence the name of this method. Upon firm pressure, the tacky adhesive mass actually flows and bonds itself both mechanically and chemically to the other surface.

However, to be functional, a pressure-sensitive adhesive must be more than just very tacky. Flypaper has a very high degree of tack but lacks internal strength or cohesiveness. A functional pressure-sensitive material will have high tack or adhesive strength combined with high cohesive strength, the basis requirement for any adhesive.

Pressure-sensitive adhesives are used in many tape areas such as packaging, masking, electrical mending, medical, and mounting. The graphic arts area includes vinyl decals and special decorative films such as clear and metalized polyester films. These areas generally require an outstanding balance of properties that must be retained under severe outdoor exposure and temperature extremes. Such properties as outstanding resistance to ultraviolet light degradation and plasticizer migration are required. Adhesives for this area must produce clear and colorless films, dictating the use of acrylics exclusively.

Solvent-based polystyrene acrylics (PSAs) have been traditionally used in these areas. More recently, they have been replaced in varying degrees with their emulsion counterparts.

However, there still exist technical areas in which the solvent systems prevail. These include applications requiring high levels of heat, water, and solvent resistance.

Inks

The use of acrylic polymers to formulate waterborne inks has grown steadily in the past 10 years. The major cause for this growth has been government regulations imposing limits on solvent emissions by printers. The same reason, of course, sponsored the growth of waterborne coatings and adhesives. The choices left to the printers are as follows:

- Install solvent recovery–incineration systems.
- Develop compliant high solids solvent inks.
- Develop adequately performing waterborne inks.

Because many printing companies have selected the third option, our discussion will focus on it. In addition to the pigment, waterborne acrylic inks are usually composed of two types of acrylic polymer, an emulsion-polymerized acrylic and an alkali-soluble solid acrylic type of resin. The emulsion acrylic contributes to finished ink properties such as film formation, adhesion to plastic films, and scuff, heat, and alkali resistance. However, emulsion polymers are very poor pigment wetters. In addition, they have poor flow characteristics compared to their solvent-borne counterparts. The alkali-soluble resins are good pigment wetters, display good rheology, and prevent drying of the ink in the cylinders when the line is stopped during printing runs.

Many different types of alkali-soluble resin can be used in combination with the emulsion acrylics. Table 1.3 compares the properties of acrylic polymers to four other generic classes: maleic rosinates, shellac, protein, and styrenated maleic copolymers.

By combining the best properties of emulsion-polymerized and alkali-soluble acrylics, reasonably performing waterborne inks are gradually replacing the traditional solvent types. These waterborne inks are being used in such diverse applications as stamps, cigarette cartons, beer cans, milk cartons, food packaging, wallpaper, grocery bags, newspapers, and corrugated boxes.

VINYL ETHER POLYMERS

General

The general formula for vinyl ether polymers used in the production of adhesives and coatings is as follows.

The consistency of these polymers depends on their molar mass and ranges from viscous oils to rubbery solids.

Vinyl ether was first converted into a resinous polymer more than a century ago. Between 1920 and 1930, it became readily accessible by the techniques of Reppe chemistry and thus attracted industrial interest. Means were then investigated for polymerizing it.

Monomers

The Reppe reaction between acetylene and an alcohol gives rise to vinyl ether:

$$HC{\equiv}CH + H{-}OR \rightarrow CH_2{=}CH{-}OR$$

It is still the only method of producing vinyl ethers that has acquired industrial significance. The properties of the constituent monomers are listed in Table .1.

Production of Polymers

Vinyl ethers can be easily polymerized in bulk or in solution, by batch or continuous techniques. Owing to the considerable heat of reaction, careful control and elaborate equipment are essential. The monomers and the initiator are metered continuously into the reactor, and the polymerization reaction sets in within a few minutes. After the reactor has been completely charged, it is closed, and polymerization proceeds further under pressure. The boiling point of the monomer or the solvent governs the rate at which the temperature rises during the reaction. The heat of reaction is removed by means of a reflux condenser and/or the reactor cooling system.

Table 1 — Monomers Used for Producing Vinyl Ether Polymers

Product	Stucture	Molecular weight	Density at 20°C(g/cm3)	Flash point (°C)	Melt Point (°C)	Boiling Point (°C)
Vinyl methyl ether	CH2=CH-O-CH3	58.1	0.747	-60	-133	6
Vinyl ethyl ether	CH2= CH-O-CH2-CH3	72.1	0.754	-45	-115	36
Vinyl isobutyl ether	CH2=CH-O-CH2-CH-CH3-CH3	100.2	0.769	-15	-112	83

Products on the Market

Table 2.2 lists vinyl ether polymers on the market. As suggested in the table, it is common practice to describe the degree of polymerization by the K value, which is a measure for the average molar mass. The form in which the polymers are offered depends on the production method and the demands imposed in application. Thus, they can be supplied in the solvent-free form or, to facilitate handling, as solutions in common solvents. Lutonal I 60 D and I 65 D are highly viscous 55% secondary aqueous dispersions and are produced from an 80% solution of Lutonal I 60 in mineral spirit (bp 60 to 140°C) with different protective colloids.

Table 2 — Conventional Vinyl Ether Polymers

Polymer Based on	Commercial Name	Manufacturer	K value	Nature of Polymer	Glass Transition Temperature
Vinyl methyl ether	Lutonal M 40	BASF	~50	Soft resin	~25
Vinyl ethyl ether	Gantrez M Lutonal A 25	GAF BASF	~12 ~60	Soft resin Viscous oil	~25 ~45
	Lutonal A 50 Lutonal A 100	BASF BASF	~105 ~25	Soft resin Tacky rubber	~30 |~25
Vinyl isobutyl ether	Lutonal I 30 lutonal I 60	BASF BASF	~60 ~60	Viscous oil Soft resin	~25 ~25
	Lumina' I 60 D Latonal I 65 D	BASF BASF	~60 ~60	Soft resin Soft resin	~20 ~15

Properties

The degree of polymerization governs the physical form of polyvinyl ethers (i.e., whether they are viscous oils, tacky plasticizing resins, or rubbery substances). The products on the market are normally colorless. They may occasionally display a slight yellow to brown discoloration, due to side reactions of the initiator system, but this seldom affects their performance in any way. The polymers offer good resistance to hydrolysis. Products with a low to medium molecular mass (i.e., K 65) may have a slight odor that emanates from residual monomers or oligomers.

The solubility of vinyl ether polymers is shown in Table .3. It depends on the alkyl group. The fact that polyvinyl methyl ether is listed as a water-soluble polymer can be explained by the hydrogen bond between the liquid medium, water, and the oxygen atom in the ether group. Thus, the polymer, which is itself insoluble, is solvated by water to form a soluble associate. On the application of energy, the hydrogen bonds are again loosened — that is, the polyvinyl methyl ether loses its hydrate envelope and is thus precipitated. This occurs, for instance, in Lutonal M 40 solutions at about 28°C. The precipitation point is influenced by the concentration of the aqueous solution and the presence of any solvents.

Table 3 — Solubility of Vinyl Ether Polymers

Polymer	Solvents[a]								
	Al H	Ar H	Chl H	LA	HA	E	Lk	Hk	Water
Polyvinyl methyl ether	-	+	+	+	+	+	+	+	+
Polyvinyl ethyl ether	+	+	+	+	+	+	+	+	-
Polyvinyl isobutyl ether	+	+	+	-	+	+	-	+	-

Applications

Vinyl ether polymers constitute a classical group of starting materials for the production of adhesives and coatings. They are primarily used in combination with other raw materials or — in one case in the pressure-sensitive adhesive sector — with polymers of the same type but different molecular mass. Blending with vinyl ether polymers improves a number of the properties of the unblended products, including anchorage, adhesion or tack on difficult substrates, resistance to plasticizers, and resistance to aging.

The main applications in the adhesives sector are blends of polyvinyl methyl ether with starch or dextrin for label adhesives or blends of medium molecular weight polyvinyl ethyl ether with resin solutions for bonding carpets. Secondary dispersions of medium molecular weight polyvinyl isobutyl ether lag somewhat behind in terms of volume but not in importance. They are blended with acrylic dispersions in the production of pressure-sensitive adhesives. Blends of vinyl ethyl ether polymers of various molecular weights are used as pressure-sensitive adhesives in the medicinal sector.

In the field of surface coatings, polyvinyl methyl ether and medium molecular weight polyvinyl ethyl ether are formulated together with cellulose nitrate, chlorinated binders, and styrene copolymers for coating metal foil, plastics, film, paper, and other flexible substrates, and for antifouling paints.

POLY(STYRENE-BUTADIENE)

Introduction

Styrene-butadiene (SB) polymers are used in coatings formulations primarily to improve coating strength, printability, gloss, pigment and fiber binding, and substrate bond strength. The styrene-butadiene polymers are produced primarily by emulsion polymerization to form a latex. A latex is a dispersion of finely divided spherical particles of polymer in water. The monomer ratios of styrene and butadiene can be adjusted to give the desired amount of flexibility or stiffness. Polystyrene latex contains hard polymer particles that will not form a continuous film upon drying at room temperature. Polybutadiene latex contains very soft, gummy polymer particles that will produce a weak, sticky film when dried. By copolymerization of styrene and butadiene monomers and adjustment of the monomer ratios, it is possible to obtain a copolymer with a wide range of intermediate properties. Thus, copolymer latex with high styrene content will produce a very tough, durable film but will not have the flexibility and adhesive performance of latex with high butadiene content.

The temperature at which the copolymer changes from a brittle to a rubbery state is known as the glass transition temperature (T g). Figure .1 illustrates the effect of the concentration of styrene on the T g of SB copolymers. As the concentration of styrene in the copolymer increases, the T g also increases. This property is important in coatings for determining durability of the dried film and optimizing curing conditions.

Typical latex will contain 45 to 55% polymer, with the balance being water. Table .1 illustrates the typical properties of styrene-butadiene latex. The particle size, composition, cross-linking, and molecular weight can all be controlled, and this is accomplished through free radical emulsion polymerization.

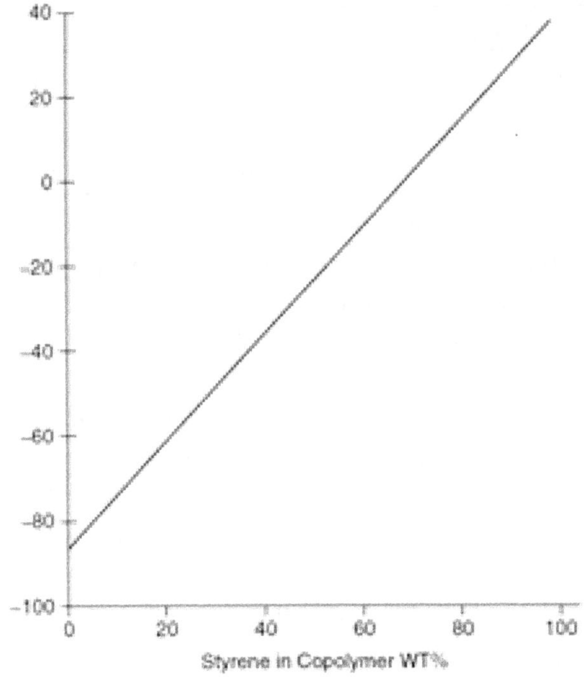

Figure 1 — The effect of copolymer styrene content on glass transition temperature.

Emulsion Polymerization

Emulsion polymerizations require an emulsifier or surfactant composed of a hydrophobic and a hydrophilic portion to give the latex colloidal stability. As low levels of surfactant are added into the aqueous solution in a reactor vessel, the surfactant will saturate the water phase, and as the concentration is increased, the surfactant molecules will aggregate to form micelles. This concentration is known as the critical micelle concentration. As the styrene and butadiene monomers are added, they will diffuse through the water phase and into the micelles until equilibrium is obtained. The majority of polymerization occurs within these monomer-swollen micelles. Figure .2 is a representation of the process.

Table 1— Typical Properties of an SB Latex

Property	Value
Color	Milky white
Solids	50% wt.
PH	7 to 9
Brookfield viscosity	<500 cp at 25ºC,
Average Particle diamerter	20,000
Suffice tension	45 dynes/cm
Specific gravity (at 25ºC)	1-01
Styrene/butadine ratio	50:50
Film properties	
Tensile strength (at break)	550psi
Elongation	520%

Polymerization begins when water-soluble free radical initiators are added. Initiators, such as persulfate salts, are used for reactions above 50 ° C, and redox systems are used for reactions at lower temperatures. The initial reactions occur in the water phase to form a free radical. The free radical will react first with monomer double bonds in the water phase, and chain growth will begin. As the molecular weight increases, the chain becomes more hydrophobic and migrates to the swollen micelles, where the majority of polymerization occurs. The polymerization occurs in random fashion, and residual double bonds have the capability of further chain growth. This gives the SB latex the ability to form a three dimensional polymer matrix. This cross-linking or branching can have a strong influence on the polymer's mechanical properties.

Chain transfer agents are added to control the molecular weight. They will lower the molecular weight of the polymer through chain termination and will result in improved wetting and binding characteristics.

The surface of the latex particle can also be modified by functional modifiers such as amides, amines, and carboxylic acids, which provide improved colloidal stability and increased adhesion properties.

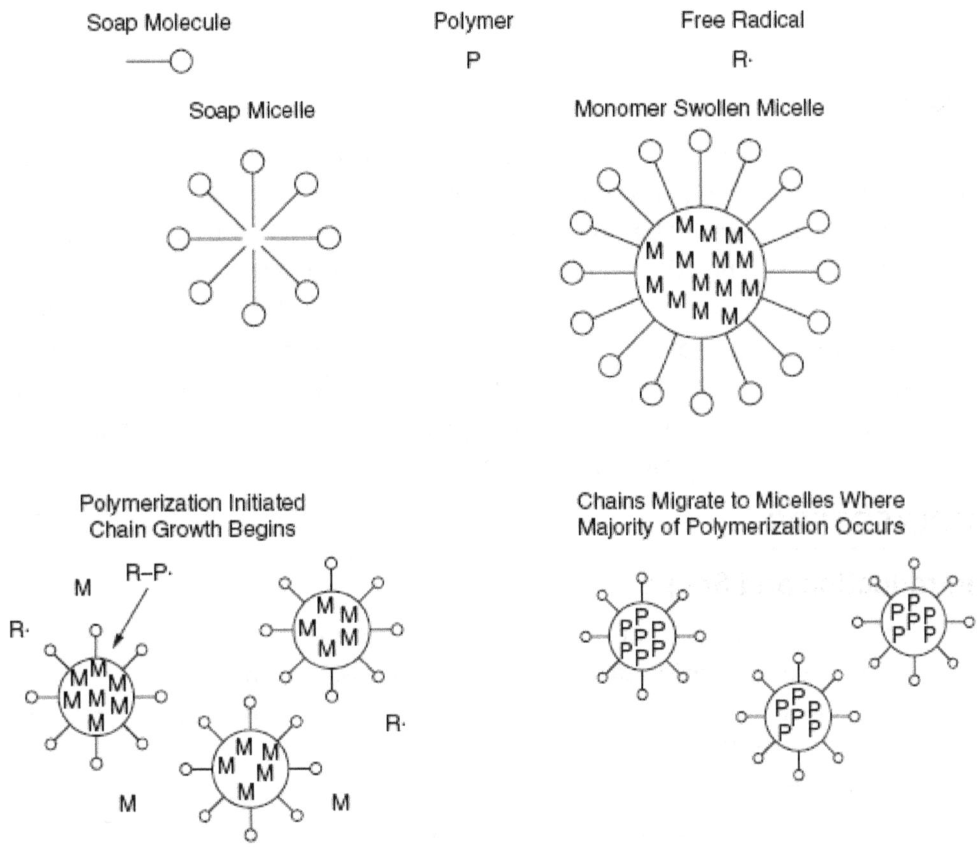

Figure 2: Free radical emulsion polymerization.

CHARACTERISTICS OF STYRENE-BUTADIENE LATEX

In coatings applications, several physical properties are key to determining performance of the latex polymer. These include glass transition temperature, particle size, minimum film formation temperature, rheology, and surface energy.

The S/B ratio and glass transition temperatures are useful as a basis for comparing performances characteristics of different SB latexes. End-use properties can be plotted against the S/B ratio and T_g to help determine the range of applicable polymers. Latex particle size and distribution affect high and low shear viscosity, gloss, and pick strength. Minimum film formation temperature (i.e., the minimum temperature at which a continuous film will form) is essential for determining process drying conditions.

The minimum film formation temperature of a typical SB latex is slightly higher than its T_g. The reactions of the latex to increased shear stress and shear rate are important rheological properties, because they will affect machine runnability. As the shear stress is increased, shear thinning, shear thickening, or Newtonian flow can occur. The surface tension or surface energy of latex will affect wettability and adhesion of the coating to substrates.

Uses

Styrene-butadiene latex is used in paper and paperboard coatings, textile coatings, as binders and coatings for flooring felts, as carpet backing, and in many other coatings applications. The material can be produced with a wide variety of properties and can be tailored to specific end uses. Advantages of SB latex in paper coatings are excellent mechanical stability, pigment hiding power, improved gloss and ink holdout, and better printability. In textiles and carpets, SB latex provides excellent fiber adhesion, high filler loading, and a degree of dimensional stability and flexibility. Many other coatings applications use these basic features of SB latexes to improve end-use performance.

POLYESTERS

Introduction and Scope

Polyesters are polymers containing the recurring ester unit ~ COO ~. They are formed by a condensation reaction between carboxyl- and hydroxyl-containing compounds. Historically, the first polyesters were condensation products of glycerol and phthalic anhydride. Those resins were used in the paint industry as early as 1910, but they were of low molecular weight and brittle. Later on, that is by 1925, they were modified by the inclusion of nonfunctional fatty acids, which compensated the third functional group of the glycerol and elastified the resins. If unsaturated fatty acids were used, they rendered the resulting resins cross-linked by air oxidation. All these resins are now generally referred to as "alkyd resins".

During the second half of this century, a large number of di- and trifunctional acids and glycols became commercially available, thus creating new possibilities for the synthesis of polyesters. It was the commercialization of terephthalic acid — or rather the dimethyl ester — that spurred the development of polyethylene terephthalate, which today plays an important role for the manufacture of fibers and films, and as an engineering plastic. In contrast to this polymer, however, which is made up of one pair of reactants only and is, therefore, frequently referred to as homo polyester, it is the combination of more than two reactants that yields copolyesters. For the manufacture of copolyesters, the chemist can today draw from a large selection of monomers. This makes copolyesters a class of polymers that can be varied over a very wide range:

- Low or high molecular weight
- Saturated or unsaturated
- Linear or branched

- Liquid or solid
- Amorphous or crystalline
- Soft/elastic or hard
- Tough or brittle

The glass transition temperature of copolyesters can be varied over a broad span. They can be modified in the main chain by other polymeric blocks (e.g., polyester units), and they can be prepared with varying terminal groups: hydroxyl, carboxyl, epoxy, acrylic functions, and others. Consequently, they can undergo further chemical reactions, especially cross-linking, to render a coating duroplastic.

The versatility of this class of polymers, combined with good stability toward light, heat, oxygen, and many chemicals, has contributed to an impressive and continuing gain in importance over the past 30 years. Unsaturated polyesters also should be mentioned. They contain maleic, fumaric, or other unsaturated acids and, thus, have double bonds in the main chain. They are used as coatings — generally dissolved in monomeric styrene — in the furniture industry. Without styrene, they may be used in conjunction with other base polymers to enhance and broaden adhesion properties. This chapter deals with saturated copolyesters only.

Classification of Saturated Polyesters

Saturated polyesters for use as binders in paints and coatings may be classified according to their molecular weight and their functionality. High molecular weight polyesters are predominantly linear, thermoplastic polymers with molecular weights from 10,000 to 30,000. Generally, they are copolyesters containing terephthalic and/or isophthalic acid and aliphatic diacids and a blend of diols. In contrast to terephthalic acid homopolyesters, they exhibit better solubility in solvents.

In coatings, they impart a high degree of flexibility paired with excellent surface hardness and stability. High molecular weight linear polyesters may be used as physically drying binder components in paints, although the majority of uses are in baking enamels for highly flexible coatings, such as coil and can coatings in combination with amino resins or other suitable hydroxyl-reactive cross-linkers.

Certain special grades of high molecular weight polyesters are ground and used as thermoplastic powder coatings. Low molecular weight polyesters range from 500 to 7000 g/mol and are, in general, not suitable as physically drying binders. Because of their low degree of polymerization, they carry a great many functional terminal groups. Low molecular weight polyesters may be linear or branched; by variation of the manufacturing process, it is possible to incorporate mostly either hydroxyl or carboxyl end groups, or both kinds.

By themselves, low molecular weight polyesters are not satisfactory film formers. They require a reaction partner capable of reacting with the end groups of the polyester and causing the formation of a cross-linked, duroplastic film. Amino resins and polyisocyanates are suitable as such cross-linking agents for hydroxyl polyesters, whereas epoxy resins and polyoxazolins may be used for carboxyl polyesters.

By proper selection of the reactants, the formulator can design products ranging from twocomponent or one-pack solvent-borne enamels with amino resins or blocked polyisocyanates to powder coatings or, via the salt formation of carboxyl polyesters, water-soluble stoving paints.

By reacting the native terminal groups of polyesters (i.e., hydroxyls or carboxyls) with at least bifunctional monomers or oligomers, saturated polyesters may be further modified in many ways. Aminoplasts may be employed to prepare thermosetting precondensates with hydroxyl polyesters. Partially blocked polyisocyanates, when blended and reacted with excess amounts of hydroxyl polyesters, will yield thermosetting binders. Similarly, one can obtain silicone-, epoxy-, or acrylic-modified polyesters.

Table 1 — Types of Polyester, Cross-linkers and Applications

Type of Polyester	Cross- linker	Field of Application
Linear high molecular weight polyesters	Melamine resins, benzoguanamine	Coil- and can-coating paints, primers, coatings for collapsible tubes
Linear low molecular weight hydroxyl polyesters	Melamine resins, benzoguanamine resin, polyisocyanates	Industrial stoving paints, two- component paints
Branched low molecular weight polyesters	Melamine resins, benzoguanamine resins, urea resins, polyisocyanates	Industrial stoving paints, two- component powder coatings
Carboxylated polyesters	Melamine resins, epoxy resins, triglycidylisocyanurate, polyoxazolin cross-linker	Waterborne coatings, powder coatings.
Silicone-modified polyesters	Siloxane melamine resins	Coil-coating paints, heat-resistant paints
Isocyanate-modified polyesters Epoxy-modified polyesters	Isocyanates, blocked polyisocyanates Polyanhydrides, carboxylated polyesters	Coil-coating paints, industrial paints Primers, industrial paints
Acrylated polyesters	Melamine resins, radiation cure	Industrial paints, adhesives, printing inks

For the purpose of further modification, polyesters with molecular weights between 1000 and 5000 are best suited, given their higher content of functional hydroxyl or carboxyl groups and their better reactivity as compared to higher molecular weight counterparts. In that respect, it is frequently desirable not to

convert all end groups, in order to leave residual cross-linkable functions for a coating or merely to improve the adhesion of the coating to certain substrates. Table .1 lists the various ways to cross-link high or low molecular weights of various modifications.

Still the most important groups of cross-linking agents are amino resins, especially melamine–formaldehyde condensation resins.

Other cross-linkers are gaining in importance, however. With the rapid growth of the market for powder coatings, polyester–epoxy hybrids and blocked isocyanate curing agents are becoming increasingly popular.

Polyester–isocyanate-based binder systems are increasingly used in solvent-borne paints, such as coil and can coatings, where a high degree of elasticity and resistance to weather or other attacks is required. Silicone-modified polyesters are known for excellent weather resistance and — with sufficiently high silicone content — good heat and chemical stability.

Epoxy-modified polyesters are suitable for cross-linking with acidic resins or acid anhydrides and offer, thus, formaldehyde- free and less toxic way to cross-link polyesters. Finally, acrylated polyesters serve as binders for radiation-cured coatings for varnishes, painting inks, and adhesives.

MANUFACTURING PROCESSES

Reaction Components

Aromatic and aliphatic polycarboxylic acids and various polyols may be used for the penetration of linear or branched polyesters of varying molecular weight. The nature of the monomers and their relative amounts determine the properties of the resulting copolyester. The most important raw materials for the preparation of copolyesters are listed in Table .2.

Whereas for the manufacture of linear polyesters, strictly bifunctional acids and alcohols must be used, for branched polyesters a certain amount of trior higher functional monomers is required. To warrant manufacture in commercial quantities on a consistently high quality level, a high degree of purity is required for the raw materials and, where applicable, a constant ratio of isomers. The polycondensation reaction is very sensitive to impurities, and even small amounts of foreign matter will affect discoloration, a change in molecular weight distribution, or even gelling.

TECHNICAL MANUFACTURING PROCESSES

Saturated polyesters are made by esterification and transesterification. Both are reversible equilibrium reactions that yield the desired product by continuing removal of condensation water. Under the conditions of the esterification reaction, transesterification occurs at the same time. Consequently, when more than two diols or decarboxylic acids are being used, a statistical distribution of the monomers must be expected. The rate of reaching equilibrium is catalytically accelerated by acids, bases, or transition metal compounds. Commonly, small amounts of titanium, tin, or antimony compounds are used as catalysts.

Table 2 — Monomers for Producing Saturated Polyesters

o-Phthalic Anhydride	Ethylene Glycol
Isophthalic acid	Propanediol 1-, 2
Terephthalic acid	Diethylene glycol
Adipic acid	Butanediol 1-, 3
Sebacic acid	Butanediol 1-, 4
Trimellitic anhydride	Hexanediol 1-, 6
Pyromellitic anhydride	Neopentyl glycol (NPG)
Trimesic acid	Trimethyl pentanediol (TMPD)
Hexahydrophthalic anhydride	Cyclohexane dimethanol (CHDM)
Hexahydrophthalic acid	Tricyclodecane dimethanol
5-t-Butyl isophthalic acid	Hydropivalic acid-NPG-ester
Dodecanoic acid	Bisphenol A bishydroxyethyl ether
Dimerized fatty acids	Trimethylol propane (TMP) Pentaerythritol

Low molecular weight linear or branched polyesters are generally prepared by a one-step process. This involves blending all monomers and reacting them at temperatures from 125 to 240°C. The condensation

products are removed either by the application of vacuum (vacuum-melt process) or by passing through a stream of inert gas such as nitrogen or carbon dioxide (driving gas-melt process) or by the addition and subsequent azeotropic distillation of a solvent (azeotrope process).

Higher molecular weight copolyesters are made in two steps. The first step involves the preparation of a polyester precondensate either by transesterification or by esterification of the acids with excess diols. The second step, referred to as the polycondensation, consists of further removal of diols from the molten precondensate in vacuum at temperatures above 250°C.

Other manufacturing processes for polyesters are known, but these are not normally used for commercial purposes.

PROPERTIES OF POLYESTERS

Morphology

Depending on the choice of components used to build up the polyester, the morphology of the polymer may be varied continuously over a wide range from completely amorphous to highly crystalline. Monomers imparting regularity — for instance, by a sequence of methylene groups or by symmetrical substituents on a ring — will lead to crystallinity in the polymer. Accordingly, polybutylene terephthalate or a copolyester consisting of terephthalic and adipic acids and butanediol will exhibit a high degree of crystallinity. Less regular molecules, such as isophthalic or orthophthalic acid or neopentyl glycol will yield only weakly crystalline or, more likely, completely amorphous polymers.

As a rule, for use in coatings, amorphous polyesters will prevail; however, there may well be instance in which crystallinity can be used to advantage — for instance, in thermolastic powder coatings, where crystallinity helps to reduce the permeability of the coating.

Solubility

Naturally, chemical composition will be the dominant factor in determining the solubility of a polyester; however, this dominance may be exhibited either directly or indirectly (by affecting the morphology).

The solubility of polyesters decreases rapidly with increasing crystallinity. Highly crystalline copolyesters are insoluble in many common solvents. They can be dissolved only in blends of phenol and o -dichlorobenzene at higher temperatures. Polyesters of medium crystallinity are soluble in methylene chloride; weakly crystalline polyesters will also dissolve in aromatic hydrocarbons, such as toluene. Amorphous polyesters will dissolve in a variety of polar solvents, such as esters, etheresters, ketones, and aromatic and chlorinated hydrocarbons.

Within the amorphous polyesters, chemical composition will have a direct effect on solubility. Thus, shifting the glycol ratio in polyester from ethylene glycol toward neopentyl glycol will enhance the solubility of the resulting polyester. Detailed information on the solubility of a variety of polyesters may be found in vendors' literature.

Polyesters are essentially insoluble in water, unless special measures are taken to impart hydrophilic sites. This may be achieved in one or more of the following ways: Incorporation of polyethylene oxide segments or anionic groups, using either carboxyls (e.g., by means of trimellitic anhydride) or salts of sulfoisophthalic acid.

Detailed treatments on water-borne polyesters may be found elsewhere. True solubility of the polymers is attained only rarely. Rather, a gradual increase in water dispersibility is observed until, by way of forming polyelectrolyte salts, a colloidal solution is obtained. Frequently, auxiliary solvents such as butyl glycol ethers are also employed.

MOLECULAR WEIGHT DISTRIBUTION

Polyesters are prepared by polycondensation, a fully reversible reaction. The desired degree of polymerization is obtained by shifting the reaction equilibrium in that excess glycol is removed. A concise treatment concerning low molecular weight linear polyesters is available in German.

Essentially, assuming comparable reactivity of end groups, the molecular weight distribution is a purely statistical one. For higher molecular weight linear polyesters, it tends to drop off fairly rapidly toward higher molecular weights and to tail off toward lower ones. For typical polyester, the following data were found: M w = 22,200; M n = 8300; and dispersity = 2.7.

By gel permeation chromatographic analysis, a content of 3% (by area) of low molecular weight (>1000) species was measured. One complication is the possible formation of cyclic oligomers — for instance, the one consisting of 2 moles of terephthalic acid and 2 moles of ethylene glycol or several others. Most of these oligomers are crystalline and tend to cause turbidity in resin solutions and, occasionally, in paint films. Much work has been devoted to the elimination of these oligomers, resulting in ways and means to avoid this interference factor.

Much more complicated is the situation when trifunctional monomers are used and branched polymers are obtained.

As Carothers and Flory have shown, the molecular weight distribution broadens rapidly with increasing degree of polymerization and soon a critical conversion point is reached, beyond which gelation of the polymerization batch occurs. This can be avoided by a two-step process that introduces additional functional groups in a controlled manner. A typical example for this technology is the preparation of branched hydroxyl polyester in a first step, followed by reaction with trimellitic anhydride as a second step.

FUNCTIONALITY AND REACTIVITY

In contrast to other polymerization processes (e.g., radical polymerization), polycondensation does not tend to cause imperfections due to branching or chain termination. Therefore, a polyester consisting of bifunctional components only will be just about perfectly linear. Only the incorporation of trifunctional monomers such as trimethylol propane or trimellitic acid causes a controlled increase in functionality above 2. As the polycondensation is generally carried out with an excess of glycol, the acid number declines more rapidly with increasing degree of polymerization than the hydroxyl number; thus, except for low molecular weight polyesters, the reactivity of the polyester is primarily linked to its terminal hydroxyl groups. Because secondary hydroxyls are less reactive than primary ones, a terminal primary hydroxyl group generally remains. This, in turn, reduces the reactivity of polyester primarily to a function of its molecular weight, in accordance with the decreasing concentration of end groups with increasing size of the polymer molecule.

Transition Temperatures

These transition temperatures are of interest in conjunction with polyesters. In order of importance for coatings, there are the following:

- Glass transition temperature, T g
- Softening point, T f
- Crystalline melting point, T m

The theory of the relationship between thermodynamic equilibrium melting points and copolymer compositions was developed by P. J. Flory. On the basis of the free volume theory, Flory and Fox proposed a general theory on glass transition theory, which led to the formula suggested by Fox, which permits the calculation of the glass transition temperature from the contributions of the individual monomers. Polyesters obey the formula quite well. Thus, from knowledge of the glass transition temperatures of the homopolyesters, copolyesters may be calculated with good accuracy. For the purpose of this chapter, however, a few basic guidelines may be helpful.

Purely aliphatic polyesters will — with increasing ratio of methylene to ester groups — asymptotically approach the glass transition temperature of polyethylene (N 160 K). Accordingly, polyester of azelaic acid and hexamethylene diol has a glass transition temperature of about 190 K, whereas one made of terephthalic acid and bisphenol A has a glass transition temperature of 480 K. More detailed data on glass transition temperatures of homopolyesters may be found in the literature.

For amorphous polyesters, the softening point — usually determined by the ring and ball method — is generally approximately 50 to 90°C above the glass transition temperature. As one would expect, this difference increases with increasing molecular weight.

Crystalline melting points depend largely on the chemical nature of the crystalline species. Thus, highly aliphatic, crystalline polyester will melt at approximately 340 K, whereas more aromatic crystalline species will melt at considerably higher temperatures (e.g., polyethylene terephthalate, 540 K). More information on melting points of polyesters is available.

COMPATIBILITY OF POLYESTERS

The compatibility of polyesters with other resins and polymers depends on many parameters, which also are partly interrelated, including molecular weight, glass transition temperature, morphology, and chemical composition of the polyester. Among the most prominent examples for the use of polyesters in combination with other polymers are blends of low molecular weight polyesters with cellulose acetobytyrate in base coats of two-coat metallic automotive finishes. High molecular weight polyesters are used as coresins for nitrocellulose lacquers and for vinyl copolymers.

Chemical Properties

Generally speaking, polyesters have excellent stability toward light, oxygen, water, and many chemicals. The weakest spot in the polymer chain is the ester group, with its potential sensitivity to hydrolysis.

Accordingly, degradation may occur, provided there is an environment that combines moisture with acids or alkali or other catalytically active materials, plus the possibility for the moisture to permeate into the polyester. The latter is largely suppressed in the case of crystalline polyesters or for cross-linked polyester paint films. Degradation of polyethylene terephthalate was investigated by Buxbaum.

Because this author was using low molecular weight model compounds, this study may, to some extent, be applicable to other polyesters as well.

Preparation of Polyester Coatings

For the preparation of paints based on saturated hydroxyl polyesters, amino resins are the most commonly chosen cross-linkers. Melakine–formaldehyde condensates are preferred over urea–formaldehyde resins in view of the superior properties of the former in weathering stability as well as the balance of surface hardness and elasticity. For further improvement of adhesion, gloss, and water resistance, benzoguannamine resins are used. Selection among the many commercially available melamine resins is primarily determined by the compatibly of the melamine resin with the type of polyester it is to be combined with.

Monomolecular, etherified condensation products such as hexamethoxymethyl melamine or only partially etherified products exhibit a very broad range of compatibility. Higher molecular weight resins containing several melamine units in the molecule are less compatible and can be used with selected polyesters only.

To meet the commonly established stoving conditions of 120 to 270 ° C with acceptably brief reaction times, the cross-linking reaction between hydroxyl polyesters and amino acid resins — especially the etherified variety — requires the use of acidic catalysts. A common choice is p -toluolsulfonic acid or its salts with volatile amines, such as morpholine, or nonionically blocked acidic compounds. Hetero- or homopolar-blocked acids are latent catalysts, which become effective only during the baking cycle.

Thus, a premature reaction between polyester and melamine resin at room temperature in the liquid paint is effectively suppressed and the shelf life of the paint appreciably extended.

The properties of polyester coatings are essentially influenced by the molecular weight and the composition of the polyester as well as by the nature and the amount of the melamine resin. With increasing amounts of amino resin, baked coatings will become harder and less elastic, whereas resistance to solvents and chemicals increases. In general, good results are obtained with 10 to 35 wt% of total binder content.

In coil and highly elastic can coating paints, the amino resin content may be reduced to as little as 5%. Of less commercial importance than melamine resin cross-linked polyesters are those with other cross linking agents such as isocyanate or epoxy cross-linked or siloxane- or radiation-cured systems. Naturally, to avoid side reactions and insufficient shelf life when designing a paint formulation, solvents, catalysts, pigments, additives, and fillers must be chosen with due regard for the chemistry involved.

Solvent-Borne Polyester Coating

For the preparation of solvent-borne polyester paints, all the binder components are first dissolved; then the pigments, fillers, and additives are added, thoroughly dispersed, and milled. For reasons of economy and to alleviate the milling procedure, it is advantageous to conduct the milling step in a fraction of the total binder. To avoid difficulties in pigment wetting when using binders of different polarity, the fraction used for milling should contain all the binder components in the proper proportion. Upon completion of the milling step, the retained binder portion is added together with leveling or flow agents or other additives and the required amounts of solvents, Paints based on saturated copolymers can be applied with all common techniques. Those most widely practiced are roller coating for can and coil coating, followed by various spray applications (air assisted, airless, or electrostatic spray) with, for instance, rotating discs or belts.

High Solids Paints

High solids paints formulated with saturated co polyesters have total solid contents between 65 and 80 wt%. They require low viscosity — that is, low molecular weight polyesters and monomolecular melamine — formaldehyde resins.

Due to the low molecular weight of the polyesters, the elasticity of paint films thus obtained is inferior to that of conventional solvent-borne polyester paints. Therefore, high solids polyesters paints are primarily used as spray-applied or dip paints. When formulating such paints, one should take into account the high polarity of the binder and use wetting agents for better pigment wetting, as well as effective flow agents and preferably nonionically blocked catalysts for improved electrostatic spray applicability. Polyester–urethane two-component paints allow the manufacture of products with higher solid content and lower emission as compared to polyester–melamine combinations.

Waterborne Paints

Waterborne polyester–melamine paints are made from saturated polyesters containing an increased amount of carboxyl groups, having acid numbers between 45 and 55 mg KOH/g and molecular weights of approximately 2000 g/mol. The polyesters are usually combined with water-soluble melamine–formaldehyde resins such as hexamethoxymethyl melamine in a ratio of 70:30 to 85:15. The use of a catalyst is not necessary, as the carboxyl groups accelerate the reaction to a sufficient degree. Before blending these polyesters with the melamine resin, the carboxyls of the polyester must be neutralized with amine, usually dimethylethanolamine. Auxiliary solvents such as butyl glycol ether may be used to reduce viscosity, improve pigment wetting, enhance shelf life, and improve the thinning characteristics of the paint. Depending on the nature of the binder, the application viscosity and the stoving cycle, 5 to 15wt% solvent based on total paint formula is used.

Lately, it has been possible to develop polyesters with acid numbers of only about 20 mg KOH/g and molecular weights of 4000 g/mol, which upon neutralization are thinnable with water. The higher molecular weight allows a further reduction in emission when curing with melamine resins. Paints thus formulated may be applied by roller; they display a high degree of flexibility and impact resistance in the cured film.

Solvent-Free Coatings

Saturated polyesters play an important role for the formulation of thermosetting powder coatings. They can be combined as carboxyl polyesters with epoxy resins or triglycidyl isocyanurate (TGIC), or as hydroxyl polyesters with blocked solid isocyanates, predominately isophorone diisocyanate (IPDI). Thermoplastic powders are used mainly to cover welding seams in cans. The preparation of powder coatings involves blending of the binder resins with pigments, additives, and catalysts, homogenizing in a kneader–extruder, then grinding and screening. Electrostatic spray is the prevailing application method.

For adhesive coatings, special, low viscosity polyesters were developed. The low viscosity in spite of comparatively high molecular weight is attributed to the special structure of these polyesters, consisting of an essentially linear main chain with a large number of alkyl side chains attached to it. By reacting the terminal groups further, the polyesters may be functionalized to carry terminal acrylic bonds, which render them reactive to high energy radiation such as ultraviolet light or accelerated electrons. Products thus formulated are solvent-free. They must be applied warm by roller or slot die coating and be radiation cured.

PROPERTIES AND APPLICATIONS OF POLYESTER COATINGS

Polyester coatings, especially cross-linked ones, exhibit excellent flexibility or even elasticity, and they show very good impact, scratch, and stain resistance. Their good adhesion properties, especially to metals, combined with good corrosion protection and weather resistance, have made them indispensable in a number of fields.

Sheet and Coil Coatings

Precoated sheet and coil have enjoyed remarkable growth over the past two decades, and polyesters have played an ever-increasing role in this industry. Polyesters are being used for primers and top coats with equal success. Polyester-based coil coating primers have been discussed concisely by Schmitthenner and Robertson. The tough requirements that coated coil stock is expected to meet — for instance, for use as façade sheeting — necessitate the use of primers, which have three main functions:

- Promotion of adhesion to pretreated metal
- Corrosion protection
- Elastification of the two-coat system consisting of primer and top coat

The two-coat system permits the optimization of adhesion of the primer to the pretreated metal and, at the same time, intercoat adhesion. High molecular weight polyesters with relatively high glass transition temperatures and in films approximately 5 µm thick have yielded excellent results on all counts. Corrosion-protective action is enhanced by the use of anticorrosive pigments. For the formulation of top coats, the end use will influence the choice of polymers. Paints with outstanding weather resistance can be formulated from medium molecular weight polyesters.

The ultraviolet (UV) absorption of polyesters used for this purpose ends at wavelengths below those at which the UV irradiation of sunlight begins. Schmitthenner has given another example of high molecular weight, highly elastified polyesters that can be used to formulate correspondingly elastic enamels for end users such as home appliances.

Can Coating

The term "can coating" embraces paints for use on a vast variety of decorated metals for packaging purposes. This includes cans of all kinds, including food cans and aerosol cans, collapsible tubes, and caps and closures of many kinds. Generally, the paint is applied to flat sheet stock, followed by printing, stacking up for storage, then stamping and forming. Especially difficult are the requirements for drawing and redrawing cans and closures, where the degree of forming is too high to be achieved in a single draw. The most common substrates are tin-plated steel, aluminum, and to a smaller extent, directly chromate steel (tin-free steel, TFS). Accordingly, the requirements for can coatings are extreme in many ways. They include printability and block and scratch resistance, yet sufficient elasticity to permit forming and drawing without damage to the paint. On the finished article, the paint must be stable with respect to its contents, it must be stain resistant, sometimes heat sealable, and for food preservation, able to withstand sterilization, nontoxic, and neutral in taste and odor. The composition of can coatings for use in conjunction with foodstuffs is regulated in many countries individually, however, the U.S. Food and Drug Administration (FDA) is recognized internationally. The FDA provides a list of components that may be used to prepare polyesters intended for use as coatings in direct contact with food.

In conclusion, polyesters offer an excellent combination of physical properties especially concerning the balance of elasticity and surface hardness, paired with excellent adhesion, resistance to yellowing, and stability against the majority of materials now packaged in cans and tubes.

Automotive Paints

In the area of automotive paints, polyesters compete with a variety of other polymers. Acrylics, epoxy esters, and even alkyds have their share of the market. As automotive paints are spray applied and low molecular weight and frequently branched polyesters tend to be used predominantly, two areas have developed as special grounds for polyesters. These are base coats for two-layered metallics and chipresistant fillers.

For the former, low molecular weight, branched polyesters are used in conjunction with cellulose acetobutyrate and polymeric melamine resins. This combination permits optimum alignment of the metallic flake pigments immediately after application. The use of polyesters for chip-resistant fillers makes use of the excellent elasticity. Special polyester fillers made of upgraded polyurethane resin are used on areas of the car body where stone chipping occurs most frequently (e.g., rocker panels).

Industrial Paints

Polyesters paints are gaining importance for the formulation of industrial baking enamels as a result of their weathering resistance, good abrasion and chemical stability, well-balanced elasticity, and surface hardness. This applies especially to high solids products and for waterborne sprayable and dip paints. Texturized paints are used for metal covers and housings for appliances, machinery, and data-processing equipment primarily because of their good abrasion resistance and favorable hardness-elasticity properties.

Automotive uses include metal parts in the engine compartment and fixtures for rearview mirrors and windshield wipers. Polyester paints are attractive here because of their corrosion protection and resistance to oil, fuel, and cleaning agents. Office furniture, such as steel desks and filing cabinets require high impact resistance, which is well met by polyester paints. Similarly, household appliances – if not made from precoated sheet – are also frequently coated with polyester-based paints.

Component Paints

Saturated polyesters play an important role as hydroxyl-functional binder components in the formulation of two-component polyurethane paints. Compared to other hydroxyl-functional materials, polyesters permit complete adjustment of the elasticity of the coating of substrate, and coating is required to avoid crack formation upon temperature fluctuations. In addition, polyesters are used in isocyanate-cured, two-component automotive repair paints because of their weathering and chip resistance.

Powder Coating

Polyesters are finding use in thermosetting and thermoplastic powders. The latter are a specialty application only. The overwhelming majority are thermosetting powders. They may be either hydroxyl polyesters cured with blocked isocyanates or carboxyl polyesters cured with di- or triepoxides.

The most important advantage of thermoplastic powders is the speed of film formation, which is accomplished merely by melting and does not require reaction time for a chemical cure. For this purpose, high molecular weight, partially crystalline, polyesters are best suited. The use is predominantly for coating the welding seams of three-piece cans.

Polyesters for use in thermosetting powders are generally amorphous, with molecular weights between 2000 and 6000, glass transition temperatures above 55°C, and softening points between 100 and 120°C. They must have good pigment compatibility. In general, they have a functionality higher than 2, which means branching in either the main chain or terminal components of higher functionality, as achieved by capping with trimellitic anhydride.

Hydroxyl polyesters are generally cured with blocked isocyanates, such as isophorone diisocyanate blocked with e-caprolactam or uretdiones.

Carboxyl polyesters may either be cured with epoxy resins (polyethylene sulfide–epoxy hybrides) or with TGIC. Polyester–epoxy hybrids enjoy the largest volume. They are very versatile and are, thus, finding use in many and diverse applications. Their advantage over hydroxyl polyesters–blocked isocyanate materials is the absence of blocking agents, which have to be driven off during the curing cycle. In turn, they require catalysts, such as quarternary ammonium compounds, to achieve satisfactory short curing times at acceptable cure temperatures. Due to the relatively high content of epoxy resins (60 to 100 parts per 100 parts polyester) they have a tendency to chalking and, hence, are limited with respect to weather resistance. The latter can be essentially improved by using TGIC at a level of approximately 10 parts per 100 parts polyester. Weather resistance of TGIC-cured carboxyl polyesters is excellent; however, powder coatings formulated on this basis have a tendency to display orange peel upon longer storage of the powder.

RADIATION-CURABLE COATINGS

Saturated polyesters, of course, are not reactive to UV light or accelerated electrons such as used generally for radiation cure. There are basically two ways to render them reactive:

- Incorporation into the polymer backbone of unsaturated acids, such as maleic or fumaric acid
- Capping with reactive, preferably acrylic or methacrylic, vinyl double bonds

The former method includes the unsaturated polyesters, such as those used as fiberglass-reinforced polymers in vehicle construction with or without the use of monomeric styrene as a reactive diluent. In the field of coatings, this material combination has found only very limited acceptance for a variety of reasons. The only type of use worth mentioning is for furniture finishes.

Vinyl-terminated polyesters, or more specifically acrylated polyesters, may be prepared by direct esterification of hydroxyl polyesters with acrylic acid or by reaction of carboxyl polyesters with glycidyl methacrylate.

The most widely practiced method of attaching acrylic double bonds to polyesters is via the reaction of hydroxyl polyesters, diisocyanates, and hydroxyalkyl acrylates leading to acrylated polyester urethanes. Such materials are finding growing acceptance either by themselves or in combination with reactive diluents for a variety of uses. The reactive diluents are predominantly di- and triacylates. The uses are diversified; overprint varnishes in the graphics and packaging sectors are presumably the largest group.

Adhesives

Adhesives should be mentioned for the sake of completeness, especially the kinds that are predominantly used in the form of continuous coatings. These include the following:

- Heat-seal lacquers and hot melt coatings
- Laminating adhesives
- Pressure-sensitive adhesives

For heat-seal lacquers and hot melt coatings, saturated, high molecular weight polyesters can be very useful. They offer good blocking resistance and swift bonding combined with attractive physiological properties (e.g., FDA conformity).

For laminating adhesives, high molecular weight polyesters may be used in conjunction with isocyanates to effect high bond strength, good temperature stability, and even sterilization resistance.

Pressure-sensitive adhesives, a totally new field for polyesters, may well gain considerable importance in the near future.

Alkyd Resins

The alkyd resin was one of the first applications of synthetic polymers in surface coatings technology; it was successful in chemically combining oil or oil-derived fatty acids into a polyester polymer structure. Benefits were the enhancement of the mechanical properties, drying speed and durability of these vehicles over and above those of the oils themselves and the oleoresinous vehicles then available. Though now surpassed by more sophisticated polymers for the more exacting applications, the alkyd, because of its partial reliance on renewable natural resource, and especially because of the enormous variety of compositions possible, still accounts for a considerable volume of total surface-coating resins produced. The long oil alkyd, optionally blended with other modified alkyds to increase toughness or alter rheology, remains the major resin used in brush-applied solvent-based decorative coatings.

Recent application has extended into higher solids and use in emulsified form, as ways of solvent reduction, in producing eco-labeled products. High solids compositions are covered later in this book. Short oil alkyds still have use in some automotive and general industrial stoving compositions where they are combined with melamine/formaldehyde hardening resins.

Oils are not directly reactive with polyester components and therefore cannot be incorporated into an alkyd structure without prior modification. We can, however, start with fatty acids saponified from the oil, though for many applications this will be an expensive route, especially if glycerol is intended to be included in the formulation. The method evolved to achieve oil modification is to carry out, as a first stage, a so-called monoglyceride preparation. In this process glycerol (or other polyol) and oil in the molar ratio of 2:1 are reacted at around 240°C in the presence of a basic catalyst (sodium hydroxide, lithium hydroxide, etc.) forming 'monoglyceride', the mono fatty acid ester of glycerol (Fig. 1).

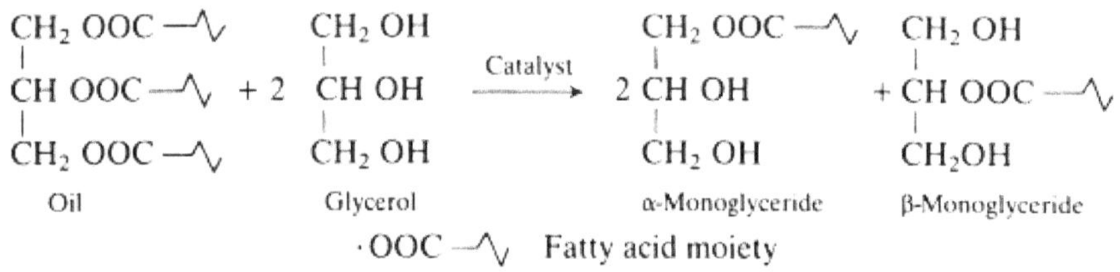

Fig. 1 — Schematic representation of monoglyceride formation.

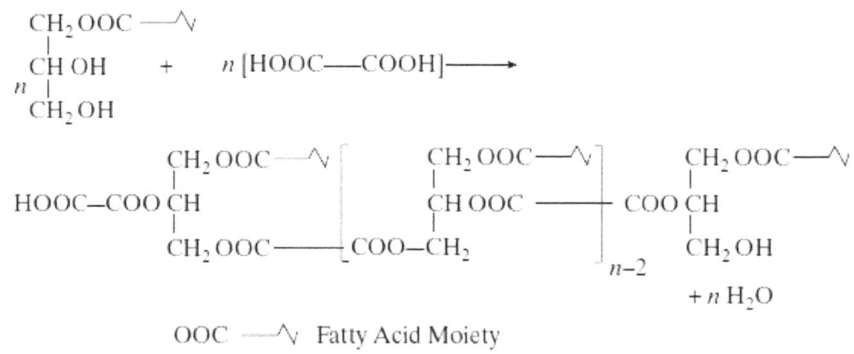

Figure 2 — Schematic representation of glycerol alkyd polymer formation

Practically, the reaction does not go to completion, and an equilibrium distribution of species is present including oil, polyol, and mono- and diglycerides. If polyols other than glycerol are to be used, such as pentaerythritol (PE) in long oil alkyds, a 'monoglyceride' stage may still be carried out using the appropriate amount of polyol to give a notional dihydroxy functional product. The satisfactory attainment of sufficient randomization of structure may be tested for by an alcohol tolerance test, when no separation of free oil should be observed. Sometimes with polyols such as PE the alcohol test is not satisfactory. In that case, reaction is carried out on a small scale in a wide-necked tube with the correct proportions of 'monoglyceride' sample and polybasic acid to test for the ability to derive a clear final product without gelation resulting. With some oils it is important not to prolong the time at temperature for this stage in order to minimize oil bodying;

for the same reason it is good practice to standardize on heat-up rate and time for the normal reaction, to ensure reproducible and predictable behaviour in the following stage.

Preparation of the final resin is carried out by the addition of further polyol and dibasic acid in the so-called bodying stage; in this stage, multiple esterification reactions occur (Fig. 2.) and water formed as a by-product of reaction must be removed.

Esterification results in the steady building-up of polymer structure and hence viscosity increases; the stage is taken to a degree of reaction at which the desired viscosity at a specified concentration (in solvent) and temperature has been attained.

While the monoglyceride method as described is used for long oil alkyds, the simpler so-called 'fatty acid process' is also practised; here, all ingredients including fatty acid (not oil) are charged together, and the process is frequently used for shorter oil alkyds where polyols other than glycerol are required. Other techniques are possible, for example an acidolysis route where oil is first equilibrated with polybasic acid prior to reaction with polyol in an apparent reversal of the monoglyceride process; this method is not widely favoured. Where it is necessary to incorporate tung oil into an alkyd, when to carry out a monoglyceride state would risk excessive bodying, a variant is the fatty acid/oil process. Here the tung oil is taken along with a good proportion of fatty acid, polyol, and dibasic acid, and a onestage reaction carried out. Providing the oil concentration is not high, transesterification of sufficient magnitude takes place during processing to react the oil into the resin and ensure a clear final product. Another technique proposed for alkyd manufacture is the 'high polymer' technique, where by stepwise addition of ingredients, more favourable molecular weight and reactive group distributions are said to be obtained and certain properties enhanced; this is claimed to be particularly effective with tall oil alkyds. This technique has recently been examined theoretically.

Composition

Each of the main ingredients of an alkyd contributes to its properties in a predictable manner. Oil content is expressed as 'oil length' where long oil alkyds contain over 60% oil, medium oil between 40 and 60%, and short oil below 40%. Long oil alkyds are generally made with drying oils and are soluble in aliphatic solvents. They are low viscosity resins that air-dry slowly to give soft flexible films with poorer gloss retention and durability. Shorter oil length generally results in the need for aromatic solvent, giving higher viscosity resins at lower solids. If formulated with drying or semi-drying oils, and of medium or long oil length, the alkyd can air-dry oxidatively at room temperature.

Otherwise it will not do so, and to form hard films it must be stoved with a cross linking resin. Medium and short oil alkyds give hard films with good gloss retention and chemical resistance; they are less flexible than long oil alkyds and are normally dried by stoving.

Aromatic acids such a phthalic anhydride or isophthalic acid, and maleic anhydride contribute hardness, chemical resistance, and durability to alkyds; in contrast long chain dibasic acids such as adipic, sebacic, and azelaic acids, are sometimes used to plasticize alkyds and provide flexibility, not hardness. Phthalic anhydride dominates in conventional alkyd formulations because of its low cost and ready availability.

It should be noted that it has added economy in use, in that by virtue of its anhydride structure, water evolution is halved compared with a dibasic acid; the first stage of its reaction is exothermic, with so-called 'half ester' formation.

Polyols used are generally at least trifunctional to permit branching or crosslinking

and can provide the alkyd with hydroxyl groups for further reaction. In general they contribute good colour and colour retention, but vary in their chemical and weathering resistance. Most general-purpose alkyds are formulated using phthalic anhydride with properties modified by polyol variation. The effects of polyol change may be readily related to structure so that in a series where glycerol is replaced by trimethylol ethane and then by trimethylolpropane, aliphatic solubility increases, viscosity falls, and flexibility increases, contributed to by increased side chain length; otherwise varying the polyol component may have a significant effect, not least because with varying polyol molecular weight the phthalic content of the final resin is affected, and hence the hardness of the final alkyd. Rosin was originally included in alkyds as a modifier because of its low cost.

However, it may also be added, as many other monobasic acids such as benzoic acid, to enable oil length to be shortened without causing increased viscosity. Rosin itself contributes to faster drying and film hardness, but degrades weathering performance; its use is restricted to alkyds for primer applications.

Where short oil length and aliphatic solubility present conflicting requirements, trimethylol propane can be used as polyol, though with some penalty of increasing softness. Para tertiary butyl benzoic acid can also be used as a monobasic acid to modify solubility, and both are effective because of the contribution of their aliphatic side groups.

Isophthalic acid gives higher molecular weight resins than orthophthalic, with better drying characteristics and harder, more durable films. Its use can narrow molecular weight distribution and enable lower viscosity/higher solids resins to be made. However, it is more difficult to incorporate, remaining solid in the reaction mixture until at least one of its acid groups has reacted. Like orthophthalic anhydride it can sublime, but unlike orthophthalic anhydride, it can lead to overhead blockages in manufacturing plant because of its lower solubility in xylene which is used as the entraining solvent for the removal of water of reaction. Special understanding is necessary to formulate high solids alkyds with longer oil length. Terephthalic acid has little use in alkyds because of its extremely low solubility in reaction mixtures and its very slow reactivity.

Mention must be made of Cardura E10 (Fig. 3; Shell Chemicals) which is the glycidyl ester of the branched Versatic Acid (Shell Chemicals), the latter being a 'tertiary' or trialkyl acetic acid with a total of ten carbon atoms. By virtue of the epoxide group, it reacts rapidly and completely with carboxyl groups at temperatures above 150°C; when incorporated into an alkyd, the effect is that which would be contributed by a non-drying oil or fatty acid component. Cardura E10 contributes special properties of good colour, gloss retention, and very good resistance to yellowing and staining. Derived alkyds can be used in stoving enamels and in nitrocellulose lacquers. Care should be taken in the manufacture of alkyds which include both Cardura E10 and phthalic anhydride since the reaction is both rapid and exothermic (around 110 kJ per epoxy equivalent). Special formulating techniques are available for this type of alkyd that also allows control of polymer architecture. Cardura E10 has special use for reducing the acid value in resins as a final treatment, for example where zero acid value is required for resins for use with photocatalyst, in metallic paints or with certain drier systems:

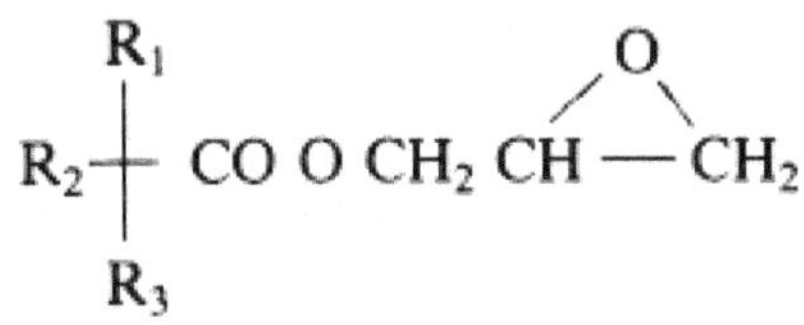

Fig. 3 — Cardura E10 (Shell Chemicals). R1, R2, R3 are alkyl, one of which is methyl.

Epoxy resin may be used as polyol in alkyd compositions in combination with the more usual polyols. Epoxy resin addition improves water and chemical resistance and contributes to hardness and toughness; however, it adds to cost and can lead to chalking on exterior exposure.

Since alkyds are soluble in either aromatic or aliphatic solvent dependent on their oil length, they are tolerant of additional stronger alcohol, ester, or ketone solvents for special applications.

ALKYD REACTIONS AND STRUCTURE

The major polymer-forming reaction in alkyd preparation is the esterification reaction between acid and alcohol, and hence the mechanism is through a type of stepwise polymerization (Fig. 2.). In alkyds this reaction is generally not catalysed. Side reactions that can occur include etherification of polyol especially in prolonged monoglyceride stages.

Pentaerythritol is particularly prone to etherification, and grades should be chosen with low dipentaerythritol content. A reason for lower molecular weight with orthophthalic alkyds compared with those prepared with iso- or terephthalic acids is the tendency to form intramolecu-lar ring structures in o-phthalate esters. Because of this, the effective functionality of phthalic anhydride is practically less than two.

It has been claimed that alkyds owe some of their special properties, including application and dry, to the presence of 'microgel' in the alkyd .While this seems to be proven, it would appear difficult to exploit in formulating for a microgel content unless a 'microgel' prepared separately were to be deliberately blended with the composition. It has been suggested that gelation of an alkyd occurs through separation of microgel particles, until particles become crowded enough to coalesce into a 'macrogel'.

Alkyd Reaction Control

Alkyd preparations must be controlled to a viscosity end point by sampling, with acid value also monitored. Viscosity measurement may be simply achieved by using Gardner bubble tubes with thinned samples; however, improved speed of measurement and higher accuracy may be achieved by using a heated cone and plate viscometer. The sampling interval will often need to be reduced as the end point is approached, owing to an increased rate of rise in viscosity, especially with short oil alkyds.

Where behaviour has been characterized from previous batches, a 'flare path' guide may be provided to follow the viscosity and acid value change; if reference to this shows deviation from normal progress, adjustment may be made by a small addition of acid or polyol. Where solvent process and oil bodying is occurring, adjustment may be made by changing the reflux rate, providing this is carried out at an early enough stage in the process. It is also possible to make temperature changes during processing to modify the bodying rate.

On attainment of end point, cooling and thinning are necessary to halt the reaction; thinning is also essential since the alkyd resin, in the absence of thinning solvent, will either be extremely viscous or solid when cold.

For new formulations, process temperature will be established by consideration of the need for higher temperatures for oil bodying and faster reaction in long oil alkyds, or for lower temperature for easier control of end point in shorter oil alkyds.

Acid value at gelation (AV gel) prediction may be possible as the preparation proceeds. The results of experiments will allow modification in both composition and processing conditions.

Describing Alkyds

In specifications for alkyds, oil type, oil length, and solvent will be stated, and sometimes the dibasic acid and its % in the composition. Oil length refers to the % weight of glyceride oil calculated on the theoretical yield of solid alkyd; fatty acid content for non-glycerol-containing compositions may alternatively be quoted. In both cases this is the major indication of the flexibility, solubility, and – where containing drying oil – drying potential to be expected for the resin. Solids, viscosity, and acid value (AV) will be determined as for other resins, and hydroxyl value (OHV) may also be measured. Both the latter are quoted in units of weight of potassium hydroxide in milligrams, equivalent to 1g of solid resin.

Calculated from the molar formula and expressed as the excess hydroxyl % over that required reacting completely with the polybasic acid moieties present. Excess hydroxyl % may be used as a formulating parameter to derive formulations with a required AV gel or average functionality.

Acid value at gelation (AV gel) may be calculated theoretically, and may also be determined practically by extrapolation of a graph of acid value against 1/viscosity as an alkyd preparation proceeds to its end point.

Alkyd Formulating

Practical alkyd formulating involves a process of calculation, application of previous practical experience, and trial preparation to arrive at the desired product. Principally oil length and oil/fatty acid type will be specified for the application, and polyol and dibasic acid dictated by a combination of other considerations including solubility, availability, and cost.

The average functionality concept is one method of formulating as first proposed by Carothers:

$$F_{\text{average}} = 2 \times \frac{\text{total acid equivalents}}{\text{total mol}} \quad \text{(hydroxyl in excess)}$$

where for gelation to occur F should be two or more. This is the basis of the well respected scheme by Patton, except that his constant k = 2/F. The basic functionality approach is recognized to require correction and Patton recommends appropriate factors for various ingredients.

An alternative method useful in arriving at a trial formulation involves calculating the extent of reaction at gelation, most usefully expressed as AV gel, which for example might be around five units below that desired in the final resin. It will be found necessary to vary the hydroxyl content, expressed as excess OH in order to alter the AV gel. Various equations have been produced as gelation theory has been developed, by for example Flory and Stockmeyer, and formulating guides are available which review the whole field of alkyd calculation.

$$p^2 = \frac{\left(\sum gB\right)^2}{\left(\sum f^2A - \sum fA\right)\left(\sum g^2 - \sum gB\right)}$$

Frequently, one of the forms of the Stockmeyer equation will be used, such as: where p = degree of reaction of acid groups at gelation, f, g = functionality of carboxyl and hydroxyl moieties respectively, and A, B = mols of carboxyl and hydroxyl moieties present. Before applying this equation to oil-containing formulations, the mols of oil should be split into constituent fatty acids and glycerol.

It is recognized that to formulate accurately, the true picture invariably lies between the two approaches. While the Flory – Stockmeyer approach is statistically correct, its inaccuracy is due to the actual formation of a number of cyclic rather than long chain polyesters through intramolecular reaction. The effect is more pronounced with the anhydrides such as phthalic anhydride, a very common ingredient.

While the simplest correction can be to consider phthalic anhydride as having a functionality of less than 2 (e.g. 1.85), Kilb developed the best theoretical solution, applied practically by Bernado and Bruins.

As a simpler alternative to some quite complicated calculations, those offered by Weiderhorn may be found a simpler and equally practical alternative, being less complex to use in the circumstances to which they apply.

In formulating short oil alkyds, monobasic acid may need to be included, dictated by solubility and hydroxyl content requirements. The use of bodied oil in long oil alkyds will be determined by the need to increase molecular weight and obtain higher viscosity, when this is not attainable by other means; the effect will be to raise the practical AV gel above that calculated.

With microcomputers being generally available, people formulating alkyds (and polyesters) regularly are likely to use computer programs to calculate both stoichiometry and gelation parameters. Programs of high functionality have been written. However, writing one's own program is quite feasible; alternatively simpler programs are available from raw material suppliers.

Alkyd Preparation Equipment

Alkyds may be prepared on both laboratory and production scales by both the socalled 'fusion' and 'solvent' process techniques. In the former, ingredients are heated in vented stirred pots; problems encountered are mainly due to variable losses of reactant, particularly phthalic anhydride which is volatilized along with the water of reaction. In the latter process, more sophisticated equipment is needed, namely reactors fitted with condensers, water separation equipment, and solvent return systems. Organic solvent, typically xylene, is added to the resin in the bodying stage (2–5%) and recycled via the overheads, water being removed by azeotropic distillation.

It can be of advantage to fit a solvent process plant with a vacuum facility which, besides providing assistance with loading, can be used, for example, to reduce the residual solvent content at the end of processing a batch.

Problems of phthalic anhydride sublimation in the solvent process may be minimized by careful plant design or avoided by using bubble cap scrubbers or fractionating equipment. Inert gas is normally supplied in all alkyd processing to improve colour and to increase safety. Process temperatures may be between 180°C and 260°C.Where consistent practice such as heating rate, gas flow, and ventilation can be achieved, it must be stressed that the fusion process can be satisfactory for almost all classes of alkyd, provided that environmental problems caused by fume can be controlled.

Water-Soluble and Dispersed Alkyds

This section specifically concerns alkyd resins. One issue is the current inability of emulsion polymer systems to provide traditional high gloss decorative coatings; for this reason, alkyd systems continue to be used. While high solids systems have been developed to satisfy eco-labeling requirements, emulsified alkyd systems have also evolved for glossy decorative applications.

The usual route to water solubility is to formulate to acid value in the region 40–60, followed by partial neutralization with alkali, ammonia, or amine. Since phthalate-based alkyds have poor hydrolysis resistance and hence poor storage stability, isophthalic acid (IPA) alkyds may be preferred. Alternatively, alkyds will be formulated using trimellitic anhydride (TMA) or dimethylol propionic acid (DMPA). The most common method in processing water-soluble alkyds containing

TMA is first to react all the ingredients except the anhydride to form a polymer with an acid number below 10; and then to add the TMA at a reduced temperature of around 175°C. Under these conditions the TMA is attached by a single ester link to hydroxyl groups on the preformed polymer, and the two remaining carboxyls are available for salt formation and subsequent water solubilization. This is termed the 'ring-opening' technique from the single stage reaction of the anhydride group of the TMA. Water-soluble alkyds of this type find application in stoving compositions with water soluble melamine/formaldehyde resins as crosslinker.

They may be supplied thinned in water-miscible solvent (e.g. butyl glycol) and water thinned at a late stage to gain maximum storage stability.

Water solubility can be conferred by incorporating water-soluble polymer. Thus, polyethylene glycols confer water solubility or dispersibility to the alkyd which may be exploited in the manufacture of gloss paints with easy brush-clean properties (as well as other use as surfactant).

Alkyd emulsions may be prepared by emulsifying using surfactants; polymeric non-ionic surfactants give advantages over normal anionic surfactants, especially when they contain autoxidizable reactivity. All resin emulsification processes produce better results when using inversion, and in the case of alkyd emulsions, special techniques using the phase inversion temperature (PIT) have been developed. The various problems additional to stability, found in the use of alkyd emulsions have been discussed in depth. These include pigmentation, drier distribution, and system rheology. In the preparation of more stable emulsions, mixed ionic/non-ionic stabilization is now favoured.

To overcome deficiencies further, techniques have evolved to prepare alkyd emulsions with core/shell-like structures. One technique has been to use as a component, fatty acid grafted with copolymer rich in methacrylic acid. Another is to use emulsified alkyd as the seed for growth of an acrylic polymer shell.

Modified Alkyds

Urethane alkyds may be prepared by replacing part of the dibasic acid in the alkyd formulation with a diisocyanate such as toluene diisocyanate (TDI); the formulating calculations applicable are those used for alkyds. Long oil urethane alkyds are used in decorative paint formulations to impart greater toughness and quicker drying characteristics. While total replacement of dibasic acid is possible (thus producing a urethane oil), it is more normal to formulate an alkyd with a reduced dibasic acid content. This is processed to a low acid value, but high residual hydroxyl content and the thinned resin is then treated with diisocyanate. The isocyanate groups react with hydroxyl groups to form urethane links, but, unlike acids in ester formation, do so exothermically at lower temperatures and without by-product evolution. It is essential to complete the reaction of all isocyanate groups present because of the high toxicity associated with free isocyanates. In-process testing may be by titration with amine, with final verification being by infrared spectroscopy.

Small remaining residues of isocyanate may be removed by deliberate addition of low molecular weight alcohol. For the preparation of urethane alkyds, a monoglyceride catalyst should be chosen for the alkyd stage that does not promote unwanted isocyanate reactions, e.g. allophanate formation, and such catalysts as calcium oxide or soluble calcium soap should be used.

Another modification of long oil alkyd is that with polyamide resin, used to impart thixotropy in the final resin solution. Polyamide resins used are typically those derived from dimer fatty acid and a diamine. The physical properties of these alkyds are very delicately balanced. Such characteristics as gel strength, gel recovery rate, and non-drip properties of derived paints are intimately connected with both the composition and the processing of the alkyd, along with the proportion and type of polyamide and degree of reaction. If this type of alkyd is being prepared, or when formulating paint compositions containing such an alkyd, care should be taken to avoid the use of, or contamination with, polar, e.g. alcoholic, diluents; these destroy the rheological structure to which hydrogen bonding almost certainly provides a considerable contribution. Typical levels of polyamide modification are around 5%, and care in processing is necessary to stop the reaction in its final stage at such a point as to achieve clarity and the optimum and desired gel properties.

Silicone modification of long oil alkyds imparts much improved gloss retention for their use in structural maintenance and marine paints. The same hydroxy or methoxy functional silicone resins are applicable as used for modification of polyesters. In general, the degree of enhancement in properties is in proportion to the level of modification with these expensive materials, and formulators need to weigh carefully the balance of cost versus durability enhancement achieved. Phenol formaldehyde resins of both reactive and non-reactive oil-soluble types can be used as alkyd modifiers. They increase film hardness and improve resistance to water and chemicals, but cause yellowing. Such modified alkyds are useful in primer systems as cheaper but less chemically resistant alternatives to epoxy resins.

Vinyl modification of alkyds is used to impart faster dry, increase hardness, and give better colour, water, and alkali resistance. However, decreased gloss retention and poorer resistance to solvents result, the latter making formulation for early recoatability difficult to achieve. Vinylated alkyds are typically used for metal finishes. Styrene and vinyl toluene are the usual modifiers, methyl methacrylate being much less common. The modification stage involves free radical chain addition polymerization of the monomer in the presence of the alkyd under conditions where grafting to the unsaturated oil component of the alkyd is encouraged. This is typically by using at least some dehydrated castor oil in the alkyd composition since the conjugated fatty acid moieties then present encourage the necessary coreaction; the typical use of peroxide catalysts such as di-tert-butyl peroxide also encourage grafting by hydrogen abstraction.

A number of techniques for preparing vinylated alkyds are possible; most common is post-vinylation of the thinned alkyd by feeding monomer and initiator into the resin held at 150–160°C. Less common alternatives are to vinylate the oil prior to alkyd manufacture or to vinylate the monoglyceride before esterification with dibasic acid. These latter techniques can allow final dilution in low boiling or reactive solvents in which the vinylation reaction would have been impossible. Where difficulties in achieving adequate grafting are experienced, resulting in cloudy resin preparations, possibilities exist of including maleic anhydride or methacrylic acid in the alkyd polyester structure to provide additional grafting sites.

To the list of modifications must be included the possibility of precondensation with melamine/formaldehyde (MF) resins. Normally for a coating composition that

is multicomponent, viscosity will be minimized by keeping the components unreacted until the curing stages; however, for some systems, advantages of increased stability, compatibility, or improvement in cure properties are seen by partial prereaction, even as here where the two polymers are compatible. For low bake curing alkyd/MF systems, precondensation of alkyd and MF can enhance final cure properties, while reducing the tendency of the MF component to self-condense when highly acidic catalyst systems are used.

PHENOLIC RESINS

Synthetic phenolic resins were developed and commercialized in the early 1900s by Leo Baekland. The reaction of phenol and formaldehyde produces a product that forms a highly cross-linked three-dimensional polymer when cured. The resins have found use in various applications in the coatings industry because of their excellent heat resistance, chemical resistance, and electrical properties. They also offer good adhesion to many substrates and have good compatibility with other polymers.

Phenolic resins have two basic classifications: resoles and novolaks. Resoles, or heat-reactive resins, are made using an excess of formaldehyde and a base catalyst. The polymer that is produced has reactive methylol groups that form a thermoset structure when heat is applied.

Novolaks are made using an excess of phenol and an acid catalyst. Reaction occurs by the protonation of the formaldehyde, and the intermediate is characterized by methylene linkages rather than methylol groups. These products are not heat reactive, and they require additional cross-linking agents such as hexamethylenetetramine to become thermosetting. These reactions can be thought of as nucleophilic attack of phenoxide ion on the formaldehyde, or electrophilic substitution by protonated formaldehyde on the aromatic ring.

Both polymerization reactions evolve water during cure. This condensation reaction serves to limit film thickness to approximately 3 mils, because the volatiles will cause blistering while curing takes place. Baking temperatures are generally 300 to 400° F with a bake time of 10 to 30 min.

Phenolic resin polymers have been used in the coatings industry for many years because of their excellent performance properties and their relatively low cost. Some of the major applications are in rigid packaging, maintenance primers, printing inks, and epoxy hardeners. A brief discussion of these applications follows.

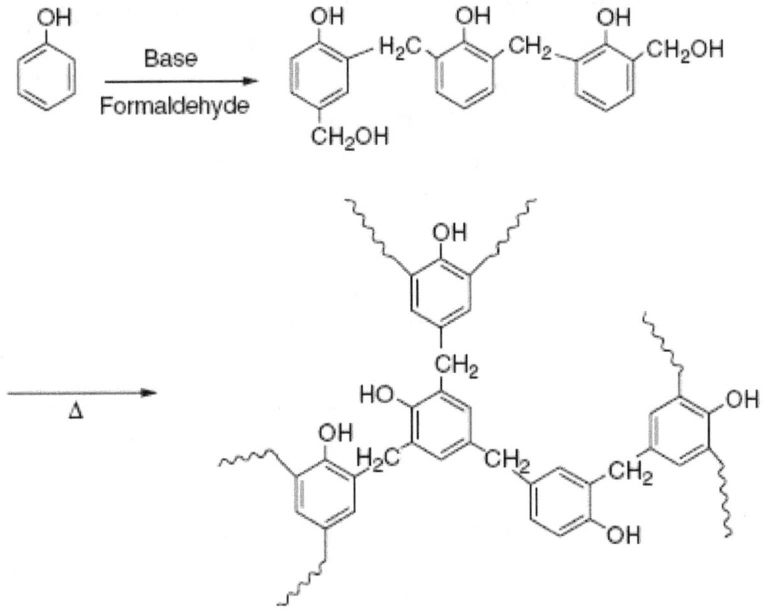

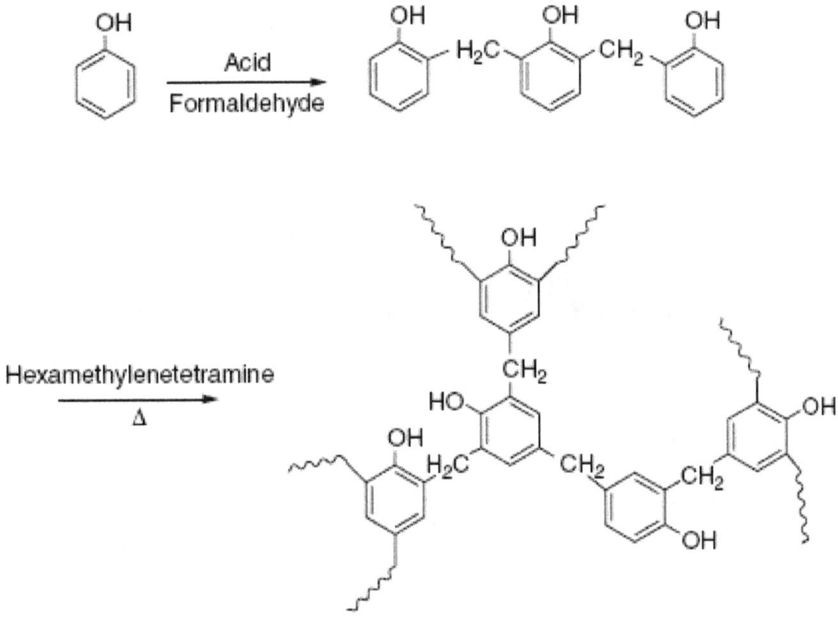

Rigid Packaging

Phenolic resins have found use in metal containers because they offer outstanding corrosion protection and excellent adhesion, along with chemical and solvent resistance. Typical end-use applications are food container coatings, drum and pail lining, pipe coatings, tank linings, and coil coating primers, or certain polyester resins.

Unmodified phenolics are often used in drum, pail, and pipe coatings. Since the substrate is usually rigid, the brittleness associated with these phenolic formulations does not present a problem. The excellent solvent resistance of phenolic resins makes them ideal for this application. A film thickness of 0.2 to 0.3 mils often provides sufficient protection. Thicker films can also be used if coatings are applied in several layers to prevent blistering during cure.

Bake schedules generally range from 300 to 400 ° F with times varying from 10 min to several hours. The coatings may be clear or pigmented with iron oxide. A typical formulation is listed in Table .1. Formulations may be applied by spray, brush, or roller coating. The requirements of food can linings are quite different from those of drum coatings. Food can coatings must have sufficient flexibility to survive any impact that could fracture the coating and thus expose the metal surface to the contents of the can.

The phenolic resins are based on blends of phenol and cresols that offer varying degrees of flexibility and chemical resistance. Resins based on bisphenol are used in critical taste applications. High molecular weight epoxies are used as modifiers to further enhance the flexibility of the cured films.

Another consideration for food can formulations is sulfur staining. Foods high in protein such as meats and fish contain sulfur, which may permeate to the substrate and cause discoloration. Phenolics, because of their high degree of cross-linking, offer outstanding resistance to sulfur-containing foods. A food container formulation is listed in Table .2.

Table 1 — Drum and Pail Lining

Ingredients	Parts by Weight
Phenolic resin	73.30
Acetoneb	7.00
Methyl ethyl ketone	9.00
Ethanol	55.00
Propyl Propasol solvent	14.00
Cyclohexanone	15.00
Leveling agent	0.05

Table 2 — Food Container Formulation

Ingredients	Parts by Weight
Phenolic resin*	22.90
Epoxy resin	34.00
n-Butanol	11.30
Methyl Propasol acetate	11.30
85% Phosphoric acid	0.25
Leveling agents	1.00

Maintenance Primers

Non-heat-reactive phenolics are used in combination with drying oils or alkyds to produce air drying maintenance coatings. The phenolic resins supply solvent and moisture resistance, while the oils or alkyds impart flexibility and good filming characteristics.

In the past, parasubstituted phenolic novolak resins were actually cooked with unsaturated oils such as linseed or castor oil. The mixtures were boiled at 400 to 500° F until a thick resinous solution was produced. The reaction mechanism is not completely understood, but it is believed that quinone methylide structures in the phenolic resins react with the double bond in the oil through a Diels-Alder type of reaction.

The resin solution is then let down in mineral spirits or aromatic solvents along with metal driers such as cobalt or manganese. Usually a combination of driers gives the best overall performance.

The result is an air dry finish that has excellent weathering resistance. Today, it is increasingly popular to simply cold-blend phenolic resins based on p-tert -butyl phenol along with solvents, driers, and highly reactive oils such as tung oil. Coating properties are excellent, and the procedure takes less energy and time.

Printing Inks

Phenolic resins are used in numerous printing ink formulations. Lithographic, letterpress, and gravure inks are oleoresinous and have phenolic resins added to the ink systems to improve film hardness and gloss. Parasubstituted phenolic novolak resins that have been modified with rosin make excellent ink additives. Rosin modification of the phenolic resin raises the melting point and gives the phenolic excellent oil solubility. The formulations consist of a phenolic resin, oils, pigments or dyes, driers, and lubricants or plasticizers. The proper balance of ingredients gives the desired combination of hardness, viscosity, penetration, and drying rate.

EPOXY HARDENERS

Phenolic resins may be combined with epoxy resins for use in protective coatings. Phenolic–epoxy products are also used in laminates, prepreg manufacturing, molding materials, and electrical insulation coatings. The phenolic resin is used as a coreactant to produce thermoset systems with improved heat and chemical resistance. Non-heat-reactive (novolak) resins are used to cross-link the epoxies. The epoxy resins are typically epoxy–phenolics or bisphenol A-based resins.

The reaction mechanism is different from the resole–epoxy reactions. The coating is heat-activated and uses a base catalyst such as an amine, dicyandiamide, or an imidazole. The phenolic hydroxyl group reacts with the epoxy group to form a polyether structure, which has an advantage, because no volatiles are released during cure. This allows for thick films to be produced with low shrinkage and no voids from volatile emission. The phenolic- epoxy reaction using a base catalyst can be demonstrated as follows:

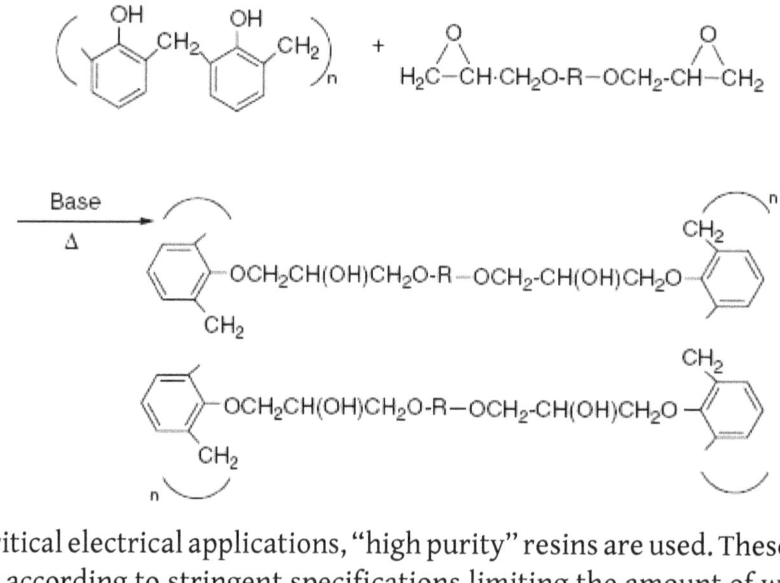

For critical electrical applications, "high purity" resins are used. These products are made according to stringent specifications limiting the amount of water, ions, and free monomers present in the resin.

Summary

Phenolic resins were one of the first synthetic polymers to have widespread commercial importance. Their outstanding performance properties have given them a permanent role in the coatings industry. They are used in applications ranging from railroad tank cars to carbonless copy paper. Phenolic resins are also used on a wide variety of substrates including metal, wood, paper, and ceramics. There are so many types of phenolic resin available to the marketplace that a particular resin can be selected for virtually any application.

Phenolic resins should not be overlooked when choosing a high performance polymer. Phenolics currently do not receive the same attention as some of the more recently developed polymers, but for many applications, there is simply no substitute for phenolic resins. Phenolics will continue to have an important role in the coating industry because of their versatility, coatings properties, and reasonable price.

ETHYLENE VINYL ALCOHOL COPOLYMER (EVOH) RESINS

Polymer

Ethylene vinyl alcohol (EVOH) resins are hydrolyzed copolymers of vinyl acetate and ethylene. The vinyl alcohol base has exceptionally high gas barrier properties, but it is water soluble and difficult to process. By copolymerizing ethylene with vinyl alcohol, the high gas barrier properties are retained and significant improvements are achieved in moisture resistance and processibility. Under proper conditions, this reaction yields a copolymer that is more than 99% hydrolyzed. Table .1 lists the range of EVOH resins presently available.

Barrier Properties

EVOH copolymers are highly crystalline, and their properties are highly dependent on the relative concentration of the comonomers. Generally speaking, as the ethylene content increases, the gas barrier properties decreases, the moisture barrier properties improve, and the resins are processed more easily (see Table .2).

Table 1 — Properties of Ethylene Vinyl Alcohol Resins

Property	Range
Melt index	0.7–15.0 g/10 min
Density	1.12–1.21 g/cm^3
Crystalline temperature	134–167°C
Ethylene content	27–48 mol%

Table 2 — Barrier Properties

EVOH (mol% Ethylene)	Gasesa				
	O_2	CO_2	N_2	He	Moisture[b]
27	0.01	0.04		7.7	6.8
32	0.02	0.05	0.002	15.5	3.8
38	0.03	0.10	0.003	25.5	2.1
44	0.07	0.20	0.005	40.0	1.4
48	0.11	0.32	0.007	52.0	1.4

Table 3 — Oxygen Transmission (O 2 TR) Rate versus Barrier Layer Relative Humidity

Film Structure				Inside Pachage			
				Outside (65% RH)		Outside (85% RH)	
Outside: 0.8 mil	Barrier: 0.6 mil	Inside 2 mils	Temperature (°C)	RH of Barrier (%)	O_2 TR[b] (Barrier %)	RH of Barrier (%)	O_2 TR[b] (Barrier %)
PP	27 mol% EVOH	LLDPE	5	79	0.03	88	0.05
PP	27 mol% EVOH	PP	20 5	75	0.04 0.02	86	0.10 0.04
PET	27 mol% EVOH	PP	20 5	72	0.03 0.02	84	0.07 0.04
Nylon	32 mol% EVOH	LDPE	20 5	67	0.03 0.01	81	0.07 0.02
PP	32 mol% EVOH	LDPE	20 5	79	0.03 0.05	88	0.05 0.05
PS	44 mol% EVOH	LDPE	20 5	70	0.04 0.05	83	0.12 0.05
			20		0.17		0.27

[a] PP, polypropylene; PET, polyethlene terephthalate; PS, polystyrene.

Table 4 — Flavor Barrier Data (Citrus and Tropical Flavors)

Film[b]	Thickness (μm)	Duration					
		1 h	2 h	15 h	1 Day	4 Days	35 Days
EVOH-32	15	A	A	A	A	A	A
EVOH-44	15	A	A	A	A	A	A
BO EVHO	1	A	A	A	A	A	A
BO Nylon	515	A	A	B	B	B	B
OPP	20	B	B	C	C	C	C
PVDC	25	A	A	B	B	B	B
LDPE	50	D	D	D	D	D	D

[a] Key : A, no detection; B, faint flavor; C, partial flavor; D, flavor clearly distinguished.

Table 5 — Flavor, Aroma, and Solvent Barrier Properties

Resin[b]	Allyl Sulfide: Garlic-Food Type (Croutons, Snacks, Salad Dressings)	Acetic Acid Vinegar-Food Type (Cheddar Cheese, Snacks, Condiments)	Ethylene Acetate: Laminating Adhesive Solvent Residual	Toluene: Printing Ink Solvent Residual	Methyl Ethyl Ketone Printing Ink Solvent Residual
HDPE-Nylon-Eva	0.00008	0.92	0.03	0.02	0.005
HDPE-EVOH-EVA	0.000075	0.035	0.0043	0.002	0.035
PVDC-PP-PVDC	0.0068	1.98	0.34	0.22	3.09
OPP-HDPE-EVOH-EVA	0.00076	1.40	0.15	0.00003	0.09
EVA-Glassine-PVDC	0.50	4.18	0.47	3.15	15.1

In addition to being outstanding as barrier properties, EVOH resins offer excellent barriers to a variety of flavors, aromas, and solvents (see Table .4 and Table .5).

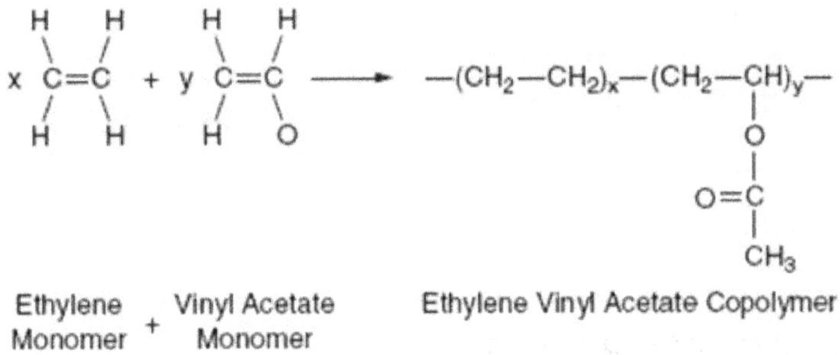

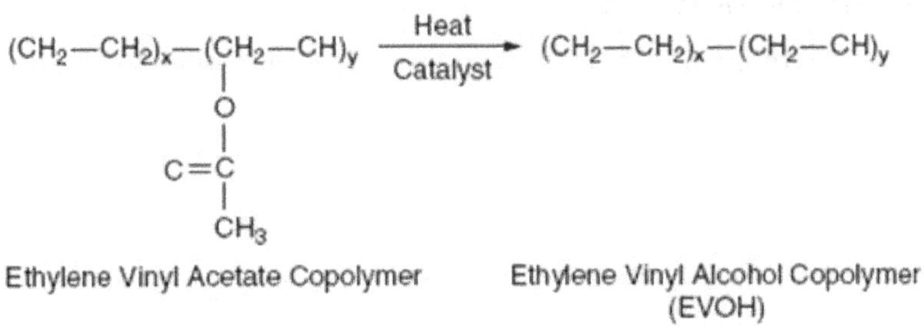

It is these exceptional barrier properties that permit the use of extremely thin layers of EVOH resins in coextrusion-coated structures for gable top drink cartons, single-serve drink cartons, dry mixes, etc. In addition to excellent barrier properties, EVOH resins have very good mechanical and thermal properties (see Table .6).

Table 6 — Properties of Three Commerical Resins

Property	Eval EVOH Resion Grade		
	F101	H101	E101
Maximum service temperature °C	240	240	250
Melting point °C	181	175	164
Glass transition temperature °C	69	62	55
Tensile strength, psi	11,220	9,400	8,520
Tensile yield strength, psi	10,365	8,300	7,385
Elongation at break, %	230	280	280
Young's modulus, psi	385,000	345,000	298,000
Ethylene content, mol%	32	38	44

Table 7 — Barrier Applications

Febrication Process	Application	Structure[1]
Lamination or extrusion coating	Food kits, pouches, dry mixes, snacks	BOB/barrier/LLDPE Nylon/PET/barrier/LDPE
Coextrusion coating	Aseptic packaging Juices and other drink products	OPP/barrier/LDPE Paper/HDPE/barrier/EVA LDPE/Paper/barrier/PP/Surlyn LDPE/Paper/LDPE/BARRIER LDPE/Paper/LDPE/barrier LDPE/Barrier

POLYVINYL ACETAL RESINS

Chemistry and Manufacture

The polyvinyl acetals are a family of high molecular weight polymers prepared by the mineral acid catalyzed acetalization of polyvinyl alcohol. The chemistry of preparation is outlined in Equation .1,where R can be H or any of a wide variety of organic radicals.

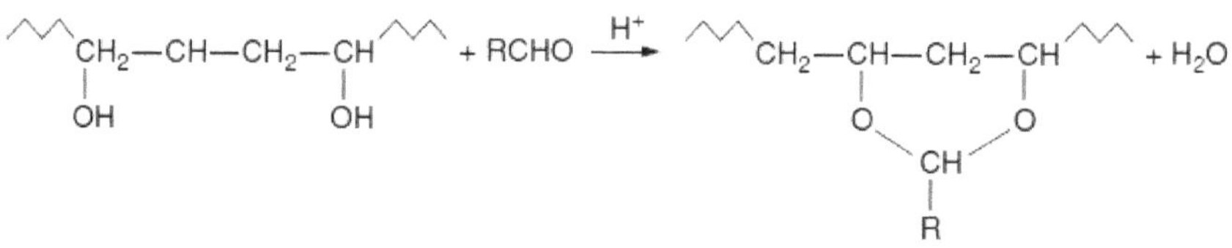

While many polyvinyl acetals have been prepared, the only commercially important polymers are the formal (R = H) and the butyral (R = n – C 3 H 7), derived from the acetalization of polyvinyl alcohols with formaldehyde and n

-butyraldehyde, respectively. The generalized structure of the polyvinyl acetal molecule is shown in Figure .1. Molecular weights, weight average molecular weight (Mw), range from about 25,000 to 150,000 for the formal and from 40,000 to 250,000 K for the butyral, depending on grade. The three chemical moieties are randomly distributed along the polymer chain. While simple in concept, the actual commercial preparation of these polymers is quite complex, requiring many processing steps. For the formal and the majority of the butyral resins currently of importance in the coating industry, these steps include extensive solvent and by-product purification and recovery operations. A schematic process flow diagram for the manufacture of a typical polyvinyl acetal, polyvinyl butyral (PVB), is shown in Figure .2.

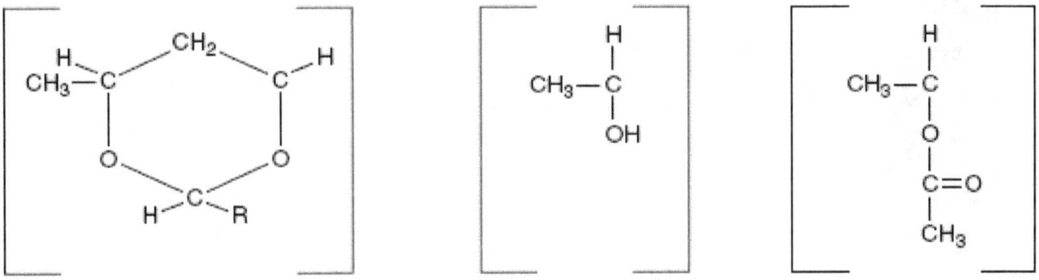

Figure — 1 Generalized structure of the polyvinyl acetyl molecule.

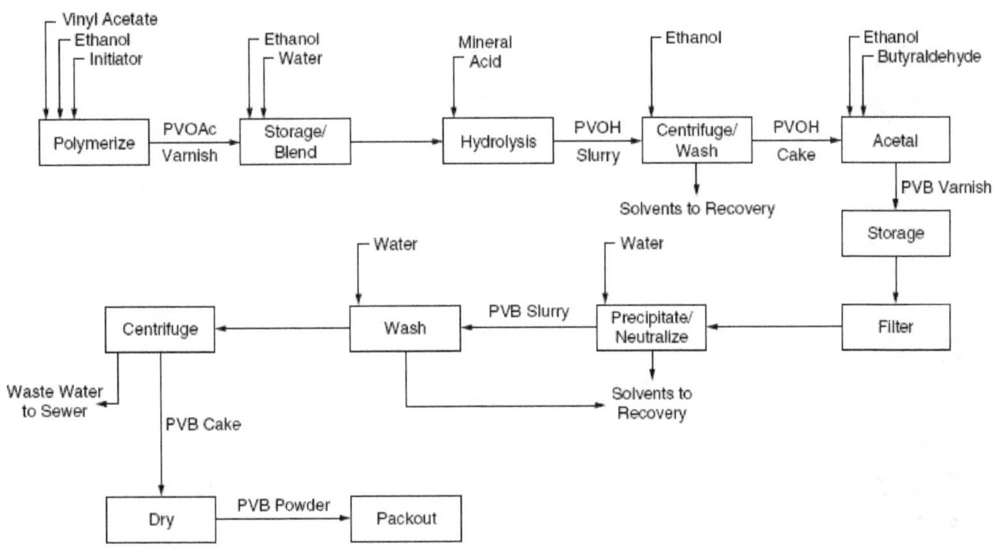

Figure 2 — Process flow diagram of a typical polyvinyl acetyl, polyvinyl butyral molecule.

Table 1 — Typical Physical and Chemical Properties of Butvar Polyvinyl Butyral Resin

	Butvar Resins							
	ASTM	White, Free- Flowing Powder						
Properties	Method[2]	B-72	B-74	B-73	B-73	B-79	B-90	B-98
Volatiles, maximum as packed,%		3.0	3.0	3.0	5.0	5.0	5.0	5.0
Molecular weight (weight average $\times 10^{-3}$)	(1)	170-250	120-150	90-120	90-120	50-80	70-100	40-70
Solution viscosity15% by wt, mPa.s	(2)	7000-14000	3000-7000	1000-4000	500-1000	100-400	600-1200	200-400
10% by wt, mPa.s	(2)	1600-2500	800-1300	400-1000	200-450	75-200	200-400	75-200
Specific gravity 23°/23°(±0.002)	D792-50	1.100	1.100	1.100	1.083	1.083	1.100	1.100
Burning rate, cm/min	D635-56T	2.5	2.5	0.5	2.5	2.3	0.9	2.0
Refractive index (±0.0005)	D542-50	1.490	1.490	1.490	1.485	1.485	1.490	1.490
Water absorption (24 h), %	D570-59aT	0.5	0.5	0.5	0.3	0.3	0.5	0.5
Hydroxyl content, as % polyvinyl alcohol	D1396-58	17.5-20.0	17.5-20.0	17.5-20.0	11.0-13.0	10.5-13.0	18.0-20.0	18.0-20.0
Acetate content, as % plyvinyl acetate	D1396-58	0-2.0	0-2.5	0-2.5	0-1.5	0-1.5	0-1.5	0-2.5
Butyral content, as % polyvinyl butyral, approx	80	80	80	80	80	80	80	80
Chemical[b]								
Resistance to								
weak acids	D543-56T	E	E	E	E	E	E	E
Strong acids	D543-56T	E	E	E	E	E	E	E
Waak bases	D543-56T	E	E	E	E	E	E	E
Strong bases	D543-56T	E	E	E	E	E	E	E

	Butvar Resins							
	ASTM	White, Free- Flowing Powder						
Properties	Method[2]	B-72	B-74	B-73	B-73	B-79	B-90	B-98
Organic solvents								
Alcohols	D543-56T	P	P	P	P	P	P	P
Chlorinated	D543-56T	G	G	G	F	F	G	G
Aliphatic	D543-56T	E	E	E	F	F	G	G
Aromatic	D543-56T	F	F	F	P	P	F	F
Esters	D543-56T	F	F	F	P	P	F	F
Ketones	D543-56T	F	F	F	P	P	F	F

Properties

Reactivity and Compatibility

In general, the polyvinyl acetals are characterized by toughness and excellent adhesion to a wide variety of surfaces. They are resistant to most nonpolar solvents and to attack by both acids and bases, although they are slowly attacked by strong aqueous acids. Their resistance to hydrocarbons and mineral oils is outstanding. Because of the residual hydroxyl groups in the polymers (Figure .2), they are readily cross linked with a variety of widely available cross-linking reagents that react with hydroxyl.

Examples of the more common cross-linking reagents are dialdehydes and phenolic, amino, isocyanate, and epoxy functional resins. The butyral shows limited compatibility with other resins (e.g., alkyd, polyvinyl chloride, silicones, urea- and melamine-formaldehyde, cellulose acetate butyrate, and ethyl cellulose) and excellent compatibility with nitrocellulose, epoxy, phenol-formaldehyde, isocyanate, and some rosin ester derivatives. The formal, being more polar than the butyral, shows more limited compatibility with other resins. It is fully compatible with most common isocyanate and epoxy resins, and shows limited compatibility with alkyd, phenolic, melamine- and urea-formaldehyde resins, and silicones.

Physical and Chemical Properties

The physical and chemical properties of Butvar resins are shown in Table .1, and the mechanical, thermal, and electrical properties are given in Table .2. Table .3 gives solubility data for the Butvar resins in a variety of solvents. The physical and chemical properties of the Formvar resins are shown in Table .4;the mechanical, thermal, and electrical properties are listed in Table .5. Solubility data for the Formvar resins are given in Table .6. Properties of Butvar dispersion BR are shown in Table .7. The data given in Table .1 through Table .7 are typical properties, not the manufacturer's specifications. For actual product specifications, the manufacturer should be consulted.

The polyvinyl formal grades are all chemically identical; they differ only in molecular weight. The polyvinyl butyrals fall into two chemical groups, high (17.5 to 21 wt%) residual polyvinyl alcohol, and low (10.5 to 3 wt%) polyvinyl alcohol. Within each of these groups, molecular weight is also varied. In general, as the percentage of polyvinyl alcohol increases, toughness, modulus, and adhesion to polar surfaces will increase, while the range of useful solvents decreases (conversely, solvent resistance increases). As the percentage of polyvinyl alcohol decreases, the resins become more broadly soluble, softer, and lower in glass transition temperature.

Table 2 — Typical Mechanical, Thermal, and Electrical Properties of Butvar Polyvinyl Butyral Resin

Properties	Method[2]	Butvar Resins						
		B-72	B-74	B-73	B-73	B-79	B-90	B-98
Mechanical[c]								
Tensile Strenght, MPa								
Yeild	D638-58T	47-54	47-5452	45-52	40-47	40-47	43-50	43-50
Break	D638-58T	48-55	48-55	41-48	32-39	32-39	49-46	39-46
Elongation,%								
Yeild	D638-58T	8	8	8	8	8	8	8
Break	D638-58T	70	75	80	110	110	100	110
Modulus of elasticity (apparent)	D638-58T	2.3-2.4	2.3-2.4	2.2-2.3	1.9-2.0	1.9-2.0	2.1-2.2	2.1-2.2
GPa								
Flexural strenght, yeild, MPa	D790-59T	83-90	83-90	79-86	72-79	72-79	76-86	76-83
Hardness , Rockwell								
M	D785-51	115	115	115	100	100	115	115
E	D785-51	20	20	20	5	5	20	20
Notched Izod Impact strenght(1.25×1.25cm), j/m	D256-56	59	59	55	43	43	48	37
Thermal (°C)								
Flow temperature	D569-59	145-155	135-145	125-130	110-115	110-115	125-130	105-110
Glass temperature		78-78	72-78	72-78	62-72	62-78	72-78	72-78
Heat distortion temperature	D648-56	56-60	56-60	55-59	50-54	52-54	52-56	45-55
Heat-sealing temeprature[b]		105	105	99	93	93	205	200

Electrical								
Dielectric contant								
50Hz	D150-59T	3.2	3.2	3.0	2.7	2.7	3.2	3.3
1kHz	D150-59T	3.0	3.0	2.7	2.6	2.6	3.0	3.0
1MHz	D150-59T	2.8	2.8	2.6	2.6	2.6	2.8	2.8
10MHz	D150-59T	2.7	2.7	2.5	2.5	2.5	2.7	2.8
Dissipation factor $\times 10^3$								
50Hz	D150-59T	6.4	6.4	5.8	5.0	5.0	6.6	6.4
1kHz	D150-59T	6.2	6.2	5.5	3.9	3.9	5.9	6.1
1MHz	D150-59T	27	27	22	13	13	22	23
Dielectric strenght, V/m (3.2mm thickness)								
Short time	D149-59	17	17	19	19	19	18	16
Step by step	D149-59	16	16	16	15	15	15	15

Table 3 — Solubility of Butvar Resins

Solvant	Butwar solution Agitated at Room Temperature for 24 h		
	5% Solid Solution	10% Solid Solution	
	B–72, B–73, B–74	B–76, B–79,	B–90, B–98
Acetic acid (glacial)	S	S	S
Aceton	I	S	SW
Butyl acetate	I	S	PS
n-Butyl alcohol	S	S	S
Butyl cellusolve	S	S	S
Cyclohexanone	S	S	S
Diacetone alcohol	PS	S	S
Diisobutyl ketone	I	SW	I
N,N-Dimethylacetamide	S	S	S
N,N-Dimethylformamide	S	S	S
Dimethyl sulfoxide	S	S	S
Ethyl acetate, 99%	I	S	PS
Ethyl acetate, 85%	S	S	S
Ethyl alcohol, 95%, or anhydrous	S	S	S
Ethylene dichloride	SW	S	SW
Ethylene glycol	I	I	I
Isophorone	PS	S	S
Isopropyl alcohol, 95%, or anhydrous	S	S	S
Isopropyl acetate	I	S	I
Methyl acetate	I	S	PS
Methyl alcohol	S	SW	S
Methyl ethyl ketone	SW	S	PS
Methylene chloride	PS	S	S
Methyl isobutyl ketone	I	S	I
Naphtha (light solvent)	I	SW	I
N-Methyl-2-pyrrolidone	S	S	S
Propylene dichloride	SW	S	SW
Tetrachlorethylene	SW	SW	SW
Tetrahydrofuran	S	S	S
Toluene	I	PS	SW
Toluene/ethyl alcohol (95%) (60:40 by weight)	S	S	S
1,1,1-Trichloroethane	SW	S	SW
Xylene	I	PS	SW

[a] Key: S, Completely soluble; PS, partially soluble; I, insoluble; SW, swells (hazy, turbid).

Within a given chemical group, increasing molecular weight results in a modest increase in tensile strength and modulus, increased solution viscosity, and a modest increase in solvent and chemical resistance. In general, even the lowest molecular weight grades of these reins are high enough in molecular weight to be well within the chain entanglement region of the E' spectrum such that physical properties do not change drastically as molecular weight increases. Molecular weight variations can be used effectively to control the rheology/viscosity of coatings formulations more than the properties of the final coating.

Table 4 — Typical Physical and Chemical Properties of Formvar Polyvinyl Formal Resin

Properties	ASTM Method	Farmvar Resins 5/95E	6/95E	7/95E	15/95E
	Physical				
Form					
Volatiles, maximum, %		2.2	2.2	2.2	2.2
Molecular weight × 10^{-3} (weight average)	(1)	25-35	35-45	40-60	70-150
Solution viscosity 15% by wt, MPa +8	(2)	140-280	250-500	325-675	1800-3500
Resin viscosity	(3)	8-12	12-15	15-20	37-53
Specific gravity, 23°/23° (± 0.002)	D792-505	1.227	1.227	1.227	1.227
Burning rate, cm/min	D635-56T	2.0	2.3	2.3	2.5
Refractive index (± 0.0005)	D542-50	1.502	1.502	1.502	1.502
Water absorption (24hs), %	D570-59aT	1.2	1.2	1.2	1.2
Hydroxyl content, as % polyvinyl alcohol	D1396-58	5.0-6.5	5.0-6.5	5.0-6.5	5.0-6.5
Acetate content, as % polyvinyl acetate	D1396-58	9.5-13.0	9.5-13.0	9.5-13.0	9.5-13.0
Formal content, as % polyvinyl formal, approx		82	82	82	82
	Chemical[b]				
Resistance to	D543-56T	E	E	E	E
Weak acids	D543-56T	E	E	E	E
Strong acids	D543-56T	E	E	E	E
Weak bases	D543-56T	E	E	E	E
Wtrong bases					
Organic solvents	D543-56T	G	G	G	G
Alcohols	D543-56T	P	P	P	P
Chlorinated	D543-56T	E	E	E	E
Aliphatic	D543-56T	G	G	G	G
Aromatic	D543-56T	G	G	G	G
esters	D543-56T	G	G	G	G
Ketones	D543-56T	G	G	G	G

Solution Viscosity

Solution viscosity is a function of chemical composition, molecular weight, and solvent composition. Within a given chemical class, solution viscosity in a given solvent will increase with molecular weight. The effect of solvent composition on solution viscosity is more complex and not within the scope of this discussion.

Solvent systems are often chosen for a variety of reasons other than viscosity: drying rate, toxicity and other environmental concerns, utility with other resins needed in a formulation, etc. Usually by judicious experimentation, a compromise solvent system can be found that will give a reasonable mix of the required properties.

Plasticizers

Plasticizers are often used to soften the films prepared from the polyvinyl acetals. Among the suggested plasticizers for polyvinyl butyral are butyl benzyl phthalate, 2-ethylhexyl diphenyl phosphate, dihexyl adipate, and a verity of other phosphate, phthalate, adipate, sebacate, and ricinoleate esters. Rosin-based polyester, and blown linseed oil plasticizers are also used.

Table — 5 Typical Mechanical, Thermal, and Electrical Properties of Formvar Polyvinyl Formal Resin

Properties	ASTM Method	Farmvar Resins 5/95E	6/95E	7/95E	15/95E
	Mechanical				
Tensile strengh, MPa	D638–58T				
Yield	D638–58T	59–66	59–66	59–66	59–66
Break		52–59	52–59	52–59	52–59
Elongation, %					
Yield	D638–58T	7	7	7	7
Break	D638–58T	50	50	50	50
Modulus of elasticity (apparent), GPa	D638–58T	2.7–3.1	2.7–3.1	2.7–3.1	2.7–3.1
Flexural strenght, yield, Mpa	D790–59T	117–124	117–124	117–124	117–124
Hardness, Rockwell					
M	D-785–51	150	150	150	150
E	D-785–51	65	65	65	65
Notched Izod impact strength (1.25 cm × 1.25 cm), J/m	D256–59	70	70	70	70
Thermal (°C)					
Flow temperature, 6.9 MPa	D569–59	140–150	140–150	140–150	160–170
Glass temperature	b	103–113	103–113	103–113	103–113
Heat distortion temperature	D648–56	83–87	58–90	85–90	87–93
Heat-sealing temperature	c	96	96	99	107

Properties	ASTM Method	Farmvar Resins 5/95E	6/95E	7/95E	15/95E
	Mechanical				
	Electrical				
Dielectric constant					
50 Hz	D150–59T	3.2	3.2	3.2	3.4
1 kHz	D150–59T	3.3	3.3	3.3	3.0
1 MHz	D150–59T	3.1	3.1	3.1	2.8
10 MHz	D150–59T	3.0	3.0	3.0	2.8
Dissipation Factor $\times 10^3$					
50 Hz	D150–59T	8.1	8.1	8.1	8.7
1 kHz	D150–59T	10	10	10	10
1 MHz	D150–59T	2	21	21	21
10 MHz	D150–59T	19	19	19	18
Dielectric strengh (3.2 mm thickness), V/m					
Short time	D149–59T	24	13	13	12
Step by step	D149–59T	12	12	12	13

Table 6 — Properties of Butvar Dispersion BR

Form	% by weight
Property	Milk-white aqueous dispersion of plasticized polyvinyl butyral
Total solids	50.0-52%
Viscosity	500-1500 mPa
PH	8.0-10.5
Particle size	Most particles between 0.25 and 1.5 μ
Particle charge	Anionic
Plasticizer content	40 parts per 100 parts of resin (28.6% of solids)
Pounds per gallon at 25°C	8.4
Grams per liter	1008

The inherent properties of adhesion to a wide variety of surfaces, film toughness and chemical/solvent resistance, and film clarity of the polyvinyl butyral resins makes them the vehicle of choice in a wide variety of specialty coating applications. They adhere tenaciously to most polar surfaces: wood, glass, metals, ceramics, pigments, etc. Their high binding efficiency allows them to be used at very high pigment loadings. Ceramic films are typically cast in thicknesses from fractions of a mil to several millimeters,using 4 to 5% Butvar B-76 as the sole binder ; most permanent coatings applications will require considerably higher binder levels than this for adequate film strength. Polyvinyl butyral resins have been used as the binder of choice in thermophotographic and photographic coatings, in photoconductive coatings, in electrophotographic coatings, and as a coating on optical recording disks, all of which depend on film toughness, binder efficiency, and optical clarity. Their toughness, chemical and heat resistance, and binding efficiency render them useful in powder coatings and photothermographic coating applications.

SURFACE COATING APPLICATIONS

Polyvinyl Butyral

Polyvinyl butyrals are used extensively as wood coatings, where their resistance to natural wood oils makes them a primary choice for sealers and wash coats. An example of an application on the Western Pine Association Knot Sealer number WP578 is given in Table 15.8. A well-known application of polyvinyl butyrals is in the manufacture of wash primers for the priming of metal surfaces to be used in hostile environments (e.g., the hulls of naval vessels). There are a number of formulations available, both singleand two-package systems. Another example of a metal coating based on polyvinyl butyral is Metal Coating 2009, described in Table. Other coating applications of the polyvinyl butyrals include solder masks for printed circuit board manufacture, heat-fusible wire coatings, and zinc oxide-based photosensitive paper coatings; they also serve as the binder/vehicle for iron oxide in the production of magnetic recording tapes. They are widely used as toughness/flexibility/adhesion promoters in the production of inks for letterpress, flexographic, and gravure printing, and as a component in toners for reprography.

Table .7 — Western Pine Association Knot Sealer, WP578: Brush Application

Material	% by weight
Butvar B-90) Monsanto Chemical Co.)	3.3
Durite P-97) Borden Chemical Co.)	40.0
SDA 35A, 95% ethan	56.7 100.0

Polyvinyl Butyral Dispersions

The dispersion of plasticized polyvinyl butyral in water, marked, by Monsanto as Butvar dispersion BR, is widely used as a permanent surface size in critical textile applications, e.g., seat belt and parachute webbing, where the toughness of the butyral film lends outstanding abrasion resistance to the fabric. The relatively high surfactant levels used in the manufacture of this dispersion reduce the inherent adhesion of the resin to most highly polar surfaces such as metals and glass. This property is used to advantage in the preparation of removable coating for temporary protection of sensitive surfaces. This dispersion is typically formulated with other dispersions/latex as extenders to reduce cost. Pigments and plasticizers can be incorporated, as can small amounts of solvents to enhance adhesion properties.

Table .8 — Metal Coating 2009: Spray or Roller Application

Material	% by weight
Diacetone alcohol	17.4
n-Butanol	17.4
SDA-35A, 95% ethanol	7.7
Xylene	34.7
Methylon 75-108 (Bakelite Thermosets, Ltd.)	5.1
Epon 1007 (Shell Chemical Corp.)	13.0
Butvar B-90 (Monsanto Chemical Co.)	2.0
10% Phosphoric acid (diluted with a portion of the given solvent mixture)	2.7
	100.0
Cure cycle: 15 minute air dry + 330 minutes at 190°F and 20 minutes at 400°F	

Polyvinyl Formal

Polyvinyl formal, available in several viscosity grades from Monsanto as Formvar, is widely used as an oil-resistant insulating coating for magnetic wire. For this application, it is normally cross-linked by formulating with phenolic, epoxy, or urethane resins to enhance properties. These coatings are tough, strongly adhering, abrasion resistant, and totally impervious to hydrocarbon oils and lubricants. Polyvinyl formal serves as the base vehicle for coatings for printed circuit boards, both permanent and fugitive (e.g., solder masking), for the surface treatment of glass fibers and boron filaments (abrasion resistance), and as an anti fog coating on glass. It finds application in photo thermo graphic and electrographic coatings.

It is used as a binder in magnetic tape coatings, particularly for the high performance chromium oxide coated tapes. It finds use as an ink binder and as a lacquer to protect the surface of lithographic plates. In general, polyvinyl formal is higher in modulus and less susceptible to solvent attack than is polyvinyl butyral. Conversely, solvent choices are more limited for the formal. Films are somewhat yellow, reducing their utility in applications calling for a colorless film.

Polyimides

Polyimides are condensation polymers that contain the imide structure –CO–N–CO– as a linear or heterocyclic unit along the polymer main chain. They exhibit exceptional thermal, thermo oxidative, and chemical resistance, and good radiation resistance and dimensional stability, while maintaining an excellent balance of mechanical and electrical properties.

Chemistry and Properties

Aromatic polyimides are generally produced by a two-step polycondensation reaction of aromatic dianhydride with either aromatic diamine or aromatic diisocyanate in a suitable reaction medium. They have the following general structure:

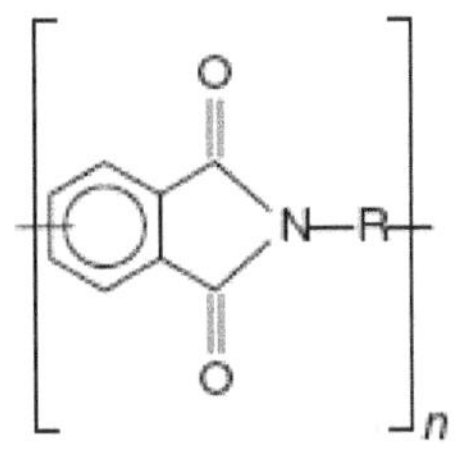

The direct production of high molecular weight aromatic polyimides in a one-step polymerization could not be accomplished because the polyimides are usually insoluble and intractable. The polymer chains precipitate from the reaction media (whether solution or melt) before high molecular weights are obtained. Therefore, processing of the aromatic polyimides can be accomplished only with the first step intermediate amic acid varnish, while it is still soluble and fusible. Were it not for the processing difficulties associated with known polymers, polyimides would be enjoying success in many more new application areas. This processing problem was partially overcome by the development of copolymers. The two major commercial polyimide copolymers are an amide-imide known as Torlon, a product of Amoco, and an ether-imide produced by General Electric under the trade name Ultem. Another approach to this processing problem was dealt with by incorporating a soft aromatic segment and/or aliphatic moiety in the polymer main chain. Among the various efforts to overcome the processing difficulties, one commercial success was achieved by incorporating a totally asymmetric diamionophenylindane isomeric mixture into the polyimide backbone; it is marketed as Matrimid 5218 from Ciba-Geigy. This material is soluble in relatively nonpolar solvents and is characterized also by exceptionally high glass transition temperature and high thermoxidative stability.

Key properties of aromatic polyimides are their outstanding high-temperature resistance, toughness, good dielectric properties, low flammability, and high resistance to radiation and structural deformation under load at elevated temperatures.

Uses

Polyimides are primarily used in the aerospace, automotive, and electronic industries, where a need exists for materials with long-term high-temperature capabilities. Molded polyimides are now widely used in load-bearing applications such as struts, chassis, and brackets in automotive and aircraft structures because of their high flexural modulus and compressive strength. Their excellent high-temperature dimensional stability, chemical resistance, and natural lubricity make them ideal bearing materials in such applications as jet engines, appliances, and office equipment. Polyimide films are used as insulation for electric motors, aircraft parts, missile wire cable, magnetic wire, and flat flexible cable.

The electrical properties of polyimides make them ideally suitable for applications in the electrical and electronic markets. They are being used in place of glass and ceramics for high heat connectors, switches, housings, and controls. They are also used in the fabrication of injection-molded printed circuit boards. Polyimide coatings are increasingly being used in microelectronic applications. Major applications are in the following areas:

- Interlayer dielectrics on silicon and gallium arsenide integrated circuits for multilevel devices
- Passivativation, thermal-mechanical buffer, and á -particle protection coatings on integrated circuits and other circuitry.

Masking for multilayered resist processing; for negative-profile lift off processing; for harsh processing such as ion implantation or dry etching. The principal features that make polyimides suitable for microelectronic applications are as follows:

- Polyamic acid varnishes and/or polyimide solutions can be easily spun on planarized layers, exposed, and etched with existing equipment.
- Polyimides have good dielectric properties.
- Polyimide coatings are tough and resilient, and have good thermal and chemical stability. They also have good radiation resistance.

AMINO RESINS

Introduction

Amino, or aminoplast, resins are an important class of cross-linkers for industrial coatings.

Amino resins are usually water-white, viscous materials that may contain added solvent to reduce viscosity for ease of handling. Most amino resins do not consist of a single chemical entity but are mixtures of monomeric and oligomeric molecules. Properties of the resin are dictated by the amount of each molecule type present in the mixture. These resins are used in thermoset and ambient cure systems.

Figure .1 shows various types of amino resin precursors. The amino resins are made by reacting these precursors with formaldehyde ($CH_2=O$) and then with primary alcohols (ROH). The resulting ether is $>NCH_2OR$. Of these, the melamine-formaldehyde (MF) and urea-formaldehyde (UF) resins are the most widely used commercially.

History

The first amino resins used in coatings were made from the reaction of urea or melamine with formaldehyde followed by butanol (either n - or iso-). They were essentially polymeric and therefore were offered at 50 to 60% solids in butanol/ xylene mixtures. These have been commercially available for nearly 70 years.

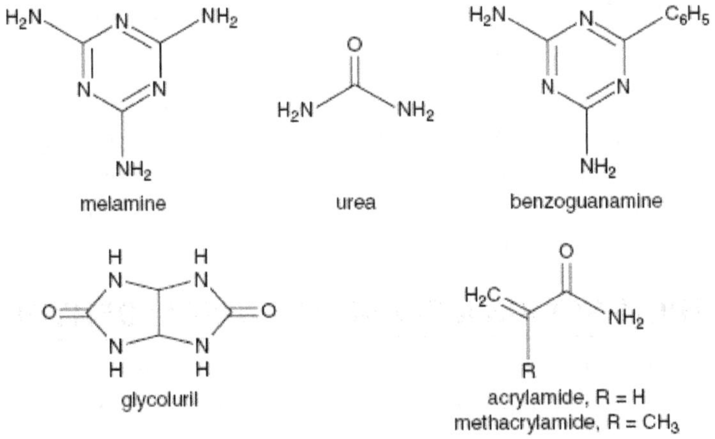

Figure .1 — Amino resin precursors.

The higher solids (80 to 100%) amino resins have been available for about 40 years. These are made with either methanol only or mixtures of methanol and other primary alcohols. These higher solids resins are now more important industrially because of the many regulations put in place, by government mandates, to reduce the level of volatile organic compounds (VOCs) emitted by paints.

$$H_2NCONH_2 + CH_2O \rightleftharpoons H_2NCONHCH_2OH$$

$$H_2NCONHCH_2OH + CH_2O \rightleftharpoons HOCH_2NHCONHCH_2OH$$

$$HOCH_2NHCONHCH_2OH + ROH \rightleftharpoons HOCH_2NHCONHCH_2OR + H_2O$$

$$HOCH_2NHCONHCH_2OR + ROH \rightleftharpoons ROCH_2NHCONHCH_2OR + H_2O$$

Figure .2 — Urea-formaldehyde resin synthesis.

AMINO RESIN SYNTHESIS AND BACKGROUND

Urea-Formaldehyde Resins

The equations shown in Figure .2 represent the stepwise synthesis of UF resins. All four reactions are equilibrium driven. The first two can be catalyzed by acid or base. The second two are acid catalyzed. The extent of reaction is dependent upon the raw material charges and catalyst ratios and, additionally, for the third and forth reactions, the extent the water is removed from the reaction.

Melamine-Formaldehyde Resins

The reactions used to generate MF resins are analogous to those illustrated above. The notable exception is that only two of the four available hydrogens on the urea molecule are easily substituted, whereas all six hydrogens on the melamine molecule are available for substitution. Figure .3 shows all the possible bonding combinations for a MF resin. Present are imino (>NH), methylol (>NCH 2 OH), alkoxymethyl (>NCH 2 OR), acetal (>NCH 2 OCH 2

OR), methylene bridge (>NCH 2 –), and methylene ether bridge (>NCH 2 –O–CH 2 –) groups. The last two bonding modes contribute to oligomeric species. Conventional MF resins consist of monomers and oligomers of melamine with a mixture of the possible functional groups. The amount of oligomers formed depends on the reaction conditions.

TYPES OF HIGH-SOLIDS MELAMINE-FORMALDEHYDE RESINS

This section will include resins with solids at ≥ 80%. One unique class of MF resins (partially acrylated) can participate in a dual-cure, ultraviolet (UV)/thermal process.

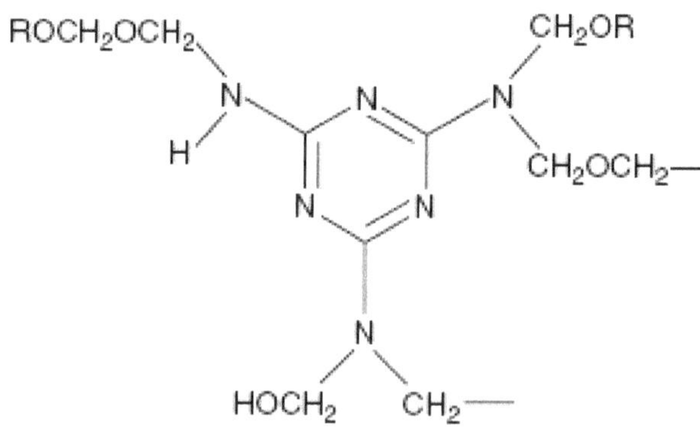

Figure .3 — Functional groups in typical MF resins.

Highly Methylated MF Resins

Formulations made with these resins, composed of methoxy ether groups only, cure faster than either those composed of methoxy–alkoxy groups or other alkoxy groups only. They can cure at temperatures down to 60°C (140°F). Some wood coatings use ambient cure conditions. Specific acid catalysts are used for cure. These resins can be used in waterborne (WB) coatings and normally have a longer shelf-life than WB formulations made with high imino-containing resins.

Imino-Containing MF Resins

These resins can be formulated to cure at temperatures of ~110°C (230°F) with no catalysts. When catalysts are required, general acid catalysts are chosen. A high level of self-condensation of the MF resin is responsible for the cure response. Because of the self-condensation, coatings made with these resins are harder and more brittle. These resins can be used in waterborne coatings.

Methoxy-Butoxy MF Resins

Formulations made with these resins show improved flow and leveling properties, and the cured systems demonstrate enhanced adhesion characteristics, and telegraphing resistance. Specific acid catalysts are used for cure.

Coatings

Urea-formaldehyde resins are used in formulations for wood (board, kitchen cabinet), paper, and some coil and general industrial coatings. Most applications are for interior finishes, because UF coatings have sensitivity to hydrolysis and have poor exterior UV durability.

Melamine-formaldehyde resins are more widely used. Coatings made with MF resins have better chemical resistance and demonstrate better exterior durability. In addition to the coating applications listed for UF resins, MF resins can be used in container, automotive, and appliance coatings.

Even though MF resins provide better stain and water resistance in coatings versus UF resins, UF resins are used in wood coatings because they can be cured with little or no heat. They also are generally less expensive than MF resins.

Film Formation – Cross-Linking

Urea-formaldehyde and MF resins can undergo self-condensation. Figure .4 shows how MF resins can self-condense. In addition to water, alcohols may also be evolved. Industrial baking finishes must possess flexibility exceeding that of self-condensed amino resins alone.

Therefore, the coreactant must introduce flexibility. The common resin types employed are acrylics, alkyds, and polyesters. These resins must possess functionality that will react with the amino resins. The principal functionalities are primary and secondary hydroxyl, carboxyl, carbamate, and amide. The reaction of the cross-linker with the coreactant is called cocondensation (Figure .5). Both self-condensation and cocondensation reactions are catalyzed by acid.

Catalysts

Many researchers have studied the role that the catalyst plays in coating performance. The type of catalysis has been divided into two categories:

General and Specific

Weak acids, such as carboxylic and phosphoric, can be used to initiate the cure, under certain conditions. When the conditions are met, this type of cure is called general acid catalyzed. When a strong acid, such as an organic sulfonic acid or hydrochloric acid, is used, the cure is designated specific acid catalyzed. The cure temperature also plays a major role. For MF resin systems, low temperature [<110°C

(230°F)] cures can only be achieved with specific acid catalysts. Examples will be given in later sections.

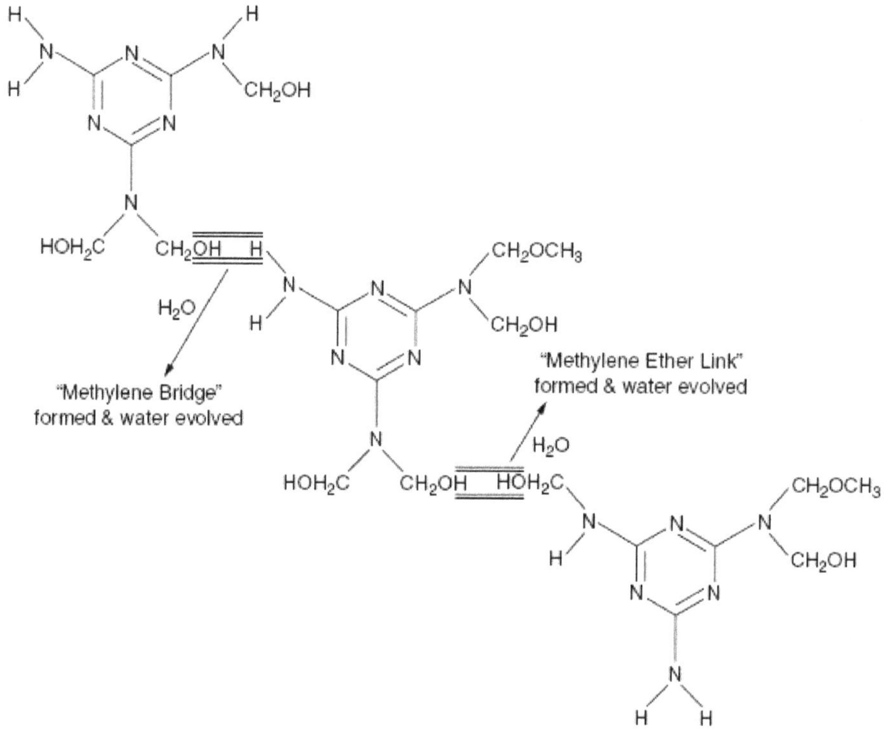

Figure — Self-condensation cure by MF resins.

Besides cure, the catalyst also influences the coating's durability. Too much residual acid will accelerate the coating's degradation, especially when exterior durability is a key property. Therefore, an optimum level of acid, and its composition, must be determined experimentally in every case.

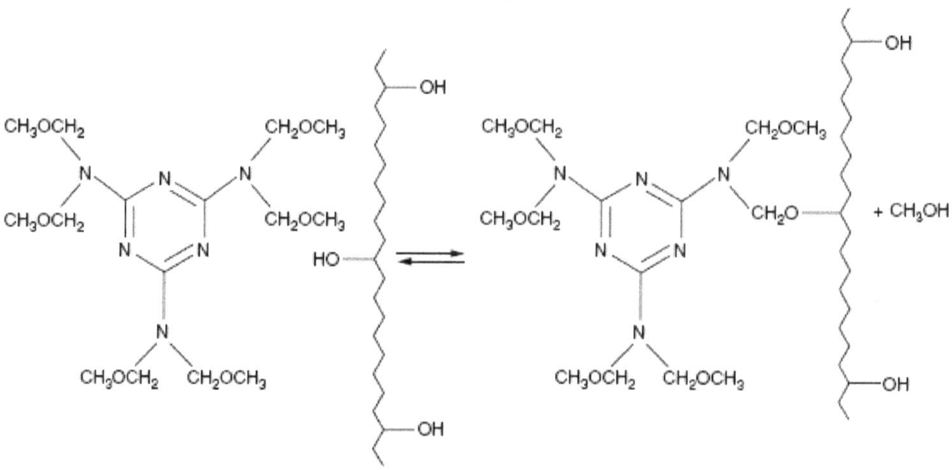

Figure .5 — Cure by transetherification of a MF resin and a hydroxyl functional coreactant.

Sulfonic Acid Catalysts

Sulfonic acids are the most widely used catalyst type. There are aliphatic and aromatic sulfonic acids commercially available. Methane sulfonic acid should be a good choice for many applications, but early versions were likely high in sulfuric acid. This caused hydrolysis problems and yellowing of coatings in which it was tested. P –Toluene sulfonic acid (p -TSA) is by far the most used acid because of cost, and because it can be easily purified (white crystalline solid). However, in the most demanding coating applications where long-term exterior exposure must be achieved, other acids are preferred. Two worth mentioning are mixed isomers of dodecylbenzene sulfonic acid (DDBSA) or dinonylnaphthalenedi sulfonic acid (DNNDSA). Both are more hydrophobic than p -TSA, and thus, they contribute less to film failures caused by water absorption.

PAINT STABILITY

Blocked Catalysts

Blocked catalysts are used so that the catalyzed paint has a reasonable shelf life. Amines are used extensively as blocking agents. At room temperature, in the paint, the amine is protonated by the acid catalyst, thus reducing the acidity of the system. When the paint is applied, the amine will volatilize, because the blocking reaction is an equilibrium process. This allows the cross-linking to begin. Oven cures are normally used when blocked catalysts have been employed. The choice of the amine is critical to coating performance. Some amine/acid combinations are not usable because the amine/acid complex may not be soluble in the formulation. Depending on the cure temperature, certain amines may volatilize too quickly and can cause severe wrinkling in the coating. This effect is due to a differential cure throughout the thickness of the coating. The concentration of the amine is lowest at the surface, so it cures faster than the interior, thus leading to a wrinkled appearance. Other amines, such as 2-amino-2-methylpropanol (AMP) can cause auxiliary cross-linking.

These amine-blocked catalysts are ionic (charged species) in character. Another type is covalently blocked.

The advantage of this type is that it can be used in electrostatic spray equipment where the paint should have low conductivity. However, covalently blocked catalysts require a cure temperature 110°C (230°F) to deblock. Certain amine blocked systems deblock at 65°C (149°F).

Primary Alcohols

Note in Figure .5, the by-product of the cross-linking is alcohol formation. A few percent of a primary alcohol in the formulated paint can prolong the stability of the system. The added alcohol effect can be explained by Le Châtelier's principle (when an equilibrium system is perturbed, the equilibrium will always be displaced in such a way as to oppose the applied change). Methanol would be the best choice because it would volatilize the fastest; however, it is also flammable. Most formulators choose butanol as the retarder. Added alcohol, along with a blocked catalyst, provides systems with the highest stability. It is worth noting that four times more n -butanol and more amine were required to stabilize a paint made with a 55% solids partially butylated MF resin than was required when a high-solids hexamethoxymethylmelamine (HMMM) was used.

Coatings — Clear Coats

This topic could fill a book in itself. Melamine-formaldehyde resins impart characteristics such as superior distinctness-of-image (DOI), high gloss, and excellent mar resistance to automotive clear coats. There has been some partial replacement of MF resins with isocyanate cross-linkers, because coatings made with MF resins do not achieve the same level of acid-etch resistance. Vaughn et al. reviewed acid-etch testing and described a straight forward accelerated test.

This accelerated test is supposed to mimic a coating exposed to acid rain. However, when mar resistance is the issue, coatings made with MF resins substantially outperform isocyanate systems.

Other Amino Resins

Benzoguanamine resins impart flexibility and superior resistance to alkali and alkaline detergent resistance versus MF resins. They also produce flexible coatings. They cost more than MF resins and have poorer UV durability, so they are used only when the above properties dictate their use. Benzoguanamine has been used significantly in can coatings. Glycoluril resin formulations require higher cure temperatures or higher catalyst levels but have excellent corrosion and humidity resistance. High cost limits their use. The acrylamide-type amino resins have had very limited use. Some applications include coil coatings for exterior use and self-cross-linkable coatings.

Future Trends

Amino resins will be with us for quite a while, because, as a class of cross-linker, they are very economical and give excellent performance. Government regulations have already forced amino resin producers to make high-solids resins. Now these resins can be supplied at lower free formaldehyde levels. A new low free formaldehyde MF resin was introduced that can cure at 82°C (180°F) with as little as 0.5% acid catalyst.

Some paint suppliers have developed new vehicle resins so that MF resins can continue to be used in high performance applications (again mainly cost driven). Carbamate- and silane-containing vehicle resins provide enhanced exterior durability benefits when cross-linked with MF resins. The desire to use amino resins in paint will keep formulators busy for years to come.

FLUOROCARBON RESINS

The commercially important fluorocarbon resins are based upon a handful of fluorinated monomers, as produced, especially when copolymerization with other, nonfluorinated monomers is a possibility.

Besides polytetrafluoro ethylene (PTFE) "coatings" — which are prepared by a high temperature sintering process and which are not considered in this chapter — the most widely used fluorocarbon coatings are based on poly(vinylidene fluoride) (PVDF) homopolymer. Typical PVDF coatings require a high temperature bake and are applied on primed metal substrates by a coil coating or spray technique. They are commonly used in high-end architectural applications such as skyscraper curtain walls and other wall panel systems, window profiles, and commercial and residential metal roofing. The first commercial grade of PVDF for coatings, KYNAR 500 ®, was introduced in 1965 by the Pennsalt Company. This same resin grade continues to be sold today by Arkema, Inc., under a worldwide licensing program.

Another commercially important class of fluorocarbon resins, fluorinated ethylene vinyl ether (FEVE) indicated in Table .1. From these basic building blocks, a wide variety of polymer products can be polymers, was introduced in the early 1980s by Asahi Glass, under the Lumiflon ® trademark. FEVE products continue to enjoy their greatest success in the Far East, especially for site-applied (air-dry) applications such as industrial maintenance topcoats. Both PVDF and FEVE coatings combine superior weatherability with excellent protective and barrier properties. They are, therefore, able to simultaneously function as decorative and functional finishes.

Among the other types of fluorocarbon resins used in coatings, fluorinated acrylics are also used in significant volumes and have a long history. They are chiefly used as surface treatment agents, for applications where low surface energy is sought, e.g., for water and oil repellency.

Poly (Vinylidene Fluoride) (PVDF) Resins for Coatings

PVDF for coatings applications is produced by a high pressure emulsion polymerization process. Vinylidene fluoride monomer normally polymerizes in a head-to-tail fashion, so the idealized structure of PVDF resin has alternating $-CH_2-$ and $-CF_2-$ groups:

Table .1 — Fluorinated Monomers Used in the Manufacture of Common Fluororesins Used in Coatings

Monomer Acronym	Common Monomer Name	Molecular Formula	Principal Use in Fluororesins for Coatings
TFE	Tetrafluoroethylene	CF_2=CF_2	Comonomer for FEVE resins and PVDF copolymers; also coating additives
VDF or VF2	Vinylidene fluoride	CF_2=CH_2	PVDF homopolymer and copolymers
CTFE	Chlorotrifluoroethylene	CFCI=CF_2	Comonomer for FEVE resins and PVDF copolymers
HFP	Hexafluoropropylene	CF3=CF=CF2	Comonomer for PVDF copolymers
–	Fluorinated decyl acrylate	*n*-$C_8F_{17}CH_2$,CH_2O-COCH=CH_2	Comonomer for fluorinated acrylics

This highly regular structure, and high fluorine content (59 wt% fluorine), contributes to the excellent chemical and photochemical stability of the PVDF resin. At the same time, because there is substantial polarity along the PVDF backbone, the resin can undergo favorable thermodynamic interactions with a variety of solvents and with certain other polymers (chiefly those with high levels of carbonyl groups).

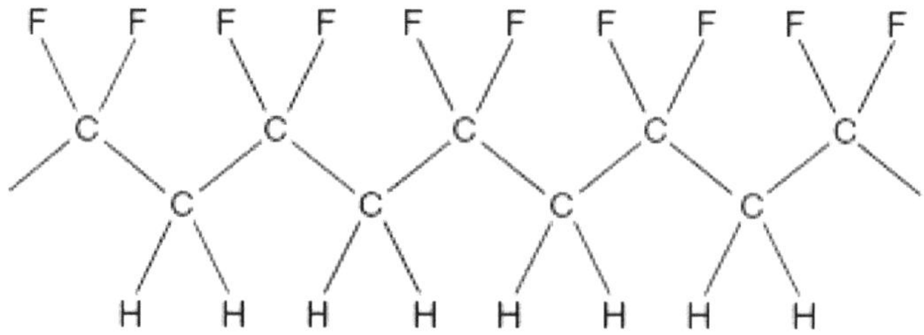

Poly(vinylidene fluoride) (PVDF)

This property is exploited by paint formulators, in the choice of co-resins and solvents that are used in PVDF formulations.

Compatible acrylic co-resins are widely used in PVDF paint formulations, to provide certain features such as pigment wetting and enhanced adhesion, which cannot be supplied by the relatively inert PVDF resin. Common acrylics used include poly(methyl methacrylate) (pMMA) grades and methyl methacrylate- ethyl acrylate copolymers. Acrylic levels of 20 to 30 wt% on total binder are generally used — these provide the desired performance enhancements, without too greatly compromising key PVDF properties such as weatherability, flexibility, and chemical resistance. While thermoplastic acrylics perform well in these systems, low levels of acrylic cross-linking are sometimes used by paint formulators to achieve specific performance enhancements, such as early MEK resistance for coil coatings.

Typical PVDF paint formulations generally contain one or more "latent solvents" that have good solvency for PVDF and the co-resin under high temperature bake conditions, and so promote an intimate resin alloy structure in the final coating film. Like pure PVDF homopolymer, PVDF-acrylic alloys at 70 to 80% PVDF are semicrystalline. The exact microscopic organization of the crystalline structures in these alloys depends on a variety of factors, including the cooling rate. However, in general, the crystallites act as labile cross-links, enhancing various coating properties, including chemical and solvent resistance, hardness, and barrier properties. To achieve the proper morphology, bake temperatures well above the PVDF 170°C melt temperature are used. This requirement limits the use of paints based on PVDF homopolymer to heat-insensitive substrates such as metals.

Most PVDF paints are formulated for application either by roll or spray coating. The substrates for spray coating are typically aluminum extrusions. Roll-coat application of PVDF paints is normally done on coil (precoated metal) lines, using a variety of metal substrates, including steel, galvanized steel, Galvalume ®, and aluminum.

Historically, PVDF coatings have been used to meet the most demanding specifications for architectural products – such as the American Architectural Manufacturers Association (AAMA) specifications. For instance, the recent AAMA 2605 specification for spray-coated aluminum extrusions and panels calls for 10-year Florida gloss and color retention. To achieve this ambitious goal, paint formulators combine proven PVDF resin technology with carefully selected weatherable pigments, typically highly stable metaloxide- type pigments.

While most PVDF coating applications continue to use solvent dispersions, a variety of technical alternatives have also been developed over the years, based on vinylidene fluoride resins. For baked applications, special PVDF homopolymer grades can be used to make PVDF powder coatings and waterborne baked coatings. When formulated at ratios like those of the standard solvent paints, the powder coating and waterborne approaches achieve performance, including weathering performance, equivalent to that of the solvent paints.

Neat PVDF resins in powdered form, and other neat fluoropolymers, are also used in limited quantities to make functional powder coatings, especially for the chemical process industry.

Copolymers of vinylidene fluoride with other fluorinated monomers have lower crystallinity and higher solubility in various solvents than PVDF homopolymer. These can be used to make solvent airdry paints, suitable for use in building restoration, and for repair and touch up of PVDF paints. [These paints, however, have high volatile organic compound (VOC) levels.] Vinylidene fluoride copolymers are also used in niche ink and printing applications. More recently, PVDF copolymer grades have been used to make water-based latex dispersions. These can be used as the base resin for waterborne paints with very low VOC levels, for instance, for industrial maintenance coatings, or for architectural coatings for objects such as PVC window frames.

Another recent trend is the use of PVDF films as paint substitutes. PVDF paint films have compositions similar to traditional PVDF paints and are prepared by solvent casting or extrusion processes. They typically have a multiple layer structure and can be used with heat-sensitive substrates, for instance, in automotive and other transportation industry applications.

FLUORINATED ETHYLENE VINYL ETHER (FEVE) RESINS FOR COATINGS

The idealized structure of an FEVE copolymer is an alternating copolymer of a highly fluorinated monomer and non fluorinated vinyl ether monomers.

Commercial FEVE products are typically based on chloro trifluoroethylene (CTFE), as shown in the above structure, or tetra fluoroethylene (TFE). Electronic induction effects from the fluorine atoms are thought to confer photochemical stability to adjacent vinyl ether portions of the FEVE backbone, greatly enhancing weatherability compared to non fluorinated ethers.

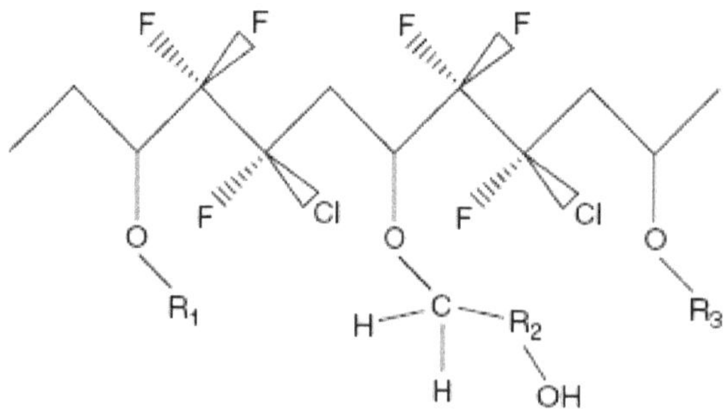

Some resin suppliers manufacture FEVE-type resins based on vinyl ester monomers rather than on vinyl ether monomers. However, the vinyl ester resins are generally understood to have inferior properties, including inferior weatherability.

Normally, a combination of vinyl ether monomers is used in order to attain a specific balance of properties. Besides using nonfunctional vinyl ether monomers such as vinyl cyclohexyl ether, vinyl ethers with functional groups on the side chain are used in measured proportions. These groups provide sites for cross-linking and can be advantageous to enhance pigment wetting, compatibility with cross-linkers, and adhesion. FEVE resins are amorphous, so unlike the case of PVDF coatings, cross-linking is normally needed for full coating property development.

Most commonly, hydroxy functional monomers such as hydroxybutyl vinyl ether are used, giving resins that can be cross-linked with isocyanate or melamine cross-linkers. However, other functional groups such as acid or epoxy groups can be incorporated into the FEVE resin as well, alone or in combination.

FEVE resins can be manufactured by a number of routes (solution polymerization, emulsion polymerization, etc.). Commercial FEVE resins are most commonly supplied in a solvent, but powder and waterborne grades are also available.

Hydroxy functional FEVE solution polymers are used with aliphatic polyisocyanate cross-linkers [e.g., hexamethylene diisocyanate (HDI) isocyanurates] to make two-component air-dry coatings. The reaction of the hydroxyl groups on the FEVE resin with cross-linker isocyanate groups makes urethane linkages, under ambient conditions. This chemistry is ideal for site-applied applications such as bridges and causeways, storage tanks, and building refinishing.

For baked systems, blocked isocyanate cross-linkers such as caprolactam-blocked isophorone diisocyanate can be used (this chemistry is often used in powder coating systems). Aminoplast cross-linkers such as melamines can also be used.

Fluorinated Acrylics

Fluorinated acrylics are typically solution polymers, prepared by copolymerizing acrylic monomers having long, perfluorinated side chains, along with other, non fluorinated acrylic monomers. By careful adjustment of the composition, fluorinated acrylics form films with –CF3 groups from the side chains aligned on the surface, through side chain crystallization. The surface –CF3 groups give coatings with an extremely low surface energy, down to 8 dyn/cm. Such surfaces are not only water repellent but are also oleophobic (oil-repellent). As such, they are useful for fabric treatments, as anti-graffiti coatings, and in kitchen and industrial workplace applications.

To maintain the favorable surface properties of fluorinated acrylic coatings, some kind of cross-linking chemistry is often incorporated into the resin backbone. Nevertheless, the durability of fluorinated acrylics is normally not equal to that of other common highly fluorinated coatings.

Other Fluorinated Resins for Coatings and Inks

As the various chapters in the bibliography reference works attest, there is a wide variety of fluoropolymers commercially available, each with a unique combination of chemical and physical properties. Besides the larger volume products described above, many other fluoropolymers have been considered for specialized niche applications. Among these are the following:

- Perfluorinated elastomeric resins (e.g., tetra fluoroethylene-hexa fluoropropylene copolymers) for situations where excellent chemical resistance, including alkali resistance, is required.
- Low molecular weight, mobile perfluoroether resins for the protection of stone, concrete, and other porous substrates.
- High fluorine content resins of various types for optical fiber cladding and anti-fouling applications.
- Functional resins for powder and liquid coating applications based on the copolymerization of fluorinated monomers and low levels of non fluorinated monomers with cross-linkable functional groups.

Epoxy resins

The epoxide or oxirane group has many reactions useful in resin chemistry, in particular those with carboxyl, hydroxyl, phenol, and amine (Fig. 1).

These reactions, which do not require high temperatures, are exothermic (70–80 kJ/epoxy equivalent) and are often readily catalysed. They can be exploited in the assembly of polymers and in curing reactions; in certain circumstances novel multistage reaction routes can be devised. The amine–epoxide reaction is particularly used in the cure of epoxy resins. The reaction with dicarboxylic anhydrides, as well as occurring in the manufacture of alkyds containing Cardura E10 (Shell Chemicals), is also used for curing powder coatings; in this reaction, requiring both initiation and catalysis, polyesters are produced both rapidly and exothermically without any water of reaction being evolved. Since hydroxyl groups may always be present when epoxy resins are being cured, even when other reactive groups predominate, it should be noted that conditions of temperature and catalysis may enhance or suppress the possible etherification reaction with hydroxyls in relation to the other epoxy group reactions taking place. For example, in the acid–epoxy reaction, base catalysis suppresses the etherification reactions that would otherwise occur.

The compound glycidyl methacrylate that has both vinyl unsaturation and an epoxy group is particularly useful as a bridge between condensation and addition stages in polymerizations. Possibilities occur to react either group initially, the second then reacting in a subsequent stage.

The best-known epoxide containing materials are the range of preformed epoxy resins based on the reaction between diphenylol propane (bisphenol A) and epichlorhydrin; their general structure is shown in Fig. 2.

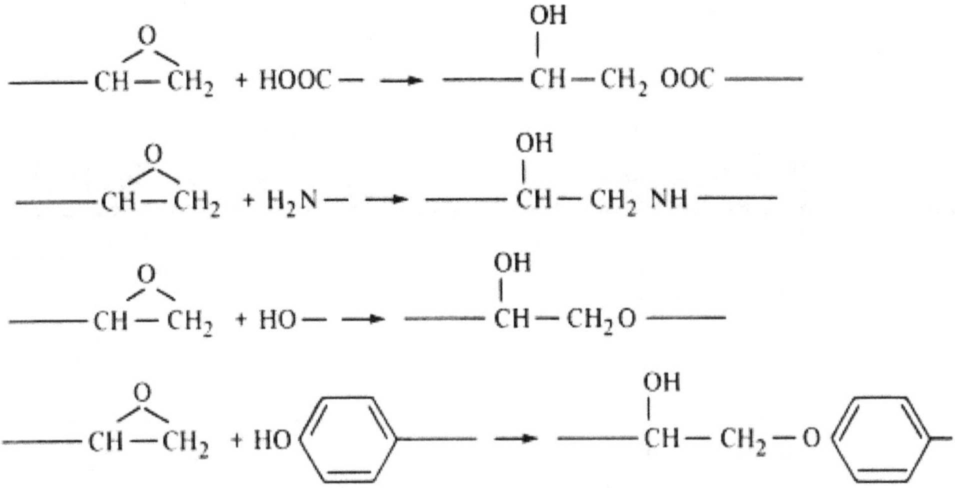

Fig. 2.25 — Typical reactions of the epoxide group.

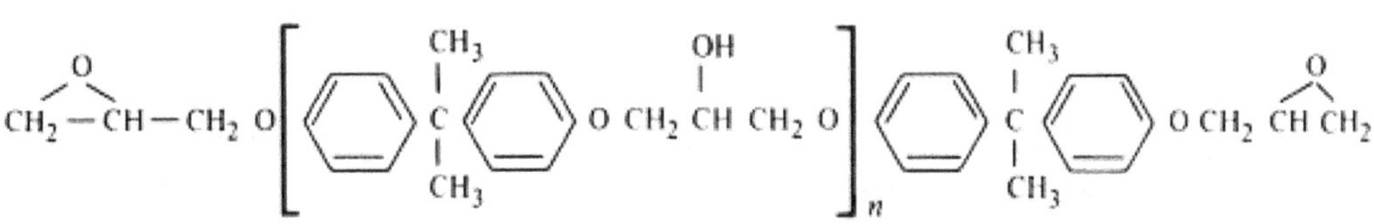

Figure 2 — Generalized structure of bisphenol epoxide resins.

Where n approaches 0, the resin approximates to the diglycidyl ether of diphenylol propane, and the product is liquid. As n increases from 2 up to about 13, solid forms with increasing melting points are encountered (Table 1). The lower melting grades are often modified by pre-reaction before use in coatings, though the higher melting grades can be used in unmodified form for can or drum lining applications.

Because of problems including handling toxic epichlorhydrin, it is normal for coatings manufacturers to purchase epoxy resins of this type; it is, however, now common practice to 'chain-extend' liquid grades to higher molecular weight by reaction with diphenylol propane. Precatalysed liquid epoxy resin is available for use with this technique, which can reduce stocking of a range of grades and can provide access to intermediate solid grades; cost savings are also possible.

The formula above would indicate that all epoxy resins possess two terminal epoxide groups and all, except where n = 0, possess in-chain hydroxy groups. Higher molecular weight epoxy resins frequently depart from this linear structure owing to the incidence of side reactions in their preparation.

Considerable care is now taken in epoxy resin manufacture to eliminate free epichlorhydrin from the product because of its carcinogenic nature. Low molecular weight di-epoxides may also be carcinogenic, and consequently the available range of low molecular weight epoxies is now limited, with certain low molecular weight aliphatic di-epoxides now no longer available.

Table 1 — Typical grades of bisphenol epoxide resins and properties

Number of repeat units (*n*)	Melting point (°C)	Expoxide equivalent
0.5	Viscous liquid	22–290
2	64–76	450–525
4	95–105	850–950
9	125–132	1650–2050
12	140–155	2400–4000

The outstanding properties of cured epoxy resins may be explained by their structure. The very stable carbon–carbon and ether links in the backbone contribute to chemical resistance, while a factor in their toughness is the wide spacing between the reactive epoxide groups and in turn the hydroxyl groups. The polar hydroxyl groups, some of which may always remain, also help adhesion by hydrogen bonding.

The aromatic ring structure enhances thermal stability and rigidity. Though these properties are attractive, aromatic epoxy resins yellow, and for this reason their major application is in primer and undercoat compositions where adhesion and corrosion resistance are particularly valuable. They are also used in can coatings for their good one-coat performance. Epoxy resins may be formulated from phenols other than bisphenol A, for example

diphenylol methane (bisphenol F). Epoxy-type materials may also be glycidyl ethers of other resins such as PF novolacs . Epoxide compounds are characterized by melting point and by their epoxide group content or epoxy equivalent which may be expressed in a number of ways.

Epoxy esters

Both the terminal epoxide groups, and the secondary hydroxy groups of solid epoxy resins, can be reacted with fatty acids to produce the so-called epoxy ester. In esterification reactions each epoxide group is equivalent to two hydroxy groups.

Epoxy esters are prepared by heating the fatty acid and epoxy resin in an inert atmosphere, preferably under azeotropic conditions to remove water of reaction, with temperatures of between 240 and 260°C normally being used. The reaction may optionally be accelerated by the addition of, for example, calcium or zinc soaps at 0.1 or 0.2% weight on total charge.

Typically the epoxy resin where n = 4 is used, and fatty acid content chosen to esterify between 40 and 80% of the available groups including hydroxyl. Medium

(50 ~ 70% modified) and long (over 70%) oil epoxy esters of drying oil fatty acids are used in air-drying brushing finishes. Short (30–50%) oil-drying or non-drying fatty acid esters are used in industrial stoving primers and finishes. Stoved epoxy resin ester films, especially when cured with MF resin, are harder and of superior adhesion, flexibility, and chemical resistance than similar alkyd/MF formulations. Increased fatty acid content, as would be expected, imparts better aliphatic solubility, better exterior durability, and decreased hardness, gloss, and chemical resistance. Linseed, tall, and DCO esters are most usual, though all fatty acids and rosin can be used.

Other Epoxy Applications

Epoxy resins may be cooked into alkyd formulations, replacing part of the polyol. They also find a place unmodified as a third component in alkyd/MF compositions to upgrade the resistance properties of the films.

Epoxy resins react with PF resins to form insoluble coatings, and well-formulated high molecular weight epoxy/PF coatings meet the highest standards of chemical resistance. These products are suitable for the linings of food cans and collapsible tubes, coatings for steel and aluminium containers, and wire enamels. Curing probably involves the formation of polyether links between the hydroxyl groups of the epoxy, and methylol groups present in PF resins of the resole type; the epoxy also reacts with phenolic hydroxyl groups on the PF. With some PF resins, compatibility problems on cold blending may be solved by pre-condensation, involving refluxing the epoxy and PF resins together in solution, when some reactive groups combine, leaving the remainder free to react in the curing process.

Urea/formaldehyde resins or melamine/formaldehyde resins may be used to cure epoxy resins, giving stoved films of paler colour, but with a reduced level of chemical resistance compared with phenol/formaldehyde resins. Again, the higher molecular weight epoxy resins are preferred.

Two-pack epoxy/isocyanate finishes require separate solutions of high molecular weight epoxy resin and polyisocyanate adduct as the two components; the epoxy resin must be in alcohol-free solvent since the curing reaction is predominantly with in-chain hydroxyl groups on the epoxy resin. One-pack finishes can be formulated with blocked isocyanates.

Epoxy resins may also be used in two-pack compositions with polyamines or polyamides .The films obtained possess outstanding chemical resistance, hardness, abrasion resistance, flexibility, and adhesion. Low molecular weight solid epoxy resins are used most. Though primary or secondary amines such as triethylene tetramine may be used, in order to avoid the toxic hazards involved in handling amines, amine adducts with low molecular weight solid epoxy resin will nowadays be used as hardeners. These adducts are prepared by the reaction of excess of an amine such as diethylene triamine with an epoxy resin to produce fully amine-terminated adduct. Reactive polyamide resins formed from dimerized fatty acids with diamine are also used.

Reactive coal tar pitches may be incorporated into an epoxy resin base for curing with amine, amine adduct, or with a polyamide resin. The derived coatings have excellent chemical resistance and are not brittle; they hence find use as pipeline, tank and marine coatings.

Epoxy resin modification with a silicone resin is possible to enhance water resistance; epoxy/silicone combinations are used in blends with other polymers with the cure mechanisms mentioned above.

Isocyanates

The isocyanate group (Table 1) is highly reactive with itself, and a number of other groups, and may be catalysed in its reactions. It reacts exothermically (40 kJ/mol) with many groups with active hydrogen atoms, in particular alcohols, amines, phenols, and water at room or moderately elevated temperatures. Amine groups are most reactive, so that hydroxy groups are the most common reactive groups on resins cured by isocyanates. However amine groups are used with, for example, cathodic electrocoat resins cured with blocked isocyanates. Secondary alcohols and secondary amines react in a similarly way to primary but with less vigour. Both aromatic and aliphatic polyisocyanates are available, with the former generally more reactive than the latter.

Table 1 — Typical reactions and reaction conditions for isocyanates

Reaction	Temperature (°C)	Catalysts
RANCO + Alkyl OH → RNHCOO Alkyl	25–50	Varied
RNCO + Alkyl NH2 → RNHCONH Alky	10–35	Not required
RNCO + Alkyl OH → RNHCONH Alky	50–75	Teritiary amine

Water reacts with the isocyanate group at similar rates to secondary alcohols; the initial product is an unstable carbamic acid (RNHCOOH) which decomposes to a primary amine and carbon dioxide, and this amine group is of course then available to react with additional isocyanate. This reaction is normally to be avoided in coating formulations, though it is exploited in the production of solid and flexible foams. For this reason, systems to be reacted with isocyanates should be dehydrated, and low moisture content urethane grade solvents used.

Isocyanates also react with themselves under certain conditions and in the presence of certain catalysts; these reactions are generally undesirable. Tin compounds are strong catalysts for isocyanate reactions and are also generally 'clean' in not promoting side reactions. It will be evident that all reaction products of isocyanates contain active hydrogen groups attached to the nitrogen atom of the urethane or substituted urea reaction product; hence, steric factors permitting, reaction can continue with these products if additional isocyanate is present, to give allophanates and biurets. All of the groups mentioned above may be present on simple molecules or attached to polymer molecules.

Low molecular weight diisocyanates are used in coatings manufacture in preparing urethane alkyds and blocked isocyanate crosslinkers. Toluene diisocyanate (TDI) is the main diisocyanate used for urethane alkyds (normally the mixed 2, 4 and 2,6 isomers). Products derived from aromatic isocyanates such as TDI and diphenyl methane diisocyanate (MDI) tend to yellow; hence aliphatic isocyanates such as hexamethylene diisocyanate (HDI) and isophorone diisocyanate (IPDI) now find higher usage (Fig. 1). Trimethyl hexamethylene diisocyanate (TMDI) and hydrogenated diphenyl methane diisocyanate (H12MDI), not illustrated, also find application. Great care is necessary in the handling of all volatile isocyanates since their vapours are irritant and they are sensitizers at extremely low concentrations.

Isocyanate Adducts and Blocked Isocyanates

Owing to the toxicity of low molecular weight (volatile) isocyanates, polyfunctional (3+) isocyanate adducts of higher molecular weight are invariably used in two-pack applications of isocyanates. Three different adduct types have been used, namely the triol adduct, the biuret and the isocyanurate. An example of the first is the reaction product of 3 mol of TDI with 1 mol of trimethylol propane. The biuret is prepared by reaction of 3 mol of diisocyanate with 1 mol of water, and the isocyanurate is the trimer formed by the reaction together of three isocyanate groups. Isocyanurates from HDI and IPDI are those most frequently used in 2K refinish paints. Isocyanurates may also be mixed with uretdione (isocyanate dimer). The structures of the HDI biuret, uretdione, and isocyanurate are illustrated in Fig. 2.

The reaction of the isocyanate group with low molecular weight alcohols, phenols, and certain other compounds such as lactams, may be considered reversible. If an isocyanate that has been 'blocked' or masked in this way is heated with a polymer with reactive groups, the low molecular weight blocking material is released. The urethane or substituted urea bond is then remade with the reactive group on the polymer.Whether free isocyanate groups are fully released in the reaction with a polymer is a matter of conjecture. Blocked isocyanate curing agents are used in automotive and various other solvent applied finishes, and powder coatings.

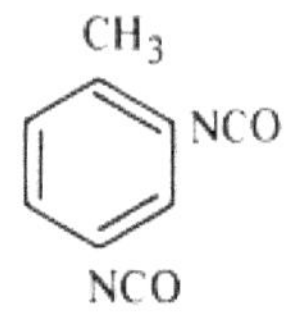

2, 4TDI

OCN $(CH_2)_6$ NCO

HDI

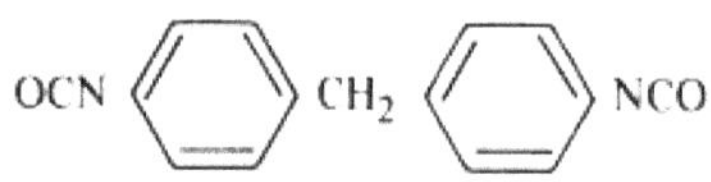

MDI

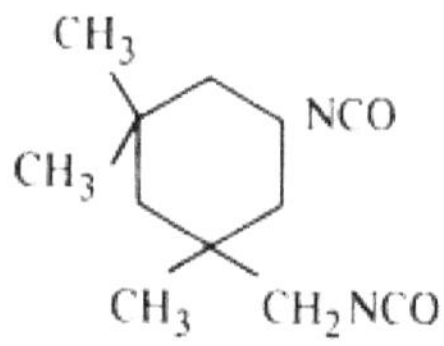

IPDI

Figure 1 — Typical diisocyanates used in coatings resins.

Depending on the blocking agent, curing temperatures may be between 100 and 175°C Automotive cathodic electrocoat formulations cure using blocked isocyanate; for powder coatings caprolactam blocked IPDI based crosslinker is used.

Polyurethane Coatings

Two-pack (2K) finishes based on hydroxy functional resins and isocyanate adducts find application for tough high solvent-resistant coatings curing at atmospheric temperature or under moderate stoving conditions. Resins that may be used are alkyd, polyester, polyether, epoxy, and acrylic. Alkyds used are frequently castor oilbased.

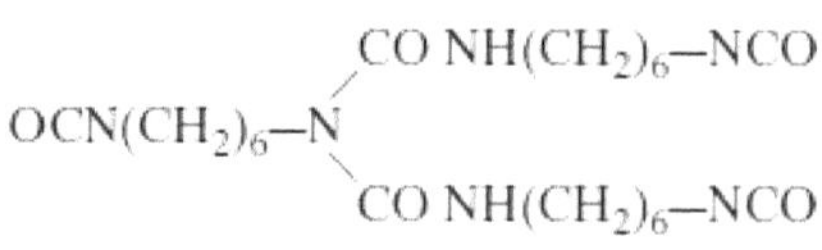

HDI biuret

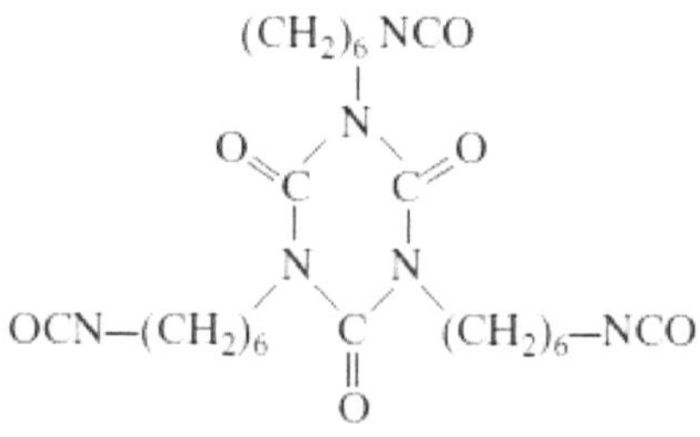

HDI isocyanurate

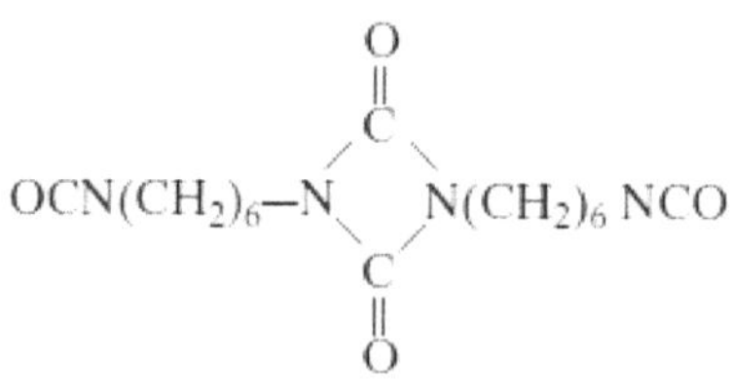

Figure 2 — HDI biuret HDI isocyanurate HDI uretdione.

Two-pack alkyd finishes are used for wood finishing. Two-pack polyester finishes find application for high-durability transport finishes including marine and aircraft and by automotive manufacturers for painting plastic parts fitted after the main metal body has been painted and stoved at high temperature.

Two-pack systems with hydroxy functional acrylic and polyester resins (2K) have high usage in car refinishing, and in most European countries this type of finish now predominates. However, because of the toxicity of isocyanates and the danger of inhaling droplets while spraying, producers have in the past striven to develop systems avoiding the use of isocyanate, but with similar properties. For these, the isocyanate component could be replaced by a crosslinker such as a melamine/ formaldehyde resin. Masked isocyanate polyurethane (PU) curing systems, sometimes mixed curing with MF, dominate in automotive OEM primers, and both solvent and water-borne systems are in use.

Two-pack systems such as the hydroxy polymer/isocyanate one may be difficult to formulate because of 'pot-life' difficulties, where conflict occurs between the need to accelerate curing but to retard bulk solidification to protect the application equipment. A development has been to accelerate curing of this type of system after application, by exposure to an atmosphere containing volatile amine catalyst (vapour phase curing).

Moisture-curing products may also be used, based on stable isocyanate terminated polymers obtained from the reaction of excess low molecular weight di- or polyisocyanate with a polyfunctional hydroxyl-terminated polyether. The pigmentation of moisture-curing finishes is difficult, though dehydrating agents are available to assist; the most frequent use of this type of product is for clear wood varnishes.

A special class of PU block copolymers incorporating both hydrophobic and hydrophilic segments is used in aqueous systems as one of the group of additives known as associative thickeners. HEUR (hydrophobically modified ethylene oxide urethane) associative thickeners consist of polyethylene glycol blocks linked via urethane segments. The end blocks are hydrophobic long aliphatic chains.

Polyurethane Dispersions

Fully reacted, aqueous polyurethane dispersions have long been available, mainly used in applications other than traditional coatings, but their use in coatings is now increasing. They are typically prepared from DMPA, polyesterpolyol and an excess of aliphatic isocyanate. These are chain extended with short chain diamines and solubilised with t-amine. Crosslinking is possible with added aziridine or carbodiimide crosslinkers, autoxidatively by fatty acid modification or even by radiation curing via attachment of acrylate.

A recent development is of 2K reactive PU systems, where HDI adduct is homogenized into a pre-prepared aqueous millbase, and the hydroxy functional coreactants can function as emulsifier for the isocyanate component. These protect the isocyanate groups from reacting with water, and while some reaction with hydroxyls does commence in the droplets, pot life extends to several hours.

SILICONE RESINS

Silicone–oxygen and silicon–carbon bonds are particularly stable, and this has a beneficial influence on the behaviour of the semi-organic silicone resins, so that they are exceptionally resistant to thermal decomposition and oxidation.

For the surface coatings formulator, a range of reactive silicone resins is available for use in the preparation of silicone-modified polymers, and these may be either hydroxy or methoxy functional. Typical structures are shown in Fig. 1. R may be phenyl or methyl, and both alkyl and aryl organo-silicones are currently available for resin modification. Reaction into a resin structure occurs through its available hydroxyl groups, when either water or methanol will be eliminated (Fig. 2.); catalysis of these reactions is possible but not essential.

Polymers that may be modified include alkyd, polyester, acrylic, and epoxy. Silicone modification may typically be from 15% to 40%, though higher levels of modification of alkyds, for example, are possible for special heat-resistant applica tions.

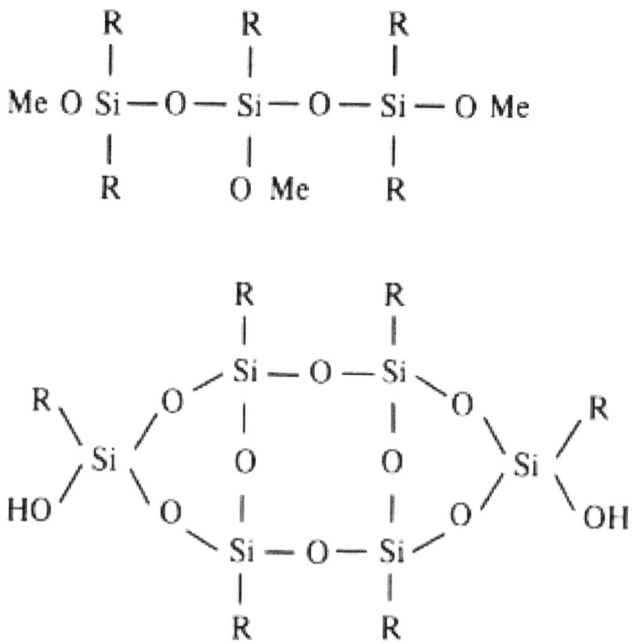

Figure 1 — Typical silicone resins for polymer modification.

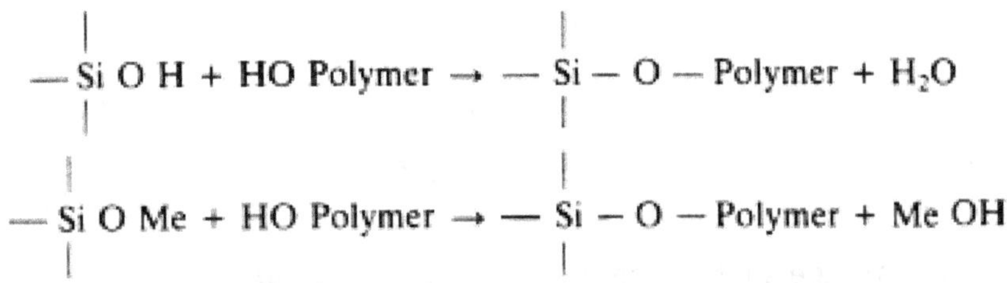

Figure 2 — Silicone resin reactions with hydroxy polymer.

The single largest use of silicone-modified resins is for coated steel coil for building structure facing. Silicone modification always considerably enhances durability. Pure silicone surface coatings are available for special application where very high heat resistance properties are required. These resins cure by the same mechanisms as for modification, but here will be catalyzed:

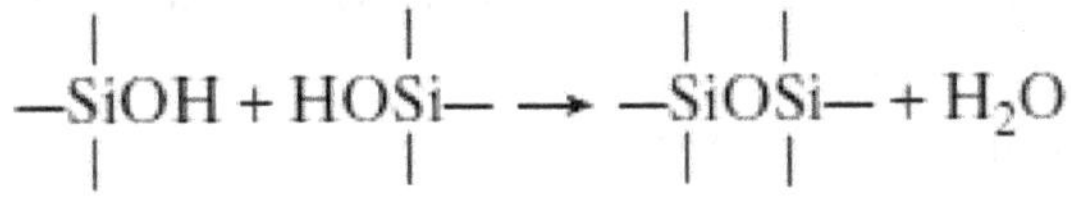

Alkoxy-silyl functional acrylic resins and emulsions were mentioned. These are moisture curing resins. Non-reactive linear silicones, e.g. poly dimethyl siloxanes of low viscosity, find frequent use in very low concentrations as flow control or marr aids in paint formulations:

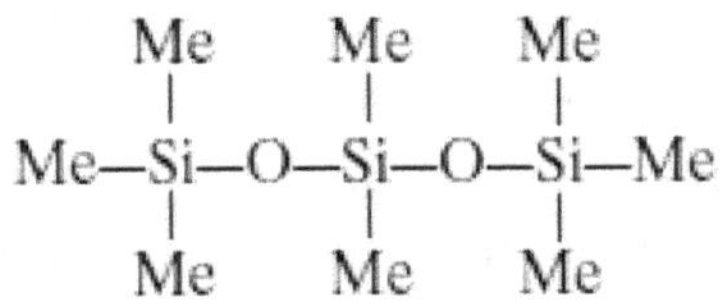

These particular silicones are generally used in solvent based coatings. For water based systems, a range of block copolymers of poly dimethyl siloxane with polyalkyleneoxides are available. Both poly ethylene oxide and poly ethylene oxide/propylene oxide di-block copolymers are used.

VINYL RESINS

The term 'vinyl resin' commonly refers to polymers and copolymers of vinyl chloride, though the term has more general meaning. Vinyl chloride is a cheap monomer whose polymers possess good colour, flexibility, and chemical resistance. Vinyl chloride is, however, gaseous and a recognized carcinogen; hence, though it is readily polymerized by emulsion or suspension polymerization techniques in an autoclave, the polymer will be purchased by most surface-coatings manufacturers. For this reason polymer preparation is not further discussed. The polymers are nevertheless quite widely used in coatings.

Both homo polymers, and copolymers usually with vinyl acetate, are used; small amounts of an acidic monomer may also be present. A serious problem with vinyl chloride-containing polymers, if heated, is dehydro chlorination, and stabilization is often necessary with, for example, organo-tin compounds or group 2 metal carboxylates.

A major advantage of PVC connected with its polar nature is its ability to accept plasticizer. Plastisols and organosols are dispersions of dried PVC particles in either plasticizer or mixed solvent/diluent. The basis of both types of coating composition is that the particulate dispersion is stable until applied and heated, when the plasticizer or solvent swells the particles, softening them and allowing coalescence; this is resisted at room temperature because of the crystalline nature of the PVC particles.

With plastisols, plasticizer is retained in the coating after coalescence; with organosols the solvent and diluent are lost through evaporation. The formulation of an organosol requires much care and a fine balance between diluent and solvent. If the mixture is too rich in diluent, the particles flocculate and give high-viscosity dispersion; if too rich in solvent the viscosity is too high, this time owing to solvent swelling of the particles.

Polyvinyl acetate/vinyl chloride copolymers are soluble in ketones or ketone/aromatic solvent blends, and this allows their use in lacquer formulations. Vinyl resins are used in coatings for strip for Venetian blinds and for bottle tops where extreme flexibility and extrusion properties are required. They are also used in heavy-duty and marine coatings where properties of toughness, elasticity, and water resistance are paramount, and have application as thick coatings on coil coated strip for building facings. Poly vinylidene fluoride dispersion is now used in the highest durability coil coating finishes in blends with acrylic resin containing up to 80% vinyl polymer. Compositions containing lower concentrations, for example 30–50% of vinyl polymer, may also be used where the acrylic resin will be a thermosetting resin of, for example, the self-reacting acrylamide-containing type.

PVC copolymer emulsions are used in emulsion paint compositions, and brief mention of formulations for this application has been made under emulsion polymerization. Other resins find use; chlorinated rubber paints are formulated using chlorinated paraffins as plasticizing component. Chlorinated rubber may also be used as an additive with resins such as alkyds and acrylics. Polyvinyl acetals (formal and butyrals) are also used, the latter particularly in resistive or active primers.

WATER-BORNE SYSTEMS

The four main technologies providing solvent reduction or elimination as so-called compliant coatings are waterborne, high solids, radiation-curing and powder coatings.

Table 1 — Comparison of the Properties of Water and Xylene

		Water	Xylene
Molecular weight		18	106
Melting point, boiling point (°C)		0,100	−25, 139
Vapour pressure (mm) 20 °C		23	8
Latent heat of vaporisation (J/g)		2259	395
Surface tension (mN/m) 25 °C		72	27.5
Solubility parameter	δ_D	15.5	20.1
	δ_P	16	1.8
	δ_D	42.3	2.5

The replacement of solvent in a coating by water also improves safety in application from the point of view of flammability and toxicity. Though water-borne systems have always found some application, legislation has accelerated the switch to waterborne systems. Electro depositable coatings are a special class of waterborne system.

Water is however very different to organic solvents (Table .1), in ways which create many problems for the formulator. To reformulate a resin system to replace organic solvent by water (Table .1), requires an increase in the polarity of the whole polymer system either by the incorporation of some water soluble groups, or by the inclusion of surfactants, or both. The polymer may finally be either in solution or in dispersed form. Unless very carefully formulated so that hydrophilic groups are destroyed or deactivated in curing, water sensitivity may be a problem in the final coating.

The high surface tension creates problems with flow and wetting, and the increased likelihood of surface defects. Replacement of solvent by water need not affect the time to dry, though a consequence of the higher heat of vaporization could be increased energy requirements for stoving. However air-flows in ovens can generally be appreciably lower. Difficulties with control of rheology can be a reason to retain some solvent in an otherwise waterborne composition.

Polymers fully water-soluble without the addition of salt-forming additives exist; examples are polyethylene oxide or glycol, polyvinyl pyrrolidone, polyacrylamide, and copolymers containing a high proportion of these materials. Polyethylene glycol, typically having one or two terminal hydroxyl groups, has a major use as a reactant to make alkyds water-soluble or dispersible; polyethylene oxide occurs as the water-soluble portion of many surfactants.

Other water-soluble polymers, including polyvinyl alcohol, acrylic polymers with a very high acid content, and modified celluloses, find application as colloid in emulsion polymerisations. Some formaldehyde condensate resins, including hexamethoxymethyl melamine and certain phenolics, are water soluble, and therefore are used in preference to other resins as cross linking resins in water-borne systems.

It will be recognized that polymers not containing soluble groups may be emulsified by added surfactant; this is not, however, common practice. Special cases of water-borne products prepared solely using surfactants were discussed in the section on emulsion polymerization, and alkyd emulsions were also covered earlier. There are cases where resins are blended and one resin acts as emulsifier for other polymer species present; this is the case, for example, where the final cure requires the presence of an insoluble curing agent. The following is concerned with systems solubilized or dispersed with the aid of hydrophilic groups on the polymer backbone. Here the polymer will normally be prepared in the absence of water, and only when at its terminal MW is it transferred into water.

The major route to water solubility/dispersibility, is by the preparation of polymers with acid or amine groups on their backbone, which may be solubilized by salt formation by the addition of an alkali/amine or acid respectively. When the resins dissolve in water, these salt groups ionize, the counter-ions being carried in the bulk solution. There is a particular advantage in using volatile amine or acid in that water sensitivity in the final film is reduced, owing to the loss of the salt-forming agent; this sensitivity can be further reduced or eliminated if the cross linking reaction can then take place with these groups. As a general point it should be noted that many successful formulations are dispersions (colloidal or micellar), not solutions, simply because this can realize the highest solids and lowest viscosity. Resins may be prepared with a higher acid or amine content than necessary, the 'degree of neutralization' with salt-forming acid or amine then being about 60% to 85%.This may be fine tuned to give the optimum viscosity, stability, freedom from settlement, and application properties.

Early water-borne vehicles were alkali neutralized maleic adducts, the simplest being maleinized oils. Maleinized fatty acids reacted with polymer backbones still play a role in water-borne systems. Epoxy resins esterified with maleinized linseed oil fatty acids alone or in combination with other fatty acids followed by neutralization can provide vehicles for both normal spray and dip and for electrodeposited application. Polybutadiene may be maleinized and can then similarly form the basis of water-borne systems. Both oil and polybutadiene-based systems may be cured by oxidation, usually stoved without the addition of further crosslinker. Water-borne alkyd and polyester systems have already been mentioned, specially formulated with a high acid value to give water solubility or to be water dispersible. Acidic acrylic systems are also preparable where, for example, maleic anhydride or acrylic acid is copolymerized with the other unsaturated monomers, and such thermosetting systems have already been described. Epoxy systems may be given acidic functionality by the ring-opening half ester reaction with a dicarboxylic anhydride, or trimellitic anhydride as described for alkyds and polyesters.

Water-borne resins may be self-cross linking (alkyds, thermosetting acrylics) or may be cross linked by added soluble or co-emulsified water-insoluble cross linking resins. Fully soluble cross linking resins may be blended into the system at any stage, and these include MF, UF, and phenolic resins, reacting as previously described for solvent-borne systems. Carboxyl groups of waterborne anionic polymers can participate in reactions with a number of added cross linkers.

b- hydroxyalkylamides were originally proposed for waterborne systems , though they now find use in the cure of powder coatings. Two other specialized cross linkers are polycarbodiimides and aziridines, where use is practically restricted to curing polyurethane dispersions. The structure of these is shown in Fig. 1.

Epoxy resins can be formulated to be amine functional by reaction with secondary amines that will then be solubilized with, for example, acetic acid (Fig. 2).

Cross linking resins must be added to these systems, for example MF, phenolic, or blocked isocyanate. A method of increasing the number of amine groups is by reacting with a diketimine, as shown later. The ketimine can be reacted with epoxy resin, without any risk of gelation, and subsequently hydrolyzed to release two primary amine groups.

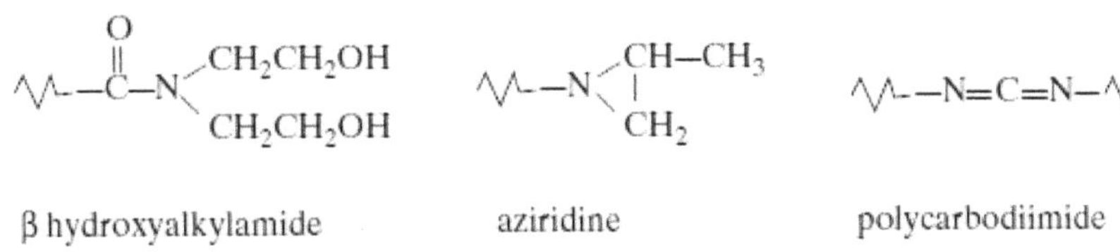

Figure 1 — Functional group of b hydroxyalkylamide, aziridine and polycarbodiimide crosslinkers.

Water-borne epoxy resins find particular application for metal primers and for can coatings, and for the latter usage water reducible acrylic grafted epoxide resins find extensive application. These are bisphenol epoxy resins grafted with styrene/ acrylic acid copolymers, using a grafting initiator such as benzoyl peroxide (Fig. 3). Specially formulated acrylic emulsions are also used in thermosetting waterborne systems, and these are prepared in the manner as emulsion polymerization. The monomers will include a hydroxy functional monomer, and the preparation will generally be carried out without the inclusion of colloid, only surfactant being present. These are usually called 'colloid-free' latexes.

Aqueous polyurethane dispersions have already been discussed. Water-soluble resins may be silicone modified, and the advantages of this have been demonstrated for alkyds and polyesters where siliconization is carried out by preacting the silicone intermediate with polyol. For acrylic latex products siliconization of the latex may be carried out as a subsequent step to the emulsion polymerization.

Certain problems are found specific to water-borne products, and require additional care in formulation. Condensation products can be prone to hydrolysis. Curing may be adversely affected by the solubilizing agent either by pH effects or, with air-drying alkyds, by neutralizing amine retarding oxidative cross linking. The presence of amines can also adversely affect colour retention. Most water-soluble compositions contain a proportion of water-miscible cosolvent which may have been a necessary component in the resin preparation state prior to water addition.

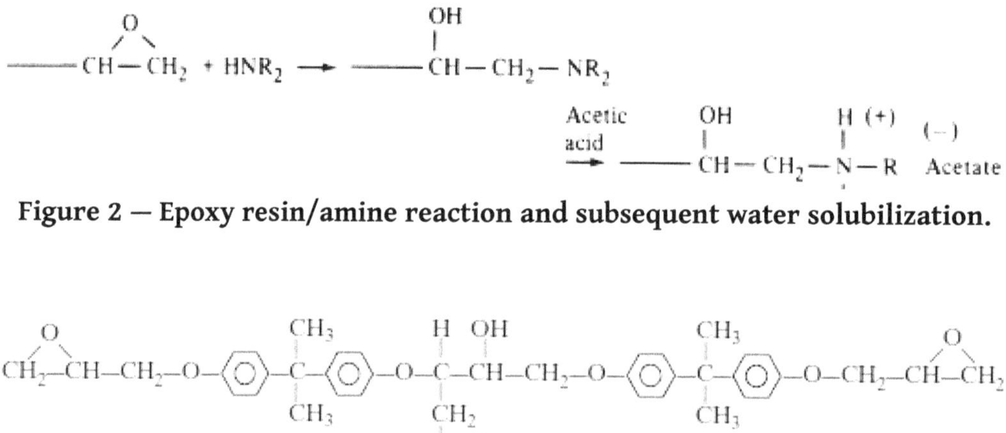

Figure 2 — Epoxy resin/amine reaction and subsequent water solubilization.

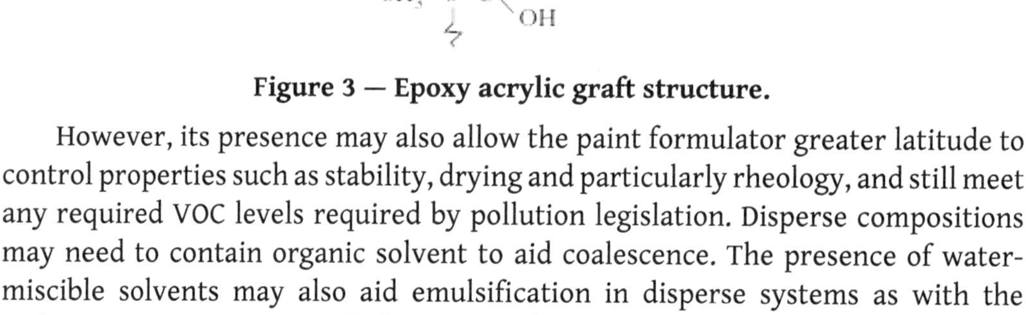

Figure 3 — Epoxy acrylic graft structure.

However, its presence may also allow the paint formulator greater latitude to control properties such as stability, drying and particularly rheology, and still meet any required VOC levels required by pollution legislation. Disperse compositions may need to contain organic solvent to aid coalescence. The presence of water-miscible solvents may also aid emulsification in disperse systems as with the techniques to prepare so-called 'microemulsions'.

RESINS FOR ELECTRODEPOSITION

Electrodepositable resins are a special class of water-borne resin. The polymer is carried in an aqueous medium and, on application of current via suitable electrodes, polymer in the vicinity of one of the electrodes is destabilized and deposited on the electrode. The deposited polymer builds up to form an insulating layer that ultimately limits further deposition. The electrodeposition process has been called electrophoretic deposition, though it is now recognized that electrophororesis plays little part in the process.

In anodic electrodeposition negatively charged polymer is deposited on the anode, and in cathodic electrodeposition positively charged polymer is deposited on the cathode. Compared with conventional painting, very uniform coverage of external surfaces can be obtained, and deposition of paint inside partially closed areas can be achieved ('throwing'). Paint utilization is high, and almost complete automation of the process is possible. These advantages are not achieved with simpler dipping processes. Since the films can only be built up on metal surfaces and to a limiting thickness, electropaints are either primers or for some industrial applications, one-coat finishes. Unpigmented clear systems are in use for coating metallic brightwork. Both anodic and cathodic commercial processes exist. Cathodic systems have now supplanted anodic systems particularly for automotive applications. Electrocoating is important in the application of can lacquers.

Almost all polymer types described in the previous section under water-borne coatings can be adapted to electrodeposition. In most cases, to be suitable for the electrodeposition process, the resin will be held in stable particulate or micellar dispersion, but not complete solution, by the action of hydrophilic ionic groups, which will provide the necessary colloidal stabilization. Non-ionically sterically stabilized dispersions, however, may also be electrodeposited.

The system should be designed such that the deposited film has high electrical resistance so that shielded areas can receive adequate coverage. The system may contain organic solvent to aid the dispersion process and dispersion stability, and act as a flow promoter during coating and curing. The role of the neutralizing acid or base is fundamental both to stability of the dispersion and to the electrocoating process. In practice, for anionic systems either alkali or amines may be used, and for cathodic systems lactic and acetic acids are usual choices for the neutralizing agent.

The earliest anodic electrodeposition vehicles were based on maleinized oils and oil derivatives, and the chemistry of these developed through vinylated and alkyd type condensates to the use of epoxy esters based on maleinized fatty acids. Since epoxy resin-based systems exhibit such good performance as metal primers, solubilized epoxy vehicles play a major role in both anodic and cathodic systems, especially for automotive applications.

Alkyd and acrylic systems have been developed for electrodepositing and have found application in industrial systems. Anodic alkyd systems are based on high acid value resins, particularly those that are trimellitic anhydride-derived, and may be drying oil-modified. They may therefore be autoxidative/stoving, or if non-drying oil-containing or oil-free, cured by co-emulsified or soluble melamine/formaldehyde resin. Acrylic systems have been proposed for both anodic and cathodic formulations, in the former case by inclusion of higher than normal concentrations, of, for example, acrylic acid. In the latter case, they have been formed by the inclusion of a copolymerized amino monomer, such as dimethylaminoethyl methacrylate or by including a glycidyl monomer that can be subsequently reacted with an amine.

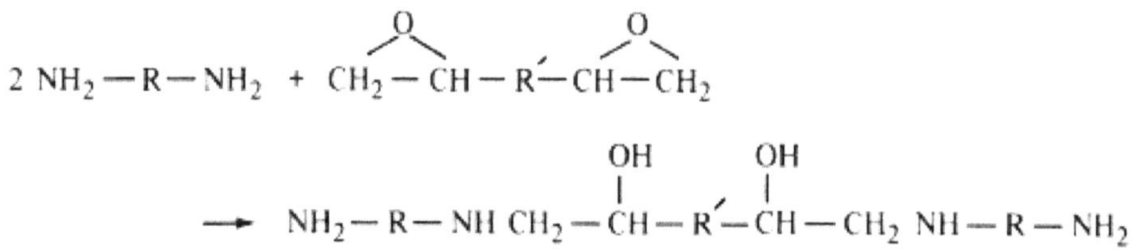

Figure 1 — Reaction of epoxy resin with excess diamine.

Cathodic epoxy systems may be either primary or secondary amino functional, reacting epoxy resins with amines or quaternary amine salts can be used to give adducts with terminal secondary, tertiary, or quaternary amines or their salts. The relatively weak nature of these amines can result in poor dispersibility of these systems, and it is more desirable to incorporate primary amine groups. This can be difficult to achieve. However, useful methods have been developed; these include reacting an excess of a di-primary amine with an epoxy resin , or blocking primary amine groups on molecules with ketones, before reaction with epoxides through other functionalities . These are illustrated in Fig. 1 and 2.

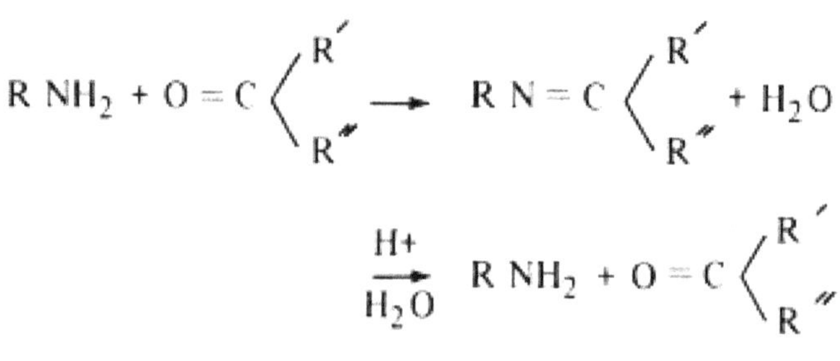

Figure 2 — Ketimine formation and hydrolysis.

Because of their extremely high film alkalinity these systems crosslink sluggishly with melamine and phenolic crosslinkers. They can be effectively crosslinked with blocked isocyanates, which are designed to be stable at bath temperatures but unblock at reasonable stoving temperatures . Acrylic cationic systems may alsobe cured with these curing agents for one-coat systems where good colour requirements are more important than corrosion resistance.

High Solids Coatings

Changing one's paint formulation to higher solids is the simplest means available to reduce actual organic solvent content to achieve lower VOC. Current expectation will be for high solids spray paint to be applied at >75% solids, and for a brush applied paint to be >85% solids. The trend to higher solids and hence lower solvent content is seen in all areas of paint application.

A number of approaches to achieving higher solids are possible. Powder coatings are of course a 100% solids system. 100% solids can be achieved where reactive diluents (e.g. unsaturated monomers) are used in place of solvent, as is the practice with unsaturated polyesters, and with radiation-curing polymers. The use of reactive diluents as a partial replacement for solvent is possible with most systems. With epoxy systems, low molecular weight or mono functional epoxides can be used. With PU systems, oxazolidines and low molecular weight polyols can be used. As described earlier, with autoxidatively drying alkyds, some solvent may be replaced by diluent which is a vinyl monomer, which copolymerizes into the structure through the free radical nature of the autoxidative drying process.

Allyl ether-based oligomers have been examined as components of high solids alkyd systems (though formation of acrolein limits their use) and specially formulated unsaturated melamine resins have also been used.

On the face of it, with all high solids systems, savings should be made in that less solvent is needed, and less energy used in the evaporation of solvent; however, the approach generally requires special formulating techniques and invariably the use of more expensive ingredients. It is possible to obtain small gains in application solids with any paint by applying at higher viscosity or by heating the paint as a means of reducing viscosity.

However, to achieve a significant increase in solids and to retain good application characteristics, it is necessary to redesign the overall system using polymer of lower molecular weight than before; the result of lower molecular weight is that less diluent is required to achieve application viscosity. However, lower molecular weight polymer always requires more cross linking reaction in order to achieve final film properties, and lowered molecular weight has the consequence that special care is necessary with the concentration of reactive functional groups and their distribution.

Polymers for high solids systems are often referred to as 'oligomeric'. This term is used for low molecular weight polymers whose molecular weight is in that part of the molecular weight spectrum where polymer properties are only just starting to become apparent; this is in particular below the molecular weight where chain entanglement is significantly affecting viscosity. An oligomer may be of molecular weight 1000–5000, where the molecular weight for an older conventional thermosetting polymer might be 10 000–40 000, and a thermoplastic polymer not requiring cross linking 80 000–100 000. Telechelic refers to linear or branched polymer molecules possessing terminal functional groups; the design of these for high solids systems has been described.

To achieve low viscosity/high solids, good understanding is necessary of the relationship between molecular weight and viscosity for the resin, and for the interaction between all the components (oligomer, crosslinker, solvent, and pigment).

Molecular weight and molecular weight distribution both require careful control. Narrowing of the molecular weight distribution is desirable; for alkyd systems narrower molecular weight for high solids systems has been achieved using iso-phthalic acid rather than o-phthalic anhydride.

In conventional coatings relatively few functional groups need react to yield cross linked films. In high solids systems, a substantially larger number of groups must react to reach the same final cross linked structure. This increase in functional group content does, however, tend to increase viscosity since these groups are generally polar; this in part offsets the viscosity decrease achieved through molecular weight reduction. The formulator must nevertheless try and ensure that each oligomer molecule has at least two functional groups attached.

Polyester formulation for high solids is relatively simple, using similar ingredients as for higher molecular weight polyester. Polyesters are usually certain to have two terminal functional groups, more if branched. For alkyds, higher functionality in reactants may be required. Thus high solids alkyds may be formulated using six functional dipentaerythritol and tri-functional trimellitic anhydride or tetra- functional pyromellitic anhydride. Higher solids alkyds will be of very long oil length in order to ensure a high overall content of oxidisable groupings and a sufficient number on each individual alkyd polymer molecule.

In the case of acrylic resins additional problems are experienced in preparation over normal formulations. High concentrations of expensive initiator may be required to reduce molecular weight; alternatively, high chain transfer agent concentrations will be required, leading to residual odour problems, difficult to tolerate for some applications. The use of high solids in preparation, which may be essential, results in higher degrees of chain transfer to polymer, leading to broadening of molecular weight distribution and consequent adverse effects on rheology.

Unlike polyesters, in acrylics functional group content from a functional monomer tends to be random ; it can hence be beneficial to include functional transfer agents such as mercapto ethanol to ensure that as many polymer molecules as possible carry at least one terminal reactive group.

Also in acrylics, the use of such bulky monomers as isobornyl methacrylate and cyclohexyl methacrylate has been found to give lower viscosity at comparable molecular weights. Higher molecular density is also achieved through branching of the polymer, so that a hyper branched polymer has a substantially lower viscosity at comparable molecular weight or solution concentration, compared to linear polymer. Formulating practice has developed techniques to provide branched, hyper branched, star and dendritic structures. Acrylic resins can be made more highly branched using ethylene glycol dimethacrylate or using multifunctional transfer agents. Techniques aiming for star or dendritic structures have been applied to obtaining both high solids alkyds and acrylics.

High solids acrylic and polyester resins are normally hydroxy functional, and hexamethoxymethyl melamine resins have found application as cross linkers in high solids thermosetting finishes. The reasons for their suitability have been discussed. With corresponding higher concentrations of functional groups in the oligomer part of the formulation, an amino resin may constitute as much as 50% or more of the composition, compared with 10–40% in more conventional finishes. Blocked isocyanate is also used. High solids 2K finishes with isocyanate adducts use the same acrylic and polyester resins.

POWDER-COATING COMPOSITIONS

Powder coatings possess advantages over conventional coatings in that no polluting solvent loss occurs on application and owing to the use of electrostatic spray, little material is lost. Electrostatic spraying is also employed with solvent-borne coatings and is a method particularly suited to automation. Powders have three distinct temperature regimes to consider and this imposes special constraints on their formulation .The ingredients, and particularly the final composition, have to be solid and glassy at room temperature in order to be ground to size and for the powder to stay free-flowing on storage, and this requires that the blend must not soften or sinter below 40°C. Most powders are produced by comminuting an extruded melt which means that the final curing composition has to withstand this melting process. (Pigmentation is also achieved in the course of the extrusion.) Finally good flowout and then cure must occur at higher temperatures still after application. In formulation and production, additional factors such as particle size control, melt flow characteristics and resistivity need careful attention.

Epoxy compositions are widely used and are based on bisphenol resins cured with amines (dicyandiamide derivatives in particular) or phenolic crosslinkers.

Optionally in so-called hybrid systems, saturated polyester may be mixed with the epoxy. Dimerized fatty acid modification may also be carried out to improve film levelling and adhesion, and epoxy resins containing branched species derived from

epoxidized novolacs may be introduced to increase hardness. Dicarboxylic acid anhydrides and acidic polyesters are also used to cure epoxy powders, and these compositions have better performance than amine cured epoxies, being non-yellowing.

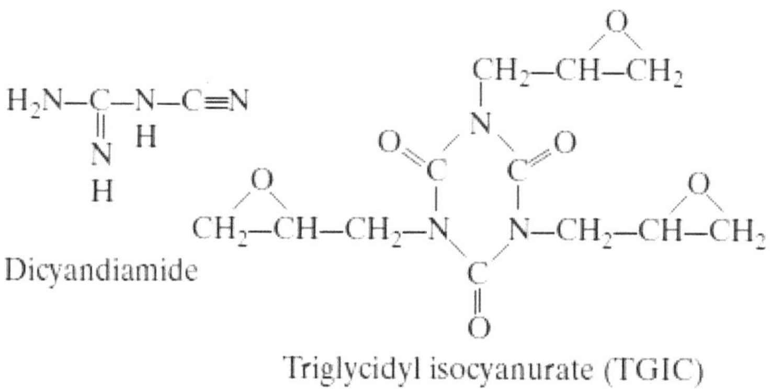

Figure 1 — Typical curing agents for powder coatings.

Polyester powder coatings may be hydroxy functional, and use melamine/ formaldehyde or masked isocyanate curing agents, or may be formulated from acidic polyesters cured with added epoxy resin or triglycidyl isocyanurate (TGIC); the latter offers excellent durability and colour stability. Caprolactam is used as a blocking agent for masked isocyanates for curing powders, though release of this agent from the curing film may be troublesome. Typical curing agents for powder coatings are shown in Fig. 1. The use of TGIC is now under threat on toxicity grounds, but b-hydroxylkylamides are finding some use as a replacement.

Acrylic systems are available but to date have found limited application; possible use in automotive clearcoats may increase use dramatically. Thermoplastic powders are also used containing, for example, nylon 11 or nylon 12, polyvinyl chloride, cellulose acetate and butyrate, or polyethylene. These are used as thicker coatings. Redispersed powders may be applied from water as the so called slurry coatings or aqueous powder suspensions, and may also be electrodeposited (electro powder coatings).

COAL TAR AND ASPHALT COATINGS

Coal Tar Types

Bituminous coal, a very complex chemical mixture, decomposes into simpler components when heated in retorts without air above 700°C (1292°F). Gas, aqueous vapor, and coal tar are driven off, leaving coke as residue. The coat tar is dehydrated and heated in stills to yield oil and coal tar pitch. Depending on the source of the coal tar and the amount of heat applied, pitches of different characteristics are obtained.

When used as bases for superior coatings, coal tar pitches are reprocessed, and any corrosion-accelerating substances are removed. Various types of coal tar pitches are then blended together.

The outstanding quality of coal tar paints is their extremely low permeability, their high electrolytic resistance, and their remarkable resistance to the disintegrating action of water. There are hardly any materials, old or new, that are as water resistant as properly compounded coal tar coatings. They will not be affected by mineral oil but may be dissolved by vegetable and animal oil, grease, and detergents, if they are in direct contact with them. Their resistance to weak mineral acids, alkalis, salts, brine solutions, and other aggressive chemicals is good. Furthermore, coal tar paints give more value per dollar than any other protective coating. This fact should not be overlooked when selecting paint for a certain job.

Coal tar paints are made by dissolving the processed pitch or blend of pitches in suitable solvents. Skill and experience are essential in compounding coal tar paints because similar physical characteristics of raw materials do not necessarily mean similar behavior of the finished product under exposure. The raw materials selected for the blending, the degree of refining, and the addition of other modifiers, often in small quantities, decide the final merit of the coating.

There are five main types of coal tar coatings:

- Thin coal tar pitch solutions without any filler
- Heavy coal tar pitch solutions with inert fillers added
- Very heavy coal tar pitch coatings containing inert fillers possessing a thixotropic gel structure but only medium inherent viscosity
- Heavy coal tar emulsions containing inert fillers and having low inherent viscosity
- Hot applied coal tar coatings

The first three types are solutions and have about the same chemical and water resistance. They vary mainly in the thickness of the film that can be laid down in a single coat. The fourth type, coal tar emulsions, consists of dispersions of coal tar pitch in water and are inferior in corrosion resistance to the solution type. This is not a fault of the pitch itself but is caused by the higher permeability of the applied film. Pitch particles dispersed in water are relatively large and do not coalesce as completely after drying as the much smaller dissolved pitch particles do. But coal tar emulsions have other very good features that will be mentioned later.

The fifth type, coal tar pitch reinforced with inert filler and applied in a molten state over a primer, is called in the trade, coal tar enamel. These enamels have all the good qualities of coal tar paints, but in higher degree, because the coating is very thick and does not depend on the evaporation of solvent to set.

Type 1 is a thin coal tar solution of low viscosity with a solids content of 60 to 70% and a spreading rate of 300 to 400 square feet per gallon, and it gives an approximate thickness of 1 to 2 miles per coat.

This thickness cannot be increased because the thin solution cannot be applied at a lower rate without sagging. Type 2 is designed to achieve a heavier coat; a filler coal tar solution must be used, which, in addition to its higher solids content, can be applied at approximately 180 square feet per gallon without sagging. This produces an approximate dry film thickness of 6 mils. To apply even heavier coatings by brush, a gel must be selected that can be applied at the low rate of 75 square feet per gallon without sagging. This will produce a dry film thickness of approximately 16 mils in one coat. Coal tar emulsions will not sag at coverage of 75 square feet per gallon and will give approximately a 12 mil dry film thickness in one coat.

Coal tar paints afford protection by the mechanical exclusion of moisture and air. If they are applied as a continuous film without holidays, they give almost perfect protection. As it is impossible to avoid pinholes and flaws in a one-coat application, more than one coat will be necessary. They dry by solvent evaporation only.

Concrete, as a rule, can be protected with thin coal tar solutions, but steel requires heavier coatings that will form an almost impervious barrier against severe corrosive influences.

All coal tar paints "alligator," more or less, in the sun. The paint will look like an alligator skin, and hence, the name alligatoring. This alligatoring is a surface defect. It is brought about by the hardening of the upper layer of the film, stimulated by the sun's rays. This causes the upper layer to contract, crack, and slip over the lower stratum which is still soft. If not enough coats are applied, these alligator marks can go right down to metal, opening the path for atmospheric corrosion. Alligatoring does not (or only to a limited degree) occur under water, where the coating is protected from the rays of the sun. Coal tar emulsions do not produce this phenomenon, probably because the pitch particles are not fused as tightly as in solution types; therefore, coal tar emulsions can be used as topcoats over badly alligatoring heavy coal tar paints. Bear in mind that these emulsions have less protection capabilities in immersion service and are not recommended for such use.

There are several popular methods to prevent alligatoring of heavy coal tar coatings, which are temporarily exposed to sun and air before submersion. The older method uses a whitewash. Add slowly and simultaneously 150 lb of processed quicklime and 1 gal of boiled linseed oil to 50 gal of water, containing 10 lb of salt dissolved therein. While being mixed and for 15 min thereafter, the mixture shall be stirred continuously and allowed to cool. It shall be free from lumps and foreign matter. The whitewash shall be aged for at least 3 days before application.

A more current approach is to use an acrylic latex paint, or similar emulsion paint applied to the surface of the coal tar coating. However, note that discoloration of this paint's film by the oils in the coal tar coating is not deemed as a cause for failure.

The solvents used in coal tar pitches are strong in odor, and adequate ventilation is necessary during applications and drying. Coal tar emulsions, which use water as a volatile thinner, are in this respect superior and should be used where proper ventilation is not possible. Coal tar paints give excellent protection at low cost in dam and flood control installations, penstocks, piers, marine work, etc.

Type 5 coatings, coal tar enamels, are used extensively for the exterior protection of steel pipelines with or without a reinforcing wrapper. They are applied by machine or hand daubing. For the interior lining of steel or concrete pipes, coal tar enamels are applied by centrifugal spinning. This process furnishes an extremely smooth surface and cuts down friction between pipe surface and moving liquid.

Coal tar enamel is furnished in different types depending on service conditions such as extremes in high and low temperatures or variations in soil stresses. These stresses are caused by certain types of soil encountered in laying underground pipelines. The modifications consist mainly of different degrees of plasticization and the amount of filler added to the coal tar enamel.

It is possible to protect numerous other surfaces with coal tar enamels, but generally, cold applied heavy coal tar coatings are preferred because they are easier to apply, stand change in temperature better than coal tar enamels, do not require a specially trained working crew, and have less overall cost.

Asphaltic Types

Asphalt is a term applied to certain bituminous substances that either occur in a natural state or are obtained as residue from the distillation of asphaltic petroleum.

The best known type of natural asphalt is gilsonite, which occurs in a number of parallel vertical veins in Utah and Colorado and is mined in a rather crude way much the same as coal. The best grade, gilsonite select, is a dark-colored, comparatively hard bitumen that has excellent chemical resistance and good weather resistance. It is one of the most valuable grades of asphaltic coatings.

Asphalts obtained from the distillation of asphaltic petroleums vary in their chemical and physical characteristics according to the nature of the crude petroleum, the length of the distillation process, the temperature at which they are subjected, and whether dry steam or air is introduced during the distillation.

Steam distilled asphalt is asphalt obtained when dry steam is introduced into the bottom of the still during distillation (which minimizes thermal decomposition of the residue). It has high stretching power but is lacking in toughness and is very susceptible to temperature changes. Air blown asphalt is asphalt obtained when air is introduced into the bottom of the still during distillation (which polymerizes the residue, forming more complex molecules). It has low stretching power but is tough and rubber-like and much more resistant to temperature changes. By taking advantage of the different characteristics of gilsonite, steam distilled asphalts, and blown asphalts, numerous asphalts can be produced by dissolving these asphalts alone or in blends of different proportions. As a rule, the higher the proportion of gilsonite, the better the coating will be.

These asphaltic paints dry by evaporation only. They are more weather resistant than corresponding coal tar paints because they do not alligator, but they are inferior in water resistance. Since the solvents used in their manufacture are mild, they do not give taste to water coming in contact with them.

Consequently, asphaltic paints are frequently used in painting steel tanks and concrete reservoirs storing drinking water. These paints may be reinforced with inert fillers to make very thick, durable coatings giving protection at low cost against severe corrosive conditions. They are, at the same time, very weather resistant, because the fillers minimize the degradation of the asphaltic components against the rays of the sun. If the inert fillers consist mainly of finely divided cork, a heavy coating that has insulating properties can be applied to cold service substrates, such as refrigeration tubing, and this will prevent condensation and sweating.

Asphaltic varnishes are made from such natural asphalts as gilsonite, with or without the addition of resin, by cooking with drying oils. The resultant varnishes are durable. Their lasting qualities depend on their oil length, that is, the more oil incorporated (up to a certain limit), the more lasting the varnish will be. Their use, of course, is limited by their black color.

Bituminous colored paints are made out of asphalt varnishes by pigmentation with strong dry colors. Since the varnish to start with is deep brown, only comparatively dark colored paints can be manufactured. But their extreme durability and low price make them very valuable coatings. They exhibit excellent resistance to industrial fumes, condensation, and the destructive action of sunlight. One drawback is the loss of gloss caused by chalking, which occurs quite soon when exposed outdoors. This, however, in no way detracts from their inherent weather resistance.

Hot asphaltic coatings reinforced with inert fillers are often used for protecting steel and concrete. If care is taken to obtain a coating free from pinholes and blisters, excellent protection for long periods can be achieved.

Bituminous coatings of all types can be applied over oil, hardened oil, and oleoresinous paints, but they themselves cannot be painted over with any coatings containing linseed or other drying oils. Oil paints and oleoresinous paints dry to an elastic but eventually very hard film. Applied over a thermoplastic bituminous coating, they cannot follow the contraction and expansion of this underlying thermoplastic material, and will crack and eventually peel off. Furthermore, bituminous substances under the heat of the sun will diffuse through a dried film of oil paint and will sooner or later stain the light colored topcoat. However, bleeding can often be stopped by separating the bituminous coating from the colored topcoat with a special sealer.

SOYBEAN, BLOOD, AND CASEIN GLUES

Soybean Glues

Preparation

Of all the available agricultural "seed meals," only soybean meal has ever developed any significant use as a raw material for wood glues. The principal reason is that its protein content, the active constituent for adhesives, is the highest available among commercial seeds and legumes. The protein assay of oil free soybean meal ranges from about 35 to 55% on a dry basis and averages 44 to 50% in commercially blended soybean meals and flours. The remaining dry meal content consists of about 30% carbohydrates, 3% fiber, and 6% ash.

In addition to protein, soybeans contain a very high percentage of triglyceride oil, which is useful in many ways. As a result of this unusually high content of both edible protein and unsaturated oil, soybeans have been a very important agricultural crop to mankind for about 5000 years. In view of this long history as a foodstuff, the use of soybeans in adhesives for wood is a very recent development, dating only from about 1920.

In practice, the oil is removed from coarse-ground soybean meal by high pressure extrusion or solvent extraction. The two products are then sold separately into their respective markets. The extracted soybeans, as meal or flour, are widely used for human nutrition around the world. For these food applications, soybean meal is deliberately heated or "toasted" during processing at temperatures above 160°F to enhance the digestibility of its proteins and carbohydrates. When soybean meal is intended for adhesive uses, the processing temperature is carefully maintained below 160°F, to preserve the alkaline solubility of its protein content. To further prepare it for adhesive applications, oil-free soybean meal is ground to an extremely find flour, mostly through a 325-mesh screen. At this fineness, measurements for process control must be made by a standard test of fine powders for specific surface, calibrated in square centimeters per gram. The normal range of specific surface for "adhesive grade" soybean flour is 3000 to 6000 cm2/g.

Table .1 — Typical High Alkali Formulation for Soybean Glue

Comonent	Parts by Weight
Water at 60-70°F	175
Adhesive-grade soybean flour	97
Pine oil or diesel oil defoamer	3[a]
Mix 3 min or until smooth	145
Water at 60-70°F	
Mix 2 min or until smooth	12
Fresh hydrated lime (as a slurry in)	24
Water at 60-70°F	
Mix 1 min	14
50% sodium hydroxide solution	
Mix 1 min	25[b]
Sodium silicate solution	5
Mix 1 min	
Orthophenyl phenol	
Mix 10 min	

[a] Normally blended with the soybean flour for dust control.

[b] 8.90% Na_2O, 28.7% SiO_2, 41° Baumé.

Wood Glues

To become an adhesive for wood or paper, soybean flour must first be wetted with plain water and then reacted with a strong alkali such as sodium hydroxide or trisodium phosphate. (The alkali should not be present during initial wetting, or permanent lumps of dry flour will form.) The chemical action of the alkali on wetted soybean flour particles is to unfold or "disperse" their complex protein structure, making all functional sites available for reaction with wood functional groups. The carbohydrate content of soybean flour becomes similarly dispersed. While this alkaline dispersion step is essential for converting soybean proteins into useful adhesives, it also starts a slow process by which the same proteins gradually lose viscosity and adhesive efficiency through hydrolytic destruction. As a result, soybean glues have a definite working life, usually 6 to 8 h at inside temperatures. Fortunately, the acids in wood neutralize most of the dispersing alkali shortly after glue application, so the dry adhesive bonds of soybean glues to wood surfaces are preserved indefinitely.

As the strong alkali in soybean glue reacts with wood, it causes a distinctive brown discoloration called alkali stain. This prominent glue line discoloration has limited the use of soybean adhesives in fine furniture and paper products, for example. As a result, soybean adhesives have found their widest application in structural and paint-grade wood products such as plywood, millwork, flush doors, and prefabricated assemblies. It is possible to prepare nonstaining soybean glues for wood and paper by using much milder alkaline dispersing agents such as calcium or ammonium hydroxide. However, the full adhesive potential of soybean protein is not developed by these moderate alkaline treatments. The resulting glues yield significantly lower bond strengths, adequate for paper and chipboard but not really satisfactory for structural wood joints.

Typical high and low alkali adhesive formulations are given in Table .1 and Table .2. The high alkali glue mix was used successfully from about 1930 to 1960 to bond interior-grade soft-wood structural plywood.

The addition of hydrated lime and sodium silicate in the high alkali formulation accomplishes two purposes: it helps maintain a level glue viscosity for a longer working life, and it improves the water resistance of the cured adhesive bond by forming water-soluble proteinates. Orthophenyl phenol imparts long-term mold and bacteria resistance to soybean adhesive bonds when glued wood products are intended for use in highly humid locations. It is currently approved by the U.S. government for this purpose. Since the alkaline pH of calcium hydroxide is so moderate, lime and soybean flour can also be blended into a single dry package, requiring only the addition of water (in two steps) to prepare the adhesive.

Table .2 — Typical Low Alkali Formulation for Soybean Glue

Component	Part by Weight
Water at 60–70 °F	225
Adhesive-grade soybean flour	97
Pine oil or diesel oil defoamer	3[a]
Mix Until smooth	
Water at 60-70 °F	150
Mix unitl smooth	
Fresh hydrated lime (as a slurry in)	30
Water at 60–70°F	50
Mix 5 min	

[a] Normally blended with the soybean flour for dust control.

This formulation has been used extensively as a briquetting binder for charcoal and other powdered materials and for paper laminating to wood.

One of the big advantages of soybean glues over many synthetic resin adhesives is their capability to be cured either hot or cold. Hot curing is accomplished in a conventional steam or oil-heat press at temperatures between 230 and about 270°F. For plywood, a pressure of 175 psi is generally employed.

Curing time is about 1.5 min per quarter inch of panel thickness, the higher press temperatures being used for the thicker panels. On dry wood, the cold curing of soybean glues is accomplished in an unheated press at about 150 to 175 psi in 15 min. During this clamping cycle, the soybean glue film develops sufficient gel strength by dehydration into dry wood to hold the plies tightly in contact when pressure is removed. Complete adhesive cure develops at room temperature over the next several days, but machining can begin in about 6 h. This widely used and patented procedure was called the No clamp Process. It is unique to soybean-based adhesives and certain low soluble blood glues.

Blends

Another unique feature of soybean proteins as adhesives is their compatibility with other protein adhesive materials to yield "blend glues" of enhanced performance properties. The two most widely used combinations have been with dried soluble animal blood and with milk casein. In each case, the somewhat granular dispersed consistency of the soybean glue adds useful working and curing properties to those of the two animal-derived proteins. The normal combining limits are between about 20:80 and 80:20 soybeans to animal protein, depending on the performance properties desired.

Table .3 — Extracted Soybean Protein Formulation

Component	Part by Weight
Water at 65–75 °F	385
Extracted soybean protein	100[a]
Chaina clay	100[a]
Powdered sodium sulfite	1[a]
Pine oil or diesel oil defoamer	1[a]
Mix 15 min	
50% sodium hydroxide solution	6
Mix 15 min	
Hexamethylenetetramine in	6
Water at 65-75°F	6
Mix 5 min	

For structural wood bonding, they are not as cost effective as soybean flour adhesives and they also tend to flow excessively in wood joints under pressure.

For applications requiring extremely smooth and fluid soybean glues, more in the nature of size coatings, soybean proteins are available as extracted and dried powders at somewhat higher cost. Typical uses would be as paper overlay adhesives, boxboard adhesives, water-based paint binders, and paper sizes.

Table .3 describes typical paper overlay glue. This glue is extremely smooth and stable. Its formulation illustrates the use of a formaldehyde donor, in this case, hexamethylenetetramine, to partially cross-link the dispersed protein for longer working life and improved water resistance. Other formaldehyde sources and additional compounds have been widely used with soybean glues for these purposes. The additions are always small, about 1%, and are usually made at the end of the mix.

For sizes and coatings, other moderately alkaline salts are frequently used to disperse fine soybean flour or extracted soybean protein. These include the sodium or potassium carbonates, phosphates, and borates, as well as ammonium hydroxide. The borates offer some advantage with soybean flour, as they also complex the carbohydrate constituents to yield increased viscosity and improved tack. None of these compositions can be considered to be adhesives in the structural wood bonding sense, however.

BLOOD GLUES

Preparation

Animal blood has a very long history of use as an adhesive in lime mortars, cements, and wood glues. From the Far East to the Baltic to the early Americas, there have been legends and reports of independently discovered adhesives applications from blood. In all cases, the raw material was fresh whole blood, which was subject to rapid spoilage. This undoubtedly limited broader historical uses for blood protein glues.

Blood adhesive technology as we know it today began in about 1900 with the development of a commercial method for drying fresh whole blood without causing it to lose water solubility. Once this had been accomplished, blood proteins could be dried and stored indefinitely for use on demand. The only preparatory step needed was the removal from freshly collected animal blood of the clotting substance fibrin, to make the blood stable for processing.

The most commonly available dried animal bloods are beef and hog, with a lesser quantity from sheep. On a worldwide basis, their primary uses at this time are as protein-rich feed supplements and edible binders for domestic animal and pet foods. In certain cultures these bloods are used extensively for human nutrition. Poultry blood, which has very high lysine content (and too low an intrinsic viscosity for adhesives), is used exclusively as a feed supplement.

For most food applications, whole blood is quickly coagulated to total insolubility with dry heat or steam. For adhesive applications, it is very carefully dried in vacuum pan ovens or spray dryers to yield controllable levels of cold water solubility. Dried bloods in the range of 80 to about 93% solubility (marketed as high soluble bloods) dissolve almost completely in cold tap water. On the addition of alkali to the water, they become extremely smooth, livery gels. Dried blood particles in the range of 25 to 40% solubility (low soluble bloods) are really wetted only in cold water. They require the addition of a fairly strong alkali to the water to become completely dissolved and dispersed into useful adhesive form. Their dispersed consistency is always quite grainy, a characteristic that is particularly useful for certain types of adhesive, described later. Dried bloods in the 40 to 70% solubility range have intermediate dispersed consistencies, tending toward smoothness as the solubility level rises.

Formulation

The adhesive constituents of dried animal blood include serum albumin and globulin and also red cell hemoglobin. Collectively, they provide nearly 100% adhesive-functional proteins in dried blood solids. Due to differences in composition and proportions of these proteins in the blood from various animals, there is wide variation in alkaline dispersed viscosity levels among soluble bloods, hog yielding the lowest and beef the highest. There are also significant viscosity variations within blood samples from a single species due to differences in age, activity, and nutrition. As a result, commercial lots of dried blood for adhesive uses are always blended in large quantities to help maintain uniformity of glue performance. In formulating adhesives from soluble animal bloods, the dry powder must be initially wetted and redissolved in plain water.

Table .4 — Typical Commercial Formulation for Blood Glue

Component	Parts by Weight
90% soluble dried animal blood	100
Water at 60-70°F	80
Mix 3 min or until smooth	
Water at 60-70°F	
Mix until smooth	60–120[a]
Ammonium hydroxide, specific gravity 0.90	
Mix 3 min	6
Powdered paraformaldehyde (sift in slowly while mixing)	15
Allow to stand 30 min	
Mix briefly until glue is fluid and smooth	

[a] Variable for viscosity control

The wetted blood is then subjected to one or more alkaline dispersing steps to unfold the protein molecules and render them fully adhesive. (If the initial water is alkaline, permanent lumps will form.) Unlike the vegetable proteins in legumes such as soybeans, high soluble bloods can be adequately dispersed with moderately alkaline compounds including hydrated lime and ammonia to become useful wood adhesives. Simple dispersions of this kind represented the earliest class of bloodbased adhesives discovered and used around the world. Much later it was learned that the addition to simple blood dispersions of chemical cross-linkers such as aldehydes greatly increased their water resistance.

Glues of this type were successfully used during and after World War I to bond aircraft propeller laminations and other structural components. In this form, blood glues represented the most water resistant wood bonding adhesives available until the advent of phenol-formaldehyde resins about 1932. Table .4 gives a commercial example.

Ammonium hydroxide yields sufficient dispersion at moderate alkalinity to expose most of the adhesive-functional polar groups on the protein structure. The paraformaldehyde actually gels the blood protein briefly, but it thins out again to yield a workable adhesive viscosity. The useful life is about 8 h. This glue delivers adequate bonds when cold pressed, but it provides the most water-resistant bonds if hot pressed to a minimum attained temperature of 160°F.

During World War II and for 20 years thereafter, blood protein glues were used extensively in the manufacture of interior and intermediate grades of structural plywood (water-resistant but not waterproof). By this time it was found that the highly alkaline multistep dispersion techniques employed with soybean glues yielded excellent consistency and adhesion when applied to dried animal bloods in the lower range of solubility. The resulting glues were used in large quantities to bond structural plywood until about 1960. (See Table .5.)

Table .5 — Typical Commercial Formulation for High Alkaline Blood Glues

Component	Parts by Weight
Water at 60-70°F	300
20% Soluble dried animal blood	75
200-Mesh wood flour	22a
Diesel oil (defoamer)	3a
Mix 3 min or until smooth	
Water at 60-70°F	330
Mix 2 min or until smooth	
Fresh hydrated lime (as a slurry in)	10
Water at 60-70°F	20
Mix 1 min	
50% Sodium hydroxide solution	16
Mix 10 min	
Sodium silicate solution	35
Mix 5 min	

[a] Normally dry blended with the dried blood for easier handling and dust control.
[b] 8.90% Na_2O, 28.70% SiO_2, 41° Baumé.

Since prior heat treatment of the blood itself has already reacted or "denatured" a significant portion of its protein content toward insolubility, glues of this type bond strongly and develop significant water resistance when cured without heat. In addition, the lime and silicate dispersion steps yield protein derivatives, which further insolubilize these blood glues upon cold curing. For optimum bond strength and weather resistance, however, hot pressing to a minimum glue line temperature of 160 ° F is still recommended. Whether cured hot or cold, blood glues of this type normally meet the published mold resistance requirements for plywood without the addition of preservatives.

Another form of blood glue employed extensively from World War II until about 1960 (and again during the petrochemical shortage of the 1970s) involves blends of high soluble blood with “adhesive grade” soybean flour. In this combination, blood protein provides the water-resistant bonds and rapid hot cure, while the soybean flour provides the desired granular consistency for machine application and assembly time tolerance and, in addition, reduced cost. A commercial example of these so-called blend glues is given in Table .6.

Table .6 — Typical Commercial Formulation of a Blood–Soybean “Blend Glue”

Component	Part by Weight
Water at 60-70°F 20% Soluble dried animal blood 200-Mesh wood flour Diesel oil (defoamer) Mix 3 min or until smooth Water at 60-70°F Parts by Weight Mix 2 min or until smooth Fresh hydrated lime (as a slurry in) Water at 60-70°F Mix 1 min 50% Sodium hydroxide solution Mix 10 min Sodium silicate solution Mix 5 min	300 75 22[a] 3[a] 330 10 20 16 35

[a] Normally dry blended with the dried blood for easier handling and dust control.
[b] 90% Na_2O, 28.70% SiO_2, 41° Baumé.

Unlike straight, low soluble blood adhesives, soybean–blood blend glues must be hot pressed to ensure adequate water resistance and must also contain a preservative. As an alternative to orthophenyl phenol, final additions of alkaline phenol-formaldehyde liquid resins (resoles) have proved effective for imparting mold resistance to these blend glues. About 15 lb of a 43% solids phenolic plywood adhesive resin would serve well in the formulation of Table 6.

Up to this point, the adhesives discussed have relied on alkaline dispersed animal or vegetable proteins as the active ingredients. Another type of protein-containing wood glue is commercially significant; namely, alkaline phenol-formaldehyde (PF) resin adhesives fortified with soluble dried blood. In this case, the PF resin solids are the principal adhesive constituents, yielding completely waterproof bonds. The blood proteins are also fully durable when cured in a matrix of phenolic solids. In addition, however, they provide the benefit of considerably shorter hot press curing times compared with straight resin adhesives. The blood proteins in this mix, fully dispersed by resin alkalinity, quickly develop a very adequate initial heat-cured bond. The phenolic resin polymers cure more slowly but completely by means of latent heating as the bonded product (plywood) is not stacked after a relatively short pressing cycle. For an example of this combined adhesive, see Table .7.

Table .7 — Typical Commercial Formulation of PF Resin Adhesive Fortified with Dried Blood

Component	Parts by Weight
Water at 70-80°F	291
Component	100
200-Mesh nutshell flour, or equivalent cellulosic extender	30
Industrial wheat flour	30
Dried 90% soluble animal blood	
Mix 2 min or until smooth	
Mix 20 min	30
50% sodium hydroxide solution	
Diesel oil (defoamer)	6
41% solids alkaline phenol formaldehyde plywood adhesive resin, added slowly Mix 5 min or until smooth	1500
Parts by Weight	

The presence of blood protein in this alkaline adhesive mix allows the normal phenolic resin hot press curing times to be reduced about 30%. The blood solids are considered to be additive to the phenolic resin solids in terms of contributing to exterior durability.

Currently, blood–phenolic resin glue similar to this one is being used as a foamable adhesive that permits glue application by extrusion onto dry veneer. After veneer assembly, plywood is produced by hot pressing in the normal manner. The only formulation change needed is a substitution of surfactant for defoamer to promote air entrainment by the dispersed blood protein. (A special intensive mixing device is employed.) This “air extension” of the phenolic adhesive permits a 25% reduction in glue application weights with no apparent loss of adhesion or durability. A few additional observations may assist in understanding the proper applications for blood protein adhesives. Virtually all blood glues are dark reddish-brown. As a result, they are generally excluded from applications in which low color glue lines are a requirement (e.g., furniture, fine millwork, paper bonding).

Alkaline-dispersed blood glues are unusually sensitive to curing with heat. Thus, their use is well worth considering when rapid hot pressing is essential.

When cured cold, alkaline low soluble blood glues perform as well as soybean glues in utilizing a patented short clamping cycle of 15 to 20 min for plywood and flush doors. The gel strength developed in this length of time is adequate to maintain the bond between the plies after pressure is released while full cure is still developing.

The working life of alkaline blood glues is about 8 h at room temperature. By this time, the dispersing alkali itself has begun to degrade the adhesive protein structure through hydrolysis. The addition of complexing chemicals such as formaldehyde donors helps to control this reaction and lengthen useful adhesive life. Blood proteins have some use in paper bonding if the colored hemoglobin is removed from the remaining serum constituents. The resulting protein mixture is similar to casein in most ways and in performance and is utilized in much the same way.

CASEIN GLUES

Preparation

According to archaeological evidence, the adhesive qualities of casein curd from milk were recognized by civilizations as early as that of the Egyptians. The record of the actual use of casein as a glue is more detailed from medieval European history, however. By that time, aqueous mixtures of casein with lime or other alkaline materials were being used for furniture gluing, paint pigment binding, and canvas sizing, to name just a few applications. In the area of furniture and musical instrument assembly, casein glues competed with gelatin glues extracted from animal bones and hides. The gelatin glues were applied hot and yielded quick bonds on cooling. However, they remained permanently sensitive to heat and moisture.

Alkaline casein glues stayed fluid and sticky at room temperature for a considerable length of time, permitting the complex assembly of parts, but they did require clamping until cured. The hardened casein glue bonds had a substantial degree of heat and moisture resistance. It is these attributes of casein glue that have characterized its performance from early history down to the present; namely, long working or assembly time tolerance, cold-clamping until cured, and significant heat and moisture resistance.

Casein protein is recovered from skim milk by acid precipitation at a pH of about 4.5. Mineral acids can be used to promote precipitation, or the milk can be cultured with a Lactobacillus that converts milk sugar to lactic acid, which in turn precipitates the casein. In either case, the precipitated casein curd is washed free of acid, dried, and ground. Casein is frequently identified for sale by the method of its precipitation (i.e., "lactic acid casein").

Because of the widespread use of milk and its by-products as foods, the price of industrial casein tends to rise and fall with the balance of production versus demand for dairy foods on world markets. Currently, the price is $2.50 per pound for edible-grade ground casein, which causes casein-based glues to be fairly expensive adhesive systems. At other times the price per pound may be a dollar or less. Regardless of current price, the unique properties of casein glues will generally justify their continued use. In preparing casein for use as wood glue, the ground product is screened to provide a particle size range of about 30 to 60 mesh. Coarse particles are recycled. Finer particles become the raw material for a variety of paper sizes and adhesives, ranging in application from foils to labels to cigarette paper gluing and bookbinding. The reason for the limited particle size range in casein woodworking glues is related to their formulation and storage as one-package dry compositions requiring only the addition of water to yield working adhesives. These dry compositions contain a substantial proportion of ground casein plus one or more alkaline salts in granular form that provide dispersing alkalinity as soon as water is added. The entire dry composition is oiled with a petroleum product, typically diesel oil, to prevent the absorption of atmospheric humidity during storage and to slow the rate of solubility of all constituents when water is added. This permits all the casein particles to become uniformly wetted in essentially neutral water. Since some alkalinity is present in the mixing water almost immediately, any finely ground casein particles present in the dry composition will quickly form permanent encapsulated lumps. Thus, casein fines simply cannot be present.

FORMULATION

In commercial practice, at least two alkaline agents are used in the formulation of modern casein woodworking glues. A carefully controlled proportion of hydrated lime is present to promote a gradual reaction with the dispersed casein protein to form calcium caseinate, which contributes significantly to water resistance. Strongly active alkali metal salts such as trisodium phosphate, sodium carbonate, and sodium fluoride (which form sodium hydroxide by double decomposition with lime) are also included to accomplish complete protein dispersion. Together, they convert the casein granules into a slick, viscous consistency, which is an excellent cold or hot press-curing adhesive for wood. Some cellulosic filler may also be included to provide glue application control and gap-filling properties. Table .8 gives a typical case in lumber laminating composition. The dry ingredients are intensively blended in an appropriate mixer while defoamer is sprayed in slowly to provide uniform distribution. The dimethylol urea, a very effective protein cross-linker and denaturant, is added in small variable amounts for glue viscosity control.

The casein materials described in Table 19.8 would be mixed as follows: 200 lb of water at 60 to 70°F and 100 lb of dry glue (as in Table 19.8) for the following steps:

- Mix 2 min or until smooth and thickening begins.
- Let stand 15 min or until thinning has occurred.
- Mix 2 min or until smooth.

Table .8 — Typical Casein Lumber Laminating Composition

Dry Glue Composition	Parts by Weight
30-60-Mesh lactic acid casein	30
30-60-Mesh sulfuric acid casein	30
200-Mesh wood flour	10
Fresh hydrated lime	13
Granular trisodium phosphate	8
Granular sodium fluoride	4
Powdered dimethylol urea	0.1
Diesel oil (defoamer)	2.9
Sodium orthophenylphenate	2
	100

Table .9 — Single-Package Casein Glue for Use in Fire Doors

Dry Glue Composition	Parts by Weight
"Adhesive-grade" soybean flour	58
60-Mesh lactic acid casein	19
Fresh hydrated lime	7
200-Mesh wood flour	5
Granular sodium carbonate	5
Granular sodium fluoride	2
Granular trisodium phosphate	1
Diesel oil (defoamer)	3
	100

The early thickened stage through which these adhesives pass requires strong agitation, preferably with counter rotating paddles and a sidewall scraper blade.

Casein adhesives of this type were widely used for lumber laminating, specialty gluing, and millwork assembly from 1930 to 1970, more or less, before being supplanted by the newer synthetic resin adhesives.

Their long assembly time tolerance and cold cure made the casein adhesives ideal for manufacturing very large curved laminated beams, for example. Similarly, their gap-filling properties, moisture resistance, and deep adhesion to almost any wood surface contributed to their reputation as one of the finest general purpose wood adhesives ever developed from a natural source.

One current application for casein glues that has not yet been preempted by synthetic resin adhesives is its use in panel and flush door assembly. For this application, casein is used in combination with adhesive-grade soybean flour to yield alkaline-dispersed protein glues of composite performance. Casein contributes deep bonds, tacky consistency, and heat resistance (particularly important for fire doors), while soybean flour contribute the granular consistency that permits quick water loss and a short clamping cycle cold cure. The fire door application (Table .9) is essentially unique among protein glues. As before, this is a single-package adhesive that requires only the addition of water. The dry ingredients are intensively blended in an appropriate mixer while the defoamer is sprayed in slowly to provide uniform distribution. From the standpoint of glue consistency, the proportion of casein shown is about the maximum at which short cycle clamping of doors and millwork is still possible.

Mixing directions are as follows: 200 lb of water at 60 to 70°F and 100 lb of dry glue (as in Table .9) for the following steps:

- Mix 2 min or until smooth and thickening begins.
- Let stand 15 min or until thinning has occurred.
- Mix 2 min or until smooth.
- Add 50 lb of water at 60 to 79°F, and mix 2 min or until smooth and fluid.

The second water addition may be varied to obtain the desired final viscosity. In general, casein glues perform best on wood in a viscosity range of about 4000 to 8000 cp at room temperature. The application rate to wood surfaces is about 70 to 90 lb per 1000 square feet of joint area.

Casein-containing adhesives for all forms of paper gluing are distinctly different in chemistry. For example, calcium and magnesium salts should not be present. Since these glues must be only mildly alkaline to avoid staining paper surfaces, a different approach to protein dispersion is employed. Typically, this involves preparing a stable casein solution containing 15 to 20% casein solids by using a mild alkali such as ammonium hydroxide or borax and then heating the solution to about 170°F for 10 to 15 min to complete the dispersion. Rapid cooling should follow to prevent degradation. Small amounts of preservatives, protein cross-linkers, and defoamers are frequently added to yield a storable product of the desired viscosity and running properties.

Casein solutions can be used alone, usually with further cross-linking and thickening, as label pastes, bag adhesives, and cigarette paper seam glues, and for a wide variety of other precision paper bonding applications. For these purposes, the same performance characteristics important to wood bonding recommend their use, namely, a sticky, wipe-resistant consistency, strong bonds to cellulose fibers, and significant water resistance.

Casein solutions are often combined with synthetic latex such as neoprene acrylic or styrene-butadiene to yield strongly adhering yet resilient paper coatings and adhesives. These combinations are specifically used for laminating metal foil to paper. The casein "glue" imparts a degree of rigidity for paper handling that the latex themselves tend to lack.

FISH GELATIN AND FISH GLUE

Introduction

Fish gelatin, or fish glue, is a proteinaceous material extracted from the skins of deep cold-water fish such as cod, haddock, and pollock. The skin and bone of all animals and fish contain collagen, which can be hydrolyzed in hot water and dilute acid to form soluble gelatin.

The difference between gelatin and glue is in the extent of processing. A gelatin processed for photographic use would be exceptionally pure, of high molecular weight, and devoid of specific chemical impurities. An edible gelatin would be free of heavy metals, microorganisms, and any impurity that would make it unsuitable for human consumption. Inedible gelating or glue does not require as much refining and is suitable for industrial or adhesive applications, where less stringent requirements are found. Fish gelatin coatings are used in all these applications.

Properties

Fish gelatin is a long chain protein molecule containing 20 different amino acids. It is amphoteric; that is, it can react as a base or acid. End groups include hydroxy, carboxy, and amino. Reactivity of each group will depend on the pH of the gelatin solution. For example, the reaction of phthalic anhydride with gelatin on the alkaline side will produce a gelatin that will precipitate at pH 4.

Fish gelatin is insolubilized by the addition of salts of polyvalent cations such as ferric sulfate or chrome alum. This principle finds use in the light sensitivity of fish gelatin coatings containing bichromates.

The action of ultraviolet light on the bichromated gelatin results in the formation of polyvalent chromium ions that insolubilize the light-struck area, leaving the unexposed areas water soluble.

Aldehydes such as formaldehyde, glutaraldehyde, and glyoxal will insolubilize gelatin. The aldehydes react with the amino end groups, and this occurs more quickly on the alkaline side.

Fish gelatin solutions are liquid at room temperature and, hence, make good coating vehicles. Many of the reactions above can be tailored so that water solutions can be coated and dried, and the dried coating then becomes insoluble.

Dried fish gelatin films are insoluble in organic solvents, but a water solution will tolerate watermiscible solvents. The following are tolerance levels of various solvents in 100 parts of 45% fish gelatin:

- Acetone, 25 parts
- Ethyl alcohol, 50 parts
- Methyl cell1usolve, 95 parts
- Dimethylforma1mide, 110 parts

Dried fish gelatin coating will not soften at elevated temperatures and can withstand temperatures up to 260°C. Baking at high temperatures (≤ 275°C) will insolubilize gelatin coatings, but they can be stripped with dilute caustic and by proteinaceous enzymes. The following are typical properties of fish gelatin and glue:

- Color — clear, light amber
- Solids — 45% in water
- Viscosity at 20°C — 6000 to 8000 cp
- Average molecular weight — 60,000 d
- Gel point — 5 to 10°C
- Ash — 0.1%
- pH — 4.6 to 5.4

Applications

Remoistenable Coatings: Gummed Tape

Fish glue coatings are excellent for gummed tape, labels, and tags. The dried coating wets very easily and develops a tackiness that allows it to be bounded quickly to another surface. It is compatible with animal glue and can be added to increase the latter's tackiness. The addition of a humectant will reduce curl of the coated paper. Glycerine, sorbitol, or glycols are suitable. Use 5 to 10% based on dry weight.

Photoresists for Photochemical Machining

Photochemical machining involves the manufacture of thin metal parts using a photoresist to image an acid-resistant stencil on the metal sheet. Photoengraving glue, a clarified fish gelatin, is used as the base for a water-soluble photoresist. The advantage over using a solvent-type organic photoresist is the ease of processing, as no organic solvents are required. Water solutions are used in the entire process.

One suggested formula consists of 100 parts by weight, photoengraving glue (45%) 10 parts by weight ammonium bichromate, and 125 to 200 parts by weight water.

The amount of water to be added can be varied to give a specific coating viscosity. A coating thickness of 0.15 to 0.25 mil (4 to 6 µm) is sufficient for etching cold rolled steel, stainless steel, and copper up to 10 mil thickness. This coating is sensitive to ultraviolet light (300 to 400 nm) and will have a shelf life of 2 to 5 days, depending on solids and storage temperatures.

Care must be taken in the use of ammonium bichromate. It is listed as a hazardous chemical, and users must conform to environmental laws. Untreated wastewater containing bichromate should not be discharged into the sanitary sewer or any other water supply.

Dyed Patterns on Glass

Bichromated photoengraving flue images also have good adhesion to glass. These images can be dyed using anionic water-soluble dyes. The same formulation given above can be used, but viscosities must be reduced to obtain a thickness of 0.10 mil (2.5 µm). Images of 10 ì m wide or less are possible, and these can be reproduced very accurately.

For very fine reticles on glass, silver wash-off emulsions can be made using fish gelatin. Silver chloride is precipitated in diluted fish gelatin by the dropwise addition of silver nitrate and sodium chloride. Upon completion, the emulsion is precipitated by the addition of 10% polystyrene sulfonate based on dry gelatin weight and adjustment to a pH 4. This precipitates the emulsion, and the remaining liquid containing dissolved salts is poured off. The emulsion is redissolved by neutralizing and sensitized by the addition of ammonium bichromate or an organic sensitizer such as 4, 4 -diazidostilbene-2,2'- disulfonic acid. Spin coating on glass will result in a 0.1 mil (2.5 µ m) coating. After exposure to ultraviolet light through a negative and washing, an image will appear on the glass. This image can be made black by using a photographic developer such as Eastman Kodak Dectol.

Ceramic Stencils

Colored pigments can be ball milled into fish glue. These inks can be stamped or stenciled onto ceramic parts and the resultant image fired to fix the pigment to the ceramic part.

Electrical Insulators

Fish gelatin, purified to have low ash, can be coated and then baked to reduce the water sensitivity. The coating will have electrical insulating properties.

Temporary Protective Coatings

Because of their water solubility, fish gelatin coatings can be used as temporary protective coating on metal parts during mechanical processing and handling. When complete, the parts are washed with a dilute alkaline cleaner to provide a chemically clean and sterile surface.

Plating Release Agents

Fish gelatin is a liquid protein solution of high molecular weight that has a good attraction to clean metal surfaces. When a clean metal plating master is dipped into a dilute solution, a monomolecular layer attaches itself to the metal surface, which acts as a releasing agent after plating. After plating on the master, the negative can be separated easily by starting at one edge and peeling the two pieces apart. The master is then recleaned, dipped into the dilute gelatin solution, and replated.

Fish Gelatin in Photographic Coatings

Fish gelatin can be used in the silver halide emulsion system. Because fish gelatin solutions are liquid at room temperature (down to 10 ° C), silver halides can be precipitated at temperatures less than required for animal gelatin. The resultant emulsion can be washed of soluble salts by using any of the common precipitating agents such as polystyrene sulfonate.

Photographic film usually requires a sublayer between the emulsion and the film to increase the adhesion of the emulsion to the film. Photographic film also requires a back coating of gelatin to balance the curling stresses of the emulsion coating. Fish gelatin can be used in both these applications.

Table .1 — Effect on Gelling Temperature of Varying Amounts of Animal and Fish Gel

Animal Gel (%)	Fish Gel (%)	Gelling Temperature (°C)
100		32
75	25	30
50	50	26
25	75	24
15	75	20
10	90	19
5	95	17
	100	8

Gelatin Capsules

Fish gelatin added to animal gelatin will reduce the softening point of animal gelatin capsules, and the resultant capsules will dissolve more readily. Table .1 gives an example of the effect on gelling temperature of the addition of fish gelatin to animal gelatin.

Conclusion

Fish gelatin and fish glue have the following features:

- Supplied as a pourable liquid
- Completely water soluble
- Excellent adhesion to metal, rubber, glass, leather, cork, wood, and paper
- Act as protective colloids to suspend small particles or monomers in solution
- Coatings can be made water resistant and insoluble in water
- Insoluble coatings will accept water-soluble dyes
- Insoluble in organic solvents
- Dry to a hard, smooth finish

WAXES

Introduction

A discussion on the use of waxes for coatings must begin with an explanation of what constitutes a wax. Because of the diverse nature of products included in the class of waxes, this is not a simple objective. There are few references that discuss waxes and their applications. In 1956 A.H. Warth published a book that covered the types and applications of waxes available.

Definition

In the medieval days (and earlier) the definition of waxes was very simple. Wax was the substance produced by bees that prevented the honey from running out the bottom of the hive. Beeswax was the only wax known. In 1848 Justus Liebig characterized beeswax as consisting predominantly of long chain fatty acids, esters, and alcohols. Several other natural waxlike products from plant and animal sources had a similar chemical makeup. The definition of a wax was still fairly vague. If the chain length of a substance was longer than the C 16 to C 18 associated with fatty acids, in a chemical sense, it was a wax.

In the 20th century, a large number of new products were introduced. Some of these were waxlike but did not conform to the chemical definition. People wanting wax defined by physical properties won out over those supporting a chemical definition. Therefore, products that have waxlike properties but are not based on long chain esters (e.g., paraffins and low molecular weight polyethylenes) are referred to today as waxes.

In 1954 the Deutsche Gesellschaft für Fettwissenschaft (DGF) established a wax definition based on physical properties and abandoned the efforts to maintain the chemical definition. A revised version was published in "DGF—Einheitsmethoden

Abteilung M—Wachse" in 1975 (Table .1). The failure to put specific limits on most of the listed properties (e.g., no upper limit on melting property and no specific viscosity range) demonstrates the diverse properties of waxes. Only five of the six DGF properties must be met for a product to fall under the definition of wax.

Table .1 — Definition of Wax from a German Technical Society

1. 20°C (68°F) kneadable solid to brittle hard
2. Macro to microcrystalline, translucent to opaque but not glasslike
3. Melting above 40°C (104°F) without decomposition
4. Relatively low viscosity slightly above the melting point
5. Consistency and solubility are strongly dependent on temperature
6. Capable of being polished (i.e., buffable) under slight pressure

Types

One generally attempts to classify waxes into one of the three basic categories of animal, vegetable, or mineral, but even this approach encounters difficulty. Mineral waxes are in reality primordial vegetation deposits subjected to pressure and millions of years of aging to form oil and coal (i.e., aged vegetable wax). A fourth classification is also needed for the numerous synthetic waxes that are derived from the three main sources. The question also arises as to how much refining and alteration of a wax can be made before it is no longer considered a natural wax but a synthetic wax.

Animal Waxes

Animal waxes are derived from two separate sources. One source is by-products produced by the animal (beeswax). The second source is from the animal itself (fatty acid derivatives).

Lesser known waxes include Chinese and shellac waxes obtained from secretions of various insects and wool wax refined from lanolin (extracted from sheep's wool). Commercially, they are found only in a few specialized areas.

Fatty Acids

Fatty acids and their derivatives come from processing tallow and lard. These products are produced in very large volume. The production is often considered to be an industry in and of itself, and these materials are not always thought of as wax products. Chemically they consist of C 16 and C 18 esters of glycerine. Typical products are stearic acid, glycerol tristearates, and glycerol monostearates.

Beeswax

Beeswax is the oldest known wax. It consists of long chain fatty acids and long chain monovalent alcohols. The exact composition of beeswax is extremely complex, and even today an exact duplicate based on synthetic materials has not been accomplished.

Vegetable Sources

Vegetable waxes are obtained from the leaves, stems, or berries of plants. These waxes consist mainly of alkyl esters. Chain length and type vary considerably depending on the surface.

Leaf Waxes

Leaf waxes are obtained by extracting wax from the leaves. They are also referred to as palm waxes. The leaf wax of greatest commercial significance, carnauba wax consists mainly of esters of alkyl alcohols (C 24 to C 34) and alkyl acids (C 18 to C 34). There are also substantial amounts of diesters and hydroxy esters. These hydroxy esters are believed to provide some of the unique paste forming properties and the ability to form microemulsions associated with carnauba wax.

Carnauba wax is extracted from the leaves of the carnauba palm (Copernica cerifera), which grows in the north and northeast regions of Brazil. The tree is also grown in other regions (Sri Lanka, West Central Africa), but only in Brazil are weather conditions suitable (torrential rains followed by a hot, dry windy period) to form significant wax deposits on the leaves.

Stem Waxes

Stem waxes include flax wax, sugarcane wax, and bark wax (Douglas fir). Efforts to develop these waxes commercially have met with limited success.

Berry or Fruit Waxes

A large number of waxlike products are derived from the fruit of various plants. Some waxes exist in the natural state as an outer coating on the fruit (bayberry and Japan waxes). Another source is refined and modified vegetable oils (i.e., hydrogenation). The cocoa fatty acid groups (C 12 to C 14) represent the largest segment, but these are marginally included in the category of waxes because most derivatives are soft and low melting.

Castor wax is derived from castor oil, which is obtained from the seeds of Ricinus communis, a plant that grows in most tropical regions. Castor oil is predominantly (90%) the triglyceride of 12-hydroxystearic acid, with a double bond in the 9–10 position. Hydrogenation of the double bond forms the castor wax.

Mineral Waxes

Mineral waxes are obtained from oil and coal deposits. The name “mineral wax” has nothing to do with the composition but derives from the source (which is beneath the earth’s surface). They are sometimes referred to as “earth waxes.”

Paraffins

Paraffins are extracted from oil. The amount and type of wax present depends on the source of the oil. The paraffin is normally separated from the oil, distilled into several fractions, and sold as commodity items by the oil refineries.

Microcrystalline Waxes

Microcrystalline waxes are normally the higher molecular weight fractions from the paraffin extractions. They vary in hardness and melt properties depending on molecular weight distribution and branching. The name "microcrystalline" evolved from the discovery that these higher melting paraffins were not amorphous but consisted, to a high degree, of microcrystalline structures.

Waxes

Montan waxes are extracted from soft coal deposits located predominantly in Germany, where the wax content in certain coal deposits is 10 to 15%. The coal is crushed and extracted with solvents to yield a dark brown to black crude montan wax with a wax content of 50 to 60%.

SYNTHETIC WAXES

Polyethylene and Polypropylene Waxes

Polyethylene waxes are made by polymerization of ethylene to form ethylene chains similar to polyethylene resin but much lower in molecular weight (2000 to 10,000). The waxes can be oxidized by a variety of methods to produce waxes that are emulsifiable. Polypropylene waxes are formed by controlled polymerization of propylene.

Modified Montan Waxes

Modified montan waxes are derived from coal. The crude montan waxes are subjected to a series of solvent extractions and oxidized with chromium trioxide ($CrO3$) generically referred to as the Gersthofen process. The $CrO3$ selectively attacks tertiary carbon structures, leaving behind only the linear montanic esters and acids. After oxidation has been completed, a light-colored product consisting mainly of montanic acid is formed. The montanic acids can be reesterified with various alcohols and polyols and partially saponified to give a series of montan ester wax products having various properties.

Miscellaneous Synthetic Waxes

Fischer-Tropsch waxes are made by reacting carbon monoxide (from the coal gasification process) with hydrogen gas to form a hydrocarbon wax. Polyglycol waxes are high molecular weight chains of ethylene oxide that form a wax-like product. Oxidized hydrocarbons (paraffins, microcystalline, and polyethelyene waxes) are formed by oxidation (Generally air blown) of hydrocarbon waxes to form a large variety of emulsifiable waxes.

COATING APPLICATIONS

Various waxes are used in different application areas for coatings to enhance certain physical properties.

Mechanism

Waxes generally function by one of two basic mechanisms: The "ball bearing" mechanism is a dispersion of discrete wax particles throughout the coating matrix. The particles that protrude above the film surface prevent the abrading media from contacting the coating surface. Instead, the other surface glides over the protruding wax particles, causing little damage to the surface.

The "migration" mechanism involves migration of the wax to the film surface, where it aids in formation of a smooth glossy film. In some applications, the plastic nature of the wax aids in filling voids formed as the solvent evaporates or the specific volume changes during curing of the resinous components.

Property Enhancement

The major properties affected by waxes are surface related.

Gloss

A high gloss is obtained from waxes predominantly by the migration mechanism. The waxes tend to fill the microvoids formed as the coating dries to provide a smooth continuous film. Often only small amounts of waxes are needed (0.5 to 2% of solids) to obtain a gloss improvement. In fact, too much wax can lead to formation of wax solid on the surface, resulting in a dull haze. A high gloss can be obtained by buffing the surface (property 6 in Table .1), but this is not practical in most situations. If the excess wax were to continue to migrate, the haze could return after several days (or weeks).

Matting

A matte (flat) surface is the opposite of high gloss, but both effects can be obtained using waxes. To obtain a matting effect, the type of wax and method of incorporation must be chosen to maximize the "ball bearing" mechanism. The wax particles that protrude above the surface diffuse the light to reduce glare. Particle size is very critical in obtaining a smooth, silky matte finish that does not feel rough or look dull or hazy. Dispersibility in the paint or varnish vehicle is important to obtain a uniform surface.

Silicas are often used as matting agents, but their high density can lead to problems with settling out. Combinations of waxes and silica can lead to optimum effects. Using agents with differences in refractive indexes can further enhance matting effects without forming a haze.

Slip

Waxes can reduce the coefficient of friction by aiding in the formation of a smooth surface, as discussed, or by reducing the number of contact points by gliding over the wax particles (ball bearing mechanism) .

Soft oily-type agents (soft, microcrystalline waxes or silicones) can also be used to form a slippery lubricating film on the surface. Mar resistance and blocking generally are very poor with this approach.

ABRASION RESISTANCE

The ability of waxes to improve abrasion resistance is generally related to their ability to increase slip.

Antiblocking

Blocking can be reduced by incorporating hard wax particles (ball bearing mechanism) to reduce surface contact. A compromise in gloss must often be accepted. Some hard waxes can reduce blocking by absorbing oils and reducing the migration to the surface of other soft tacky components present in the coating formulation.

Antisettling/Antisagging Agents

When dissolved and cooled in solvent-based systems, many waxes form a thixotropic dispersion. The formation of a gel, which will liquify when stirred, will reduce the tendency of denser materials (pigments, fillers, etc.) to settle out.

Methods

How a wax is incorporated into a coating formulation can affect the end performance. Particle size and uniformity of dispersion are critical factors. The best method depends on the type of wax and the desired end effect. Four basic methods are used.

Wax Compounds

Solvent-dispersed pastes are formed by heating the wax in a suitable solvent (ideally the same as the vehicle used in the coating formulation) and quickly cooling to room temperature under high shear conditions. The rapid cooling and high shear rates are required to prevent the formation of large seed crystals, which would appear as blemishes in the coating when applied. Minor variations in the production of wax compounds can cause changes in the particle size distribution. Because of the sensitivity related to production conditions of wax compounds, most users rely on wax compound producers instead of preparing such waxes in-house.

Micronized Powders

Air milling and classifying the wax powder gives the optimum control of particle size. Micronized powders are based mainly on very hard, high melting waxes such as polyethylene waxes, amides, and polytetrafluoroethylene (PTFE). According to the wax definition, PTFE is not a wax, but because it is used in the same area associated with waxes, it is often included in discussions on wax applications, Attempts to prepare these waxes in a compound form is difficult. Their crystallinity invariably leads to formation of larger particles if the procedure is not properly done.

The dry powder reduces the handling of flammable and hazardous solvents often associated with waxy compounds. Both solvent- and water-dispersible grades are available. On the negative side, improper handling can lead to dusty conditions and explosion hazards. Micronized powders are also more expensive on a per-pound basis (however, greater efficiency can lead to lower use levels).

Milling

The wax can be obtained as a coarse powder and milled in the same manner as pigments, or along with the pigments. This approach provides the end user with wax at the lowest cost. However, the time involved and the cost of the equipment rarely justify the savings in raw materials. Those who currently use this approach do so because they already have the equipment, which would be idle otherwise.

Emulsification

This approach is limited to water-based systems. The emulsion can be prepared in-house, provided the equipment is available, or it can be supplied by outside producers. The wax must be an emulsifiable type. Most commonly used for emulsification are oxidized polyethylene waxes and the refined montan waxes.

03 Colored Inorganic Pigments

THE COLOUR INDEX SYSTEM

The Colour Index System is a coding system developed under the joint sponsorship of the Society of Dyers and Colourists (SDC) in the United Kingdom and the Association of Textile Chemists and Colorists (AATCC) in the United States. The system is referred to as the "Colour Index." In referring to pigments using the Colour Index System, we may describe, for example, Molybdate Orange as Pigment Red 104, Colour Index number 77605.

The Colour Index names for pigments are abbreviated as follows:

- PB = Pigment Blue
- PBr = Pigment Brown
- PM = Pigment Metal
- PV = Pigment Violet
- PW = Pigment White
- PBk = Pigment Black
- PG = Pigment Green
- PO = Pigment Orange
- PR = mPigment Red
- PY = Pigment Yellow

PIGMENT SELECTION

Once a formulator has decided on the shade required for a particular application, the next most important criteria of any pigment are its fastness properties. It is pointless to formulate a coating with a pigment that will not withstand the exposure specifications of the coating's end use. Such specifications can extend to requiring as much as 5 years of outdoor exposure in states such as Rajasthan.

Due attention must be paid to the appropriate manufacturers' literature to ensure that a pigment has been chosen that will satisfy the end-use criteria for fastness to light, solvents, heat, and chemicals, as well as flocculation and crystallization. Once a pigment class has been selected that will perform adequately in the end-use application, the formulator can consider such other factors as economy.

The formulator should be aware; however, that the fastness properties of a pigment will be affected by the medium into which it is incorporated. Thus, even though intrinsically the pigment may feature the required properties, it is still necessary for the final pigmented coating to be tested in the end-use application. Use of "base coat - clear coat" formulations in the automotive and industrial marketplace in recent years has had a major effect on pigment choice in this area, since incorporation of UV absorbers in the clear coat has done much to extend the fastness properties of the pigments used in the base coat.

INORGANIC BLUES

Iron Blue

The most common and economical inorganic blue is iron blue, Pigment Blue 27, a complex ferriferrocyanide having the formula FeNH4Fe (CN)6.xH2O. Originally called Prussian blue, the pigment was discovered by Diesbach in 1704 and was characterized as having a very jet masstone, a red tint, and a very hard texture. Development of a pigment that had a softer texture and improved dispersion by the French chemist Milori led to the introduction to the marketplace of a product chemically identical to Prussian blue but known as Milori Blue.

$$Na_4Fe(CN)_6 + FeSO_4 + (NH_4)_2SO_4 \circledR Fe(NH_4)_2Fe(CN)_6 + 2Na_2SO_4$$

$$6Fe(NH_4)_2Fe(CN)_6 + 3H_2SO_4 + NaClO_3 \circledR 6FeNH_4Fe(CN)_6 + NaCl + 3(NH_4)_2SO_4 + 3H_2O$$

Iron blue is manufactured by reacting ferrous sulfate with sodium ferrocyanide in the presence of ammonium sulfite to yield a leukcoferricyanide known as Berlin White (Equation .1). This intermediate stage is then dissolved in sulfuric acid and oxidized with sodium chlorate to give iron blue (Equation .2). Various specialty grades of iron blue exist that differ in masstone, tint strength, dispersion, and oil absorption.

Chinese Blue is a fine grade that exhibits a green undertone as compared to the original Prussian blue. Bronze blue is a variety that exhibits a surface bronze that depends on the angle at which the surface is viewed.

Iron blue exhibits good resistance to weak acids but very poor resistance to even mildly alkaline systems. Iron blues also show marked tendency to lose their blueness if stored in a paint system that contains oxidizable vehicles such as linseed oil. An extension of this phenomenon occurs when the paint is applied to a surface, since the color will return as the film dries and atmospheric oxygen reverses the reaction.

Although iron blues feature good light fastness properties when used at masstone or deep tint levels, this feature rapidly becomes unacceptable if the pigment is used highly extended with white. Use of iron blues in the coatings industry has been greatly reduced because of the availability of the more economical and durable copper phthalocyanine blue.

Cobalt Blue

Cobalt blue is a mixed metal oxide or solid solution with a typical composition of CoAl2O4 that is acrystalline matrix of Al2O3 (65 to 70%) and CoO (30 to 35%). The pigment is Pigment Blue 28, and it is made by calcining cobalt oxide and aluminum oxide at temperatures approximating 2400°F. The calcined pigment is reduced to a fine powder using physical means such as hammers or air milling.

Cobalt blue is chemically inert, extremely heat stable, and lightfast; it is hard and abrasive and has a high hiding power due to its relatively high opacity. The material is not as strong or as clean as copper phthalocyanine blue and as such is relatively more expensive. The major markets the product serves are the coatings and plastics market, with a particular niche in the vinyl sidings industry.

Ultramarine Blue

Ultramarine blue is catalogued as Pigment Blue 29 and is also known under more common names such as Laundry Blue, Dolly Blue, and lapis lazuli. Chemically the pigment is a sodium aluminum sulfosilicate of inexact composition but closely related to the zeolites and having a suggested formula of Na 6 Al 6 Si 6 O 24 S4.

Two chemically distinct types of ultramarine blue, a green shade and a red shade, can be made by varying the calcinations process by which the product is manufactured. Although their basic characteristics are the same, the green shade contains more aluminum and less sulfur than does the red shade.

The major use for ultramarine blue, as one of its common names implies, is as a laundry aid in soap and detergents. Less than 5% of the production is used in the coatings industry. Its uses are in the area of interior gloss and emulsion paints that require excellent alkali resistance and light fastness.

Cobalt Chromate

Cobalt chromate, also called Cerulean Blue, is related to cobalt blue in that it too is a mixed metal oxide.

Cobalt chromate, however, contains chromium in the matrix in addition to cobalt and aluminum. The empirical formula is Co (Al,Cr) 2 O4, and the compound is catalogued as Pigment Blue 36. The material is a crystalline matrix of cobalt oxide (30 to 35%), chromium oxide (30 to 60%), and aluminum oxide (10 to 40%).

The pigment is made by calcining an appropriate mixture of these metal oxides at a temperature of 2400°F.

Cobalt chromate is used for premium coatings, where outstanding long-term durability and weather fastness are a prerequisite of the desired end use.

INORGANIC BROWNS

The major inorganic browns used in today's coatings applications are those based on oxides of iron; they are available as the naturally mined product or as the synthetically produced material. The coatings market is the largest consumer of iron oxides, irrespective of their hue. Until 1920 all the major iron oxide pigments were wholly from naturally occurring deposits with little changes other than that of physical separation from impurities found alongside the natural ores. Natural iron oxides suffer from the obvious disadvantage of non uniformity because their properties, particularly particle size, vary with their source. For this reason, natural oxides have traditionally carried names such as Raw Sienna, Burnt Sienna, Turkish Umber, and Burnt Umber.

Oxides produced synthetically offer the user a more uniform product, with minimal batch-to-batch variation and controlled physical properties.

Natural Iron Oxides

By definition, these iron oxides are obtained from naturally occurring deposits of ore. The source mines for these pigments can be classified as either "iron ore mines" or "pigment mines." Mines of the former type operate primarily as suppliers of ore as feedstock for blast furnaces with only a small offtake for the pigment industry. Nevertheless, this relatively small output represents the major source of red oxides.

The pigment mines are in existence solely to supply crude ore for pigments. The earth's crust contains massive deposits of both hematite (red iron oxide) and magnetite, the mythical lodestone (black iron oxide). Deposits of the Yellow Ochers and Brown Siennas and Umbers are relatively sparse.

A typical mining operation may involve dumping crude ore on an incline followed by slurring this ore into an aqueous suspension using high pressure water jets. The slurry is next directed through a series of crude washing stages to remove coarse sand and rock particles. This removal of suspended impurities continues as the slurry passes into a separator tank and then through a Dorr bowl rake. The ore leaving the rake overflows to a separating tank, from which it is withdrawn under controlled conditions.

The slurry is dried as a thin aqueous layer using a rotary drum drier. For a particle to be described as pigmentary, it is generally accepted that the particle size should be smaller than 25 μ m. For natural iron oxides to possess this physical property, the dried ore must be pulverized and classified to eliminate agglomerates. After classification, the particle size is defined by reference to the percentage of particles that will pass through a 325-mesh sieve — that is, a screen with a mesh size of 44 ì m. In the drying process, steam is used as the heat source, to permit the hydrated oxides to dry without loss of water of crystallization and consequent color shift. The process of calcining, where temperatures in excess of 1200°F are reached, involves the treatment of the hydrated oxides at temperatures that will result in the complete loss of water to form the corresponding Burnt Siennas and Burnt Umbers.

In today's industry, fluid energy mills and micronizing or jet mills are used to grind and de-aggregate the dried agglomerates. With many commercial pigmentary oxides, the final stage in processing involves the blending of batches showing variations outside the accepted standard specification to produce a standardized product.

Iron Oxide Browns

Pigment Brown 7 is an iron oxide brown that is available in shades ranging from light red to deep purple brown. The empirical formula is Fe2 O3 and as a natural product it contains varying amounts of clay. Offered as Metallic Brown, made from calcined hematite (PR 102) and Burnt Sienna, made from calcined limonite (PY 43), the composition of the pigment varies depending on both the source of the ore and the conditions of calcinations. Major deposits are worked in Cyprus and in Italy; in the United States, there are deposits in Georgia and Virginia.

Pigment Brown 7:x is a ferrosoferric oxide derived from ores containing 25% manganese dioxide having a very indistinct composition that may be described as Fe 2 O 3 · x MnO with varying proportions of clay. Classical names include Raw Umber, Burnt Umber, and Turkish Umber. Principal producers of the "umbers" continue to be the United States, Cyprus, and Turkey.

Synthetic brown oxide is Pigment Brown 6. Also known as brown magnetic iron oxide, the pigment is produced by controlled oxidation of Pigment Black 11. Chemically the material is Fe 2 O 3 with a proportion of FeO and Fe 2 O 3.H2O. Pigment Brown 11 is magnesium ferrite made by calcining a blend of ferric oxide and magnesium oxide. Chemically the pigment is MgO. Fe 2 O3. The volume of brown iron oxides used in coatings is generally low, since most users achieve browns by formulating blends of yellow, red, and black pigments.

Brown iron oxides have good chemical resistance and high tint strength and are permanent pigments. They are ideal pigments for use in wood stains and furniture finishes.

Mixed Metal Oxide Browns

The other browns used in the coatings industry are Pigment Browns 24, 33, and 35. Each is a mixed metal oxide of variable composition.

Pigment Brown 24 is chrome antimony titanium buff rutile made by calcining oxides of titanium, chromium, and antimony. The pigment is a yellowish brown with excellent fastness to heat, light, and weather.

Pigment Brown 33 is zinc iron chromite brown formed by calcining a mixture of the oxides of zinc, chromium, and ferrous and ferric iron. The product is a dark brown with outstanding durability, chemical resistance, and baking fastness.

Pigment Brown 35 is iron chromite brown made by calcining chromium oxide with a mixture of ferric and ferrous oxides. Again, typical of this class of pigments, the fastness properties are excellent.

GREENS

Chrome Green

Chrome greens, Pigment Green 15, are blends or mixtures of a green-shade chrome yellow (PY 34) and iron blue (PB 27). Varying the ratio of these two substituents gives greens with a light yellow shade to a deep, dark shade. An empirical formula is PbCrO 4 . x PbSO 4 . y FeNH 4 Fe (CN)6.

The pigments provide good hiding, high tint strength, and moderate chemical resistance at a relatively good price. Bake temperatures up to 300°F can be tolerated. Use of chrome greens is restricted to exterior and industrial coatings applications because of the lead content of the pigments.

Oxide Green

Chromium oxide green, Pigment Green 17, finds a special application in camouflage paints because of its ability to reflect infrared light. The pigment is pure, calcined chromium oxide, Cr 2 O 3, manufactured by reduction of sodium bichromate with sulfur or carbon.

$$Na_2Cr_2O_7 + 2C \text{ ® } Cr_2O_3 + CO\text{-} + Na_2Co_3 \quad (3)$$

The pigment is used where its price can be justified by the resultant excellent light and chemical fastness properties the pigment gives to the coating.

Hydrated Chromium Oxide Green

Also known as viridian green or Guignet's green, this material is Pigment Green 18 and is chemically a hydrated chromic oxide of formula Cr2O3 . 2H2O made by hydrolyzing the product of the calcining of sodium bichromate with boric acid. The pigment is a bright, blue-shade green with exceptional chroma and brilliance that has outstanding fastness properties in both masstone and deep tints.

Mixed Metal Oxide Greens

Cobalt chromite green, Pigment Green 26, is a mixed metal oxide of variable composition made by calcining cobalt and chromium oxide to give $CoCr_2O_4$. The pigment is a blue shade green with excellent light, weather, and heat fastness.

Cobalt titanate green, Pigment Green 50, is another mixture made by calcining cobalt and titanium oxides to produce Co_2TiO_4. The pigment offers high infrared reflectivity in camouflage paints and outstanding light, weather, chemical, and heat fastness.

INORGANIC ORANGES

Cadmium Orange

Also known as cadmium sulfoselenide orange, this pigment, Pigment Orange 20, is a solid solution produced by calcining cadmium sulfide with cadmium selenide at a temperature approximating 1000C.

Change in the stoichiometry of the cadmium sulfide–cadmium selenide mixture results in a range of pigmentOry products with a shade range from bright yellow (PY 35) to a bright red (PR 108). Barium sulfate added or formed in the process extends the pigment and gives the lithopone version of the orange designated by a suffix as Pigment Orange 20:1. Cadmium orange is used in the coatings industry in industrial color coding applications, where chemical and heat resistance are prime requisites of the applied coating.

Chrome Orange

Chrome orange, Pigment Orange 21, is a basic lead chromate formed under alkaline conditions to give a product of empirical formula $PbCrO_4 . xPbO$. Shades from a yellow-shade to a red-shade orange can be produced, depending on the alkalinity of the reaction mass.

Because of the pigment's low cost and moderate light fastness, the major coatings use of chrome orange is in the production of protective coatings. As with all lead-containing pigments, the color will rapidly darken on exposure to atmospheric sulfur. Restrictions on the use of lead-containing products have resulted in a marked reduction in consumption of chrome orange in recent years.

A variation of chrome orange is Pigment Orange 21:1, a basic lead silicochromate of formula $PbSiO_3 \cdot 3PbO$. This pigments major outlet is in road traffic paints.

Cadmium Mercury Orange

Known by the Ciba-Geigy trade name of Mercadium Orange, Pigment Orange 23 is a solid solution of mercury sulfide in cadmium sulfide. Careful control of the formation of the mixed crystal enables a product of controlled hue to be produced.

The pigment is manufactured by precipitation of the sulfides of mercury and cadmium from a solution containing their soluble salts followed by calcining this precipitate in an inert atmosphere.

$$x\mathrm{CdSO_4} + (1 - x)\mathrm{HgSO_4} + \mathrm{Na_2S} \text{ ® } x\mathrm{CdS} + (1 - x)\mathrm{HgS} + \mathrm{Na_2SO_4} + 2\mathrm{O_2} \quad (4)$$

An extended or lithopone-type pigment, Pigment Orange 23:1, can be produced by using barium sulfate as the precipitant, thus producing insoluble barium sulfate as a by-product of the reaction sequence which is retained with the pigment.

$$x\mathrm{CdSO_4} + (1 - x)\mathrm{HgSO_4} + \mathrm{BaS} \text{ ® } x\mathrm{CdS} + (1 - x)\mathrm{HgS} + \mathrm{BaSO_4} + 2\mathrm{O_2} \quad (5)$$

Cadmium mercury oranges are extremely heat stable, withstanding temperatures up to 700°F, and have excellent chemical resistance and outstanding weatherability and solvent fastness.

REDS

Oxide Reds

As with the iron oxide browns, we again see pigments that are available as both natural and synthetic products. Iron oxide red, Pigment Red 101, is available under such names as Hematite, Persian Gulf oxide, Mars red, Turkey red, Indian red, Ferrite red, and Rouge. The synthetic material is chemically Fe2O3, whereas the naturally occurring product is Fe2O3 with varying amounts of FeO and clay.

Synthetic red iron oxides represent the largest class of synthetic iron oxides manufactured. Four principal manufacturing routes are employed:

1. Copper as red oxide is formed by calcining iron sulfates.

$$12\mathrm{FeSO_4} \times \mathrm{H_2O} + 3\mathrm{O_2} \text{ ® } 2\mathrm{Fe_2O_3} + 4\mathrm{Fe(SO_4)_3} + 12\mathrm{H_2O} \quad (6)$$

2. Synthetic red oxide is produced from synthetic black oxide by calcining the material in a controlled atmosphere containing oxygen.

$$4\mathrm{FeO} \times \mathrm{Fe_2O_3} + \mathrm{O_2} \text{ ® } 6\mathrm{Fe_2O_3} \quad (7)$$

3. Precipitated red oxide is prepared in an aqueous medium by growing seed nuclei in the presence of a ferrous salt and scrap steel. Control of the pigment particle size during the process results in control of the darkness/lightness of the oxide.
4. Synthetic red oxide may be produced by calcining synthetic yellow iron oxide (Fe2O3. H2O) to give the dehydrated product, which is Fe2O3. The wide range of shades available in the red oxide class, coupled with their acid and alkali resistance, heat stability, and the economics associated with using these pigments, accounts for the large tonnage of these pigments used annually in the coatings industry.

Molybdate Orange

Although given the common name classification of an orange, this pigment is in fact Pigment Red 104 with empirical formula $PbCrO_4 \cdot xPbMoO_4 \cdot yPbSO_4$. Other common names for this pigment include Chrome red, Molybdate red, and Chrome vermillion. The pigment is in fact a solid solution of lead chromate, lead molybdate, and lead sulfate. The material is produced by adding a solution of sodium chromate, sodium molybdate, and sodium sulfate under controlled conditions into a solution of lead nitrate at a temperature between zero and 40°C to precipitate the mixed crystal. Control of the particle size distribution and crystalline shape determines the actual hue of the manufactured pigment from redshade yellow to red-shade orange.

Molybdate orange is an opaque pigment with high solvent fastness, moderate heat fastness, and good economy. The pigment has poor alkali and acid resistance, and the nontreated grades will exhibit considerable darkening on exposure to atmospheric pollutants, particularly in heavily industrialized areas.

The pigment is also sensitive to overgrinding when, if care is not exercised, the crystal will disproportionate to give a yellower, weaker looking product.

The major outlet for Molybdate orange is in the coatings industry, with an estimated 60% of the pigment produced going into coatings. In decorative and industrial paints Molybdate orange is used as the basis for many red and orange formulations, particularly in combinations with organic pigments such as quinacridone, when used for original equipment manufacturers (OEM) finishes in the automotive industry, in countries that still allow lead to be used in OEM coatings formulations.

Cadmium Red

Pigment Red 108 is cadmium sulfoselenide red, a solid solution of cadmium sulfide and cadmium selenide produced by calcining a coprecipitate of cadmium sulfide and cadmium sulfoselenide. The amount of cadmium selenide incorporated into the solid solution dictates the final hue of the pigment as does also, to a lesser extent, the temperature of calcinations. The empirical formula is $CdS \cdot xCdSe$.

A lithopone version, $CdS \cdot xCdSe \cdot yBaSO_4$, Pigment Red 108:1, is also available as an extended product coprecipitated with barium sulfate.

The pigment exhibits excellent stability to heat, alkali, organic solvents, and light when used at masstone and high tint levels. It offers clean bright shades of high chroma. Lighter shades will show poorer resistance to light and moisture than will the more intense shades.

Mercury Cadmium Red

Another inorganic pigment, known by the Ciba-Geigy trade name of Mercadium Red, is Pigment Red 113. This calcined coprecipitate of mercuric sulfide with cadmium sulfide has the empirical formula CdS . xHgS.

An extended lithopone type also exists, Pigment Orange 23:1, which can be produced by using barium sulfate as the precipitant, thus producing insoluble barium sulfate as a by-product of the reaction sequence, which is retained with the pigment. Not as heat resistant as cadmium reds but offering economy, good hiding, good solvent resistance, and excellent brightness, mercury cadmium reds do show a marked sensitivity to acids, which precludes their use in coatings that are likely to come into contact with such substances.

INORGANIC VIOLETS

Violet

This violet is Pigment Violet 15, a sodium sulfosilicate prepared by oxidation of Pigment Blue 29, Ultramarine blue. The redness of hue of the violet is determined by controlling the degree of oxidation. Chemically the pigment is $Na_4H_2Al_6Si6O_{24}S_2$.

The pigment possesses a brilliant hue and offers good lightfastness and heat stability. It will react with metals to form sulfides and finds use in cosmetic applications and in nontoxic acrylic poster paints and artists' colors.

Manganese Violet

First introduced at the turn of the century, this pigment, Pigment Violet 16, is manufactured by slurrying together a mixture of manganese dioxide and diammonium phosphate in phosphoric acid at high temperatures. The mixture is dehydrated during the reaction sequence to produce the pigment as MnNH4P2O7. The manufacture is unusual in that unlike many of the inorganic reactions , it does not proceed via a calcinations process.

The pigment does not have a bright hue; it has only moderate opacity and poor tint strength. It has excellent acid resistance and poor alkali resistance. On the positive side, the pigment has high bleed resistance and light fastness. The major outlet for the pigment is in the cosmetics industry and as a toner in the plastics industry.

Metal Oxide Violets

For completeness, the mixed metal oxide violets can be listed as follows:

1. Cobalt violet phosphate, Pigment Violet 14, is a crystalline pigment produced by calcining a mixture of cobalt and phosphorus oxides. The pigment is a blue shade violet of formula Co3 (PO4)2.

Used in artists' colors and plastics, this pigment has poor tint strength and a dirty undertone but outstanding chemical, weather, and light fastness.

2. Cobalt lithium violet phosphate, Pigment Violet 47, is produced by calcining cobalt, lithium, and phosphorus oxide to give CoLiPO4, a red-shade violet. Again this pigment has poor tint strength, lacks brightness, and has a dirty tint tone. Although the pigment has excellent chemical, light, weather, and heat stability, it finds little use in coatings and is primarily used in plastics.
3. Cobalt magnesium borate, Pigment Violet 48, is a crystalline borate manufactured by calcining a mixture of the oxides of cobalt, magnesium, and boron to give a pigment of formula (Co,Mg)2B2O5.

Another red-shade violet, this pigment features the same properties as cobalt lithium violet phosphate and finds its major use in engineering resins.

4. Cobalt ammonium phosphate, Pigment Violet 49, is produced by a coprecipitation process to give a crystalline pigment of formula CoNH4PO4 . H2O. Used in artists colors and as a toner in whites, this pigment exhibits excellent solvent and light fastness. The pigment has poor acid and alkali resistance, poor bake stability, and poor tint strength.

INORGANIC YELLOWS

Strontium Yellow

This yellow is Pigment Yellow 32 and is chemically strontium chromate, SrCrO4, prepared by precipitating a soluble chromate with a solution of an appropriate strontium salt. The pigment is used primarily in corrosion-inhibiting coatings. Its poor tint strength, low hiding power, and unsatisfactory acid and alkali resistance limit the pigment's use in the coatings industry.

Primrose Chrome Yellow

Primrose chrome yellow, Pigment Yellow 34, is a coprecipitate of lead chromate and lead sulfate of empirical formula PbCrO4 . X PbSO4.

$$2Pb(NO_3)_2 + H_2O + Na_2Cr_2O_7 \rightarrow 2PbCrO_4 + 2NaNO_3 + 2HNO_3 \quad (8)$$

Primrose chrome is made by substitution of a soluble sulfate for part of the sodium bichromate in Equation .8; 23 to 30% lead sulfate is contained in typical solid solutions, which are sold as Primrose chrome.

Primrose chrome is stabilized by the use of proprietary additives during the formation of the pigment crystals, such that the orthorhombic crystal form is produced exclusively. The pigment must not be precipitated as the monoclinic form; otherwise instability will result.

Many different grades of lead chromate based pigments are available to offer such improved properties as better chemical resistance, decreased tendency to darken on exposure to the atmosphere, and silica encapsulation to minimize the solubility of the lead in the pigment.

Primrose chrome offers a green shade pigment of good light fastness, high opacity, low rheology, and low cost. The coatings industry is the largest consumer of Primrose chrome yellow, followed closely by the printing ink and plastics industries.

Cadmium Zinc Yellow

This pigment is yet another coprecipitate or solid solution. For this pigment, Pigment Yellow 35, cadmium sulfide is coprecipitated with zinc sulfide followed by calcinations to give pure cadmium zinc yellow, CdS . xZnS. The hue of the pigment can be varied by altering the amount of zinc sulfide used in the solid solution. A primrose hue is achieved with levels of 14 to 21% of zinc sulfide, whereas 1 to 7% will give a redder shade or "golden" hue.

A lithopone version, Pigment Yellow 35:1, is available. As with all "lithopones," this pigment is merely manufactured in the presence of barium sulfate such that an intimate precipitate is produced containing barium sulfate as part of the solid solution with formula CdS xZnS yBaSO4.

The major outlet for cadmium zinc yellows is in the plastics industry. The pigments offer bright, clean, high hiding colors with outstanding resistance to heat, light, and solvents. The pigments, however, have poor fastness to mineral acids and will fade markedly when used at low tint levels or in the presence of moisture.

Zinc Chromate

Zinc chromate or zinc yellow is Pigment Yellow 36. A lithopone version, extended with barium sulfate, also exists as Pigment Yellow 36:1. The pigments are bright green-shade yellows made by precipitating hydrated zinc potassium chromate from the reaction of sodium bichromate with zinc oxide and potassium chloride.

Zinc chromate is used principally in corrosion-inhibiting coatings. It has poor tinting strength and poor resistance to mineral acids and alkalies, thus limiting its use within such systems.

Cadmium Sulfide Yellow

Simply calcinated calcium sulfide, CdS, this pigment, Pigment Yellow 37, can be produced in hues ranging from a green shade to a red shade by varying the conditions during calcinations. The pigment offers excellent stability to heat, light, mineral acids, and alkali. A major drawback is the tendency to fade in the presence of moisture. The lithopone version, extended with barium sulfate, CdS . $xBaSO_4$, is Pigment Yellow 37:1.

Oxide Yellows

As with most of the iron oxides discussed, this pigment can be obtained as the natural oxide (Pigment Yellow 43) or the synthetic version (Pigment Yellow 42).

The natural yellow iron oxide is FeO . xH_2O, containing various amounts of clay and other minor minerals. The pigment is also known as limonite and is mined in Africa (African ocher), India (Indian ocher), France (Ocher), and Italy (Sienna). The French material is reported to be of the highest quality.

The synthetic material can be produced in a variety of ways. A direct precipitation process uses an alkali such as ammonium hydroxide and ferrous sulfate. The Penniman-Zoph process uses scrap steel and a ferrous salt to grow seed particles, and the aniline process reacts nitrobenzene with metallic iron to produce iron oxide and aniline, a primary aromatic amine, as a by-product. Irrespective of the manufacturing process, the product has the empirical formula Fe_2O_3 . xH_2O. Iron oxide yellows are economical pigments with excellent light fastness, durability, opacity, and rheology.

On the negative side, they have a dull masstone, which results in dirty tints and only fair tinting strength and baking stability. Nevertheless, they find widespread use within the coatings industry because of their economics of use.

Mixed Metal Oxide Yellows

For completeness, the mixed metal oxide yellows can be listed as follows:

1. Nickel antimony yellow rutile, Pigment Yellow 53, is a product from the calcinations of the oxides of nickel, titanium, and antimony, $(Ti, Ni, Sb)O_2$. A green-shade yellow with excellent light fastness, chemical resistance, and heat fastness, this pigment offers only poor gloss and low tint strength.
2. Zinc ferrite brown spinel, Pigment Yellow 119, is made by calcining the oxides of zinc and ferrous and ferric iron, $(Zn,Fe)Fe_2O_4$. A brown-shade yellow or buff with outstanding durability, chemical fastness, and heat stability, this pigment has moderate tinting ability compared to Pigment Yellow 53 but lacks brightness and has a dirty tint tone.
3. Nickel niobium titanium yellow rutile, Pigment Yellow 161, is a crystalline solid solution prepared by calcining the oxides of titanium, niobium, and nickel, $(Ti,Ni,-Nb)O_2$. This pigment is a greenshade yellow with low tint strength and dirty tint tones. Typical of the mixed metal oxides, the pigment offers high heat resistance and outstanding chemical stability and outdoor durability.

4. Chrome niobium titanium buff rutile, Pigment Yellow 162, is a solid solution obtained from the calcinations of titanium, chromium, and niobium oxides, (Ti,Cr,Nb)O2. It gives a red-shade yellow with outstanding chemical, heat, and light fastness, coupled with excellent durability. Again its tinting strength is poor, and the pigment lacks brightness.
5. Manganese antimony titanium buff rutile, Pigment Yellow 164, is a pigment prepared by calcining the oxides of manganese, titanium, and antimony (Ti,Mn,Sb)O2. Typical oxides are manufactured by calcinations. Although it has low tint strength and lacks brightness and cleanliness, the pigment does offer excellent heatfastness and exterior durability, coupled with outstanding chemical resistance.

Bismuth Vanadate/Molybdate Yellow

The most novel inorganic yellow to appear in the pigment marketplace within the past decade is bismuth vanadate/molybdate, Pigment Yellow 184, which was introduced by BASF in 1985. The pigment is manufactured by dissolving bismuth nitrate, sodium vanadate, and sodium molybdate in nitric acid, precipitating a complex oxide of the metals, followed by calcining the complex coprecipitate to give the pigment as a polycrystalline product. The accepted formula is 4BiVO4· 3BiMoO6.

The pigment is a green-shade yellow used primarily for brilliant solid shades in both automotive and industrial coatings. The pigment has excellent weather fastness, high opacity, and good gloss retention.

04 Organic Pigments

INTRODUCTION

A definition of a pigment, as distinct from a dyestuff, has been prepared by the Dry Color Manufacturers' Association (DCMA) in response to a request from the Toxic Substances Interagency Testing Committee. This definition should clarify the term "pigment" and "dyestuff," which are often erroneously used interchangeably:

Pigments are colored, black, white, or fluorescent particulate organic and inorganic solids which usually are insoluble in, and essentially physically and chemically unaffected by, the vehicle or substrate in which they are incorporated. They alter appearance by selective absorption and/or by scattering of light.

Pigments are usually dispersed in vehicles or substrates for application, as for instance in inks, paints, plastics, or other polymeric materials. Pigments retain a crystal or particulate structure throughout the coloration process.

As a result of the physical and chemical characteristics of pigments, pigments and dyes differ in their application; when a dye is applied, it penetrates the substrate in a soluble form after which it may or may not become insoluble. When a pigment is used to color or opacify a substrate, the finely divided, insoluble solid remains throughout the coloration process.

THE COLOUR INDEX SYSTEM

The Colour Index System is a coding system as developed under the joint sponsorship of the Society of Dyers and Colourists (SDC) in the United Kingdom and the Association of Textile Chemists and Colorists (AATCC) in the United States. The system is referred to as the "Colour Index." (This system is registered as a trade name and the use of the "u" in "Colour" must be retained.) By giving a compound a "Colour Index Name" and a "Colour Index Number," a colored compound can be readily placed into a classification according to its chemical constitution and color. This description is recognized by many government bodies as adequate information for inclusion in hazard data sheets or material safety data sheets to fully identify the pigment in question and to provide accurate reference when including a pigment in any inventory listing. Thus, for example, phthalocyanine blue has a Colour Index name Pigment Blue 15 and the Colour Index number 74160.

Table .1 : Range of Colour Index Constitution Numbers. The Colour Index names for pigments are abbreviated as follows:

PB = Pigment Blue

PBr = Pigment Brown

PM = Pigment Metal

PV = Pigment Violet

PW = Pigment White

PBk = Pigment Black

PG = Pigment Green

PO = Pigment Orange

PR = Pigment Red

PY = Pigment Yellow

Additionally, the Colour Index constitution number conveys information regarding the structure of the compound as shown in Table .1.

Pigment Selection

Once a formulator has decided upon the hue that is required for a particular coating, the next most important criteria of any pigment are its fastness properties. It is useless to formulate any coating with a pigment that will not withstand the exposure specifications of the coating's end use. Such specifications can extend to requiring as much as 5 years' outdoor exposure in states such as Florida and Arizona.

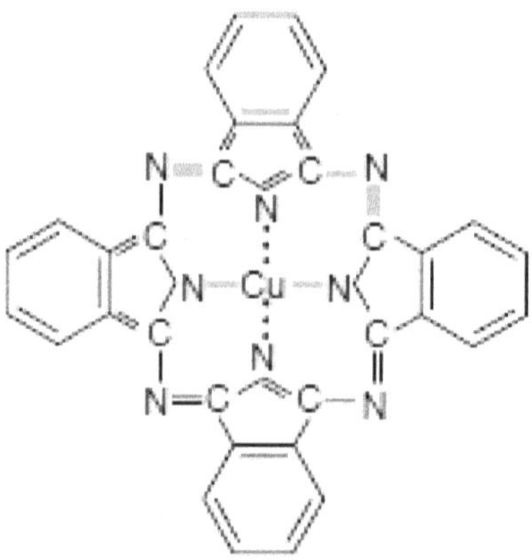

Figure .1 — Structure of copper phthalocyanine blue (pigment blue 15).

Due attention must be paid to the respective manufacturers' literature to ensure that a pigment has been chosen that will satisfy the end-use criteria for stability to flocculation and crystallization, as well as fastness to light, solvents, heat, and chemicals. Once a pigment class has been selected that will perform adequately in the end-use application, the formulator can consider such other factors as economy. The formulator should be aware; however, that the fastness properties of a pigment will be affected by the medium into which it is incorporated. Thus, even though intrinsically the pigment may feature the required properties, it is still necessary for the final pigmented coating to be tested in the end-use application.

ORGANIC BLUES

Copper Phthalocyanine Blue

The major blue used within the coatings industry is copper phthalocyanine blue (PB 15), with its usage far outweighing other blues such as Indathrone blue (PB 60). Phthalocyanines are planar molecules with a tetrabenzotetraazoporphin structure as shown in Figure .1. Manufacture is comparatively easy despite the superficial complexity of the phthalocyanine molecule. Reaction of a phthalic acid derivative at temperatures approximating 190°C with a source of nitrogen such as urea and a metal or metal salt is usually all that is required to produce the appropriate metal phthalocyanine. Molybdate, vanadates, and certain compounds of titanium have been found to be useful catalysts for this condensation reaction.

Figure .2 illustrates the chemistry behind the production of copper phthalocyanine blue. This condensation reaction results in the formation of copper phthalocyanine in a crude, nonpigmentary form. The product has thus to be finished or conditioned to give the pigment grade of choice. Typically crude phthalocyanine blue is characterized by a crystal size of the order of 50 ì m, purity in excess of 92%, and a poor pigmentary strength.

Metal-free phthalocyanine blue (PB 16) is normally manufactured via the sodium salt of phthalonitrile. Acid pasting is used to condition the crude and give the pigment. Copper phthalocyanine is commercially available in two crystal forms known as the á and â . The á form is described by the designations Pigment Blue 15, 15:1, and 15:2 and is a bright red-shade blue pigment. The â form is described as Pigment Blue 15:3 and 15:4 and is a bright green or peacock shade. The á form is meta-stable and requires special treatment to stabilize the crystal against its tendency to a Molybdate or vanadate.revert to the more stable, green-shade crystal. If either of the unstable crystal forms (PB 15 or 15:1) is used with strong solvents, conversion to the form will occur upon storage of the system. Conversionfrom the to the form is usually accompanied by an increase in crystal size with subsequent loss ofstrength and shift to a greener hue.

As stated earlier, copper phthalocyanine gives excellent service in most coatings applications, but there is considerable variation between both the chemical and crystal types available.

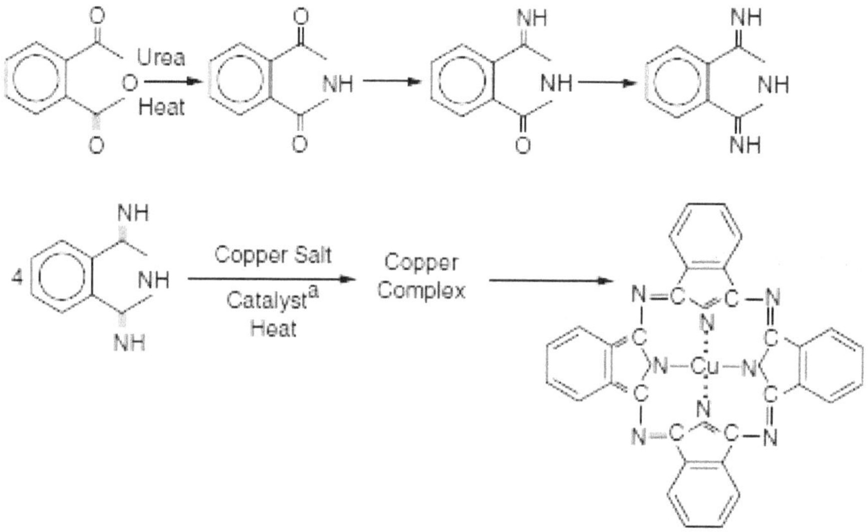

Chemistry of copper phthalocyanine

Pigment Blue 15 is a á crystal with the reddest shade of the types commonly available. It is the least stable of the family and as such is often referred to as crystallizing red-shade (CRS) blue. This crystal form cannot be used in any solvent containing systems.

Pigment Blue 15:1 is also a á crystal, but chemical modifications have been made to stabilize the structure against crystallization. Most commonly the molecule is chlorinated to the extent of introducing one chlorine molecule to give "monochlor" blue. Another technique involves the use of a substituted phthalocyanine, added to the pigment at levels approaching 10 to 15% that confers crystal stability to the system. The monochlorinated grade is, as a consequence of the introduced chlorine atom, greener than the additive-stabilized crystal.

Pigment Blue 15:2, described as "noncrystallizing nonflocculating" red-shade blue, and is widely used within the coatings industry. The product is a á crystal that is stabilized against both crystallization and flocculation using additive technology.

Pigment Blue 15:3 represents the green-shade, â crystal phthalocyanine blue and, as it exists in the stable crystal form, it is less susceptible to crystallization. Most commercial grades of Pigment Blue 15:3, however, contain from 4 to 8% of the á crystal, which will be adversely affected by strong solvent systems.

A 100% â blue is too dull, opaque, and weak to be commercially attractive; hence, a proportion of the á crystal is left in the system, contributing considerably to the attractiveness of the system. Pigment Blue 15:4 represents a â blue that has been modified with phthalocyanine-based additives to give a green-shade blue that is resistant to flocculation and can be used in strong solvent systems.

Copper phthalocyanine approximates the ideal pigment. It offers strength, brightness, economy, and all-around excellent fastness properties. Perhaps the pigment's only disadvantages are its tendencies to change to a coarse, crystalline, non pigmentary form in strong solvents and to flocculate or separate from white pigments when used in paints and lacquers.

Miscellaneous Blues

Although the organic blues used in the coatings industry are primarily copper phthalocyanines, brief mention must be made of other blue pigments that find use in the coatings marketplace.

Indanthrone blue, Pigment Blue 60, belongs to the class of pigments described as "vat pigments." This pigment is expensive relative to copper phthalocyanine, and thus economic considerations are a limitation to its widespread use. Idanthrone blue is a very red-shade pigment with outstanding fastness properties.

Carabazole violet, Pigment Violet 23, is a complex polynuclear pigment that is a very valuable redshade blue of high tinctorial strength. The pigment possesses excellent fastness properties, and only its relatively high cost and its hard nature limit its more widespread use. From an economic standpoint it costs approximately three times as much as phthalocyanine blue.

The pigment is used as a shading component in high performance coatings that call for particularly red-shade blue.

ORGANIC GREENS

Copper Phthalocyanine Green

The major green pigment used as a self shade in the coatings industry is based on halogenated copper phthalocyanine and, as such, is termed phthalocyanine green. The Colour Index names are Pigment Green 7 and Pigment Green 36.

Pigment Green 7, the blue-shade green, is based on chlorinated copper phthalocyanine with a chlorine content that varies from between 13 to 15 atoms per molecule.

Pigment Green 36, the yellower shade, is based on a structure that involves the progressive replacement of chlorine on the phthalocyanine structure with bromine. The composition of Pigment Green 36 varies with respect to the total halogen content, chlorine plus bromine, and in the ratio of bromine to chlorine.

Figure .3 illustrates the proposed structures of the phthalocyanine greens. In practice, no single pigment consists of a specific-molecular species; rather, each pigment is a complex mixture of closely related isomeric compounds.

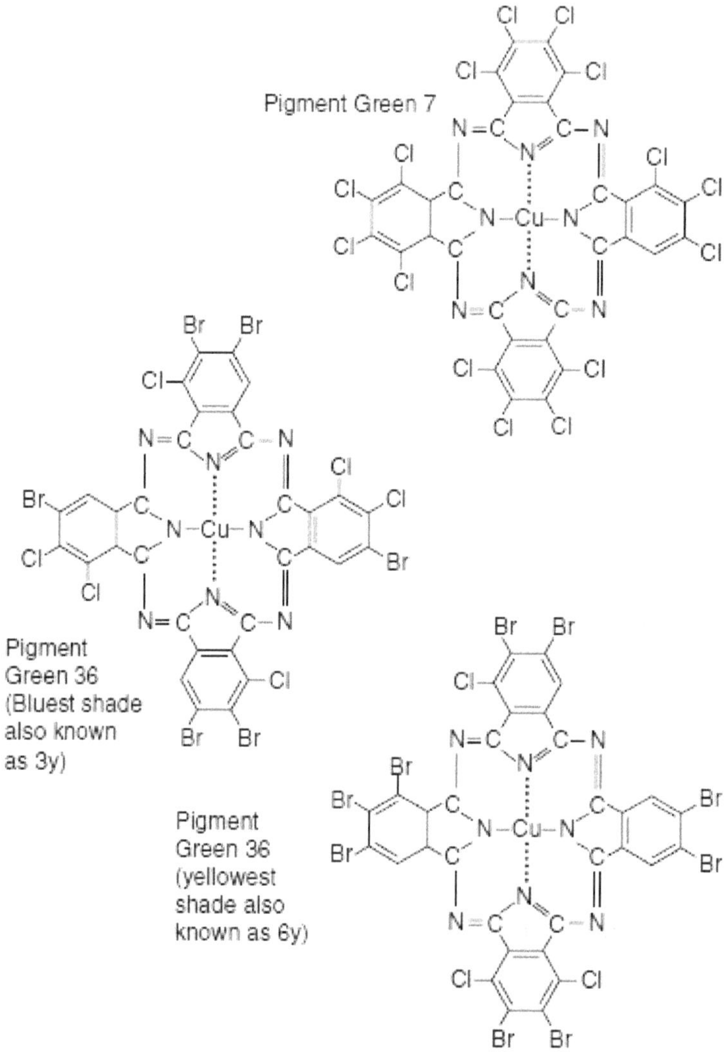

Figure .3: Structures of copper phthalocyanine greens

These pigments are ideal, since their tinctorial and fastness properties allow their use in the most severe application situations. They possess outstanding fastness to solvents, heat, light, and outdoor exposure. They can be used in masstone shades and tints down to the very palest of depths.

Phthalocyanine greens are manufactured by a three-step process: crude phthalocyanine blue is first manufactured, then halogenated to give a crude copper phthalocyanine green, and finally conditioned to give the pigmentary product.

Miscellaneous Greens

Table .2 gives a summary of the properties of some other commercially available organic greens that may find some minor application in the coatings industry.

Table .2 — Summary of Miscellaneous Green Properties

Colour Name	Chemical Name	Comments
PG 1	Brilliant Green (Triphenylmethane PTMA)	Brilliant, blue shade; poor alkali and soap resistance, solvent bleed, and lightfastness; may be used in interior finishes.
PG 2	Permanent Green (Triphenylmethane PTMA)	Blend of Pigment Green 1 and Pigment Yellow 18; bright yellow shade; poor fastness overall
PG 4	Malachite Green (Triphenylmethane PTMA)	Bright, blue shade; poor fastness properties overall
PG 8	Pigment Green B (Nitroso)	Yellow shade; dull hue; poor fastness properties; may be used in interior emulsions
PG 10	Green Gold	Yellow shade; loses metal in strong acid or alkali; good lightfastness; moderate solvent fastness; used in automotive and exterior paints

Table .3 — Organic Orange Pigments

Colour Index Name	Colour Index Number[a]	Chemical Type
PO 2	12060	Azo
PO 5	12075	Azo
PO 13	21110	Bisazo
PO 16	21160	Bisazo
PO 31	n/a	Bisazo condensation
PO 34	21115	Bisazo
PO 36	11780	Benzimidazalone (azo)
PO 38	12367	Azo
PO 43	71105	Azo
PO 46	15602	Perinone
PO 48	n/a	Quinacridone
PO 49	n/a	Quinacridone
PO 51	n/a	Pyranthrone
PO 52	n/a	Pyranthrone
PO 60	n/a	Benzimidazolone (azo)
PO 61	n/a	Tetrachloroisoindolinone
PO 62	n/a	Benzimidazolone (azo)
PO 64	n/a	Heterocyclic hydroxy
PO 67	n/a	Pyrazoloquinazolone

[a] n/a = Not assigned.

Organic Oranges

Table .3 lists the orange pigments that are available in today's marketplace. A broad subdivision can be made based on such chemical features of the molecule as azo based, benzimidazolone based, and miscellaneous oranges.

Azo-Based Oranges

The structures of the seven orange pigments that can be placed into the "azo" category are shown in Figure .4 and are listed below:

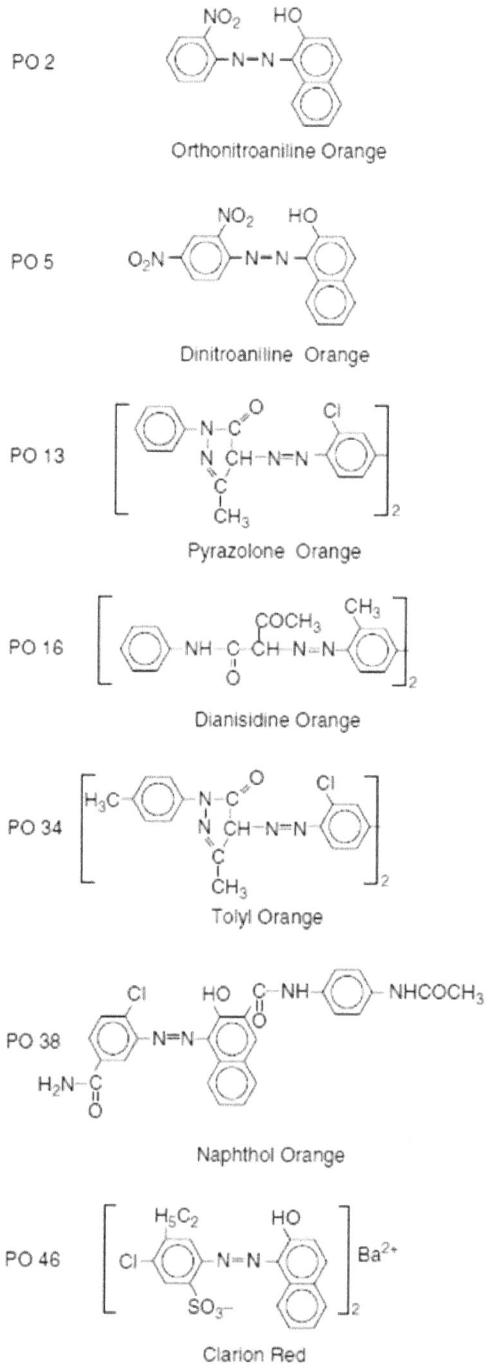

Figure .4: Structure of azo-based oranges

Pigment Orange 2, Orthonitroaniline Orange, is prepared by the classical diazotization and coupling technique in which diazotized orthonitroaniline is coupled to β -napthol. The pigment's major market is in the field of printing inks. Its use in coatings is not recommended because the pigment's solvent fastness is poor and its light fastness inadequate.

Pigment Orange 5, Dinitroaniline Orange, is manufactured by a diazotization and coupling sequence in which diazotized dinitroaniline is coupled onto â -napthol. This pigment offers good light fastness in full tone and moderate solvent fastness. As such, the pigment finds widespread use in latex-based paints and, with the exception of high bake enamels, in both architectural andindustrial coatings.

PO 36

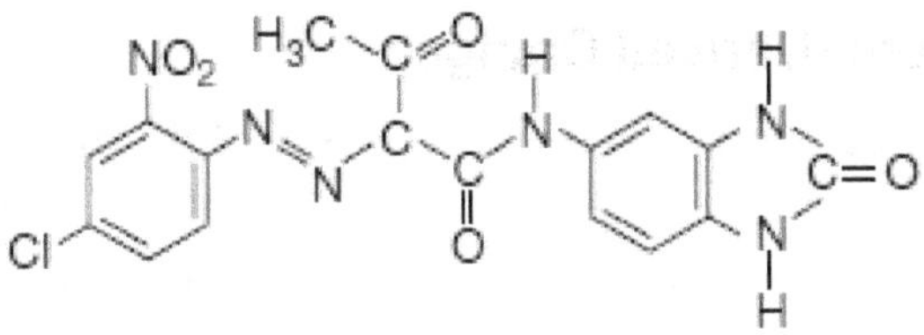

Benzimidazolone Orange

PO 60

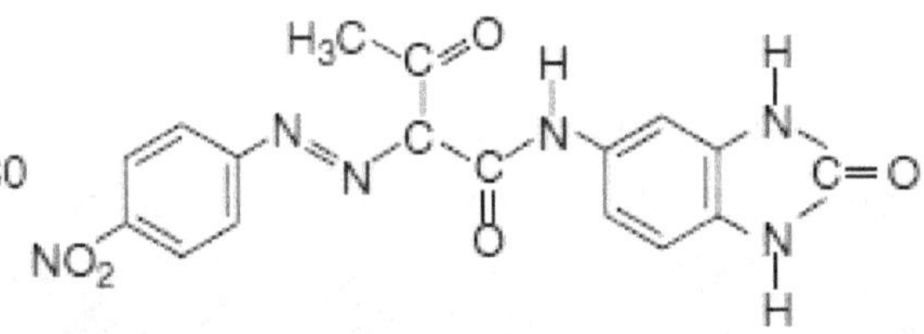

Figure .5 : Structure of the benzimidazolone oranges

Pigment Orange 13, Pyrazolone Orange, is synthesized by coupling tetrazotized 3, 3-dichlorobenzidine onto 3-methyl-1-phenyl-pyrazol-5-one. The pigment is a bright, clean yellow-shade product that is tinctorially stronger than Pigment Orange 5. It may be recommended for interior coatings, particularly as a replacement for lead-based oranges.

Pigment Orange 16, Dianisidine Orange, is a diarylide orange that is prepared by coupling tetrazotized 3, 3-dimethoxybenzidine onto acetoacetanilide. The pigment finds use in baking enamels, since its heat fastness is superior to that of other orange pigments with similar economics. Usage in interior coatings at full tone levels is also recommended.

Pigment Orange 34, Tolyl Orange, is produced by coupling tetrazotized 3,3-dichlorobenzidine onto 3- methyl-1-(4-methyl-phenyl)-pyrazo-5-one. The pigment is a bright, reddish orange that offers moderate light fastness and good alkali resistance, but poor solvent fastness. As such, the material is used in interior coatings applications, particularly where a lead-free formula is specified.

Pigment Orange 38, Naphthol Orange, is manufactured by coupling diazotized 3-amino-4-chlorobenzamide onto 4-acetamido-3-hydroxy-2-napthanilide. The pigment is a bright reddish orange that exhibits excellent alkali and acid fastness, moderate solvent fastness, and acceptable light fastness when used at full tint. As such, the pigment finds use in baking enamels, latex, and masonry paints.

Pigment Orange 46, Clarion Red, is a metallized azo pigment manufactured by coupling diazotized 2- amino-5-chloro-4-ethylbenzene-sulfonic acid onto â -napthol and metallizing the product with barium to yield the pigment. The pigment has poor light fastness, inferior alkali resistance, and inadequate solvent fastness, hence is not recommended for use in coatings.

Benzamidazolone-Derived Oranges

The three benzimidazolone-derived oranges contain the azo chromophore and are all based on the 5- acetoacetylaminobenzimidazolone as the coupling component. Pigment Orange 36 is the product of coupling diazotized 4-chloro-2-nitroaniline to the benzimidazolone.

Pigment Orange 60 is the product of the coupling of 4-nitroaniline to the benzimidazolone.

Because of the proprietary nature of this product, the structure of Pigment Orange 62 has not been fully declared :

Table .4 — Summary of the Properties of the Miscellaneous Oranges

Colour Index Name	Common Name/ Description	Properties
PO 43	Perinone	Red shade, strong, clean, vat pigment with excellent fastness properties; used in metallized finishes and high grade paints; shows slight solvent bleed
PO 48	Quinacridone Gold	Yellow shade; excellent lightfastness; lacks brightness in masstone; used in metallic finishes
PO 49 PO 51	Quinacridone Deep Gold Pyranthrone Orange	Red shade; dull masstone; excellent durability; used in metallics Medium shade; excellent fastness to solvent, light, and heat; dull in tin; exhibits slight solvent bleed; used in air dry and bake enamels
PO 52	Pyranthrone Orange, red shade	Vat pigment with excellent fastness to solvent, light, and heat; dull in tints; slight solvent bleed; used in air dry and bake enamels Medium shade; exhibits some solvent bleed; used in air dry and bake enamels

PO 61	Tetrachloroisoindolinone Orange	Medium shade; exhibits some solvent bleed; used in metallic automotive finishes
PO 64	Bright shade red	Excellent solvent and lightfastness; used in industrial coatings
PO 67	Yellow shed	Excellent brilliance in full shade; good gloss retention; very good weather- and lightfastness in full shade; used in industrial and automotive coatings

Pigment 60 is a transparent, yellow-shade orange that also exhibits excellent heat and solvent fastness with an exterior durability that allows the pigment to be used in high quality industrial and automotive finishes.

Pigment Orange 62 is also a yellow-shade orange that shares the light fastness properties of the other two oranges. Currently it is used in oil-based inks and artists' colors. Its use in the coatings industry has yet to be fully explored.

Miscellaneous Oranges

The structures of Pigment Orange 53, a pyranthrone, Pigment Orange 64, a heterocyclic hydroxy, and Pigment Orange 67, a pyrazoloquinazolone, have not been fully declared. Table .4 summarizes the properties of this class of pigments, which represents a series of oranges that are finding increased application in the coating industry.

REDS

Metallized Azo Reds

Many of the reds used in the coatings industry fall into the chemical category of azo pigments because the azo chromophore —N=N— is a feature of the molecule.

A further subdivision may be made into acid, monazo metallized pigments such as Manganese Red 2B (PR 48:4) and Calcium Lithol (PR 49:2) and nonmetallized azo reds such as the Naphthols (e.g. PR 17 and PR 23) and Toluidine Red (PR 3). Typically, each of the acid, monoazometallized pigments contains an anionic grouping such as a carboxylic (—COOH) or sulfonic acid (—SO 3 H) group, which will ionize and react with a metal cation such as calcium or manganese to form an insoluble, metallized pigment.

Nonmetallized pigments do not contain an anionic group in their structure and, as such, will not complex with a metal cation.

All azo reds contain one or more azo groups and are produced by similar reaction sequences. The initial reaction sequence, described as diazotization, involves reacting an aromatic primary amine with nitrous acid, formed in situ by reacting sodium nitrite with hydrochloric acid at low temperatures to yield a diazonium salt. Invariably the diazonium salt that is formed by this process is unstable and should be kept cold to avoid any decomposition.

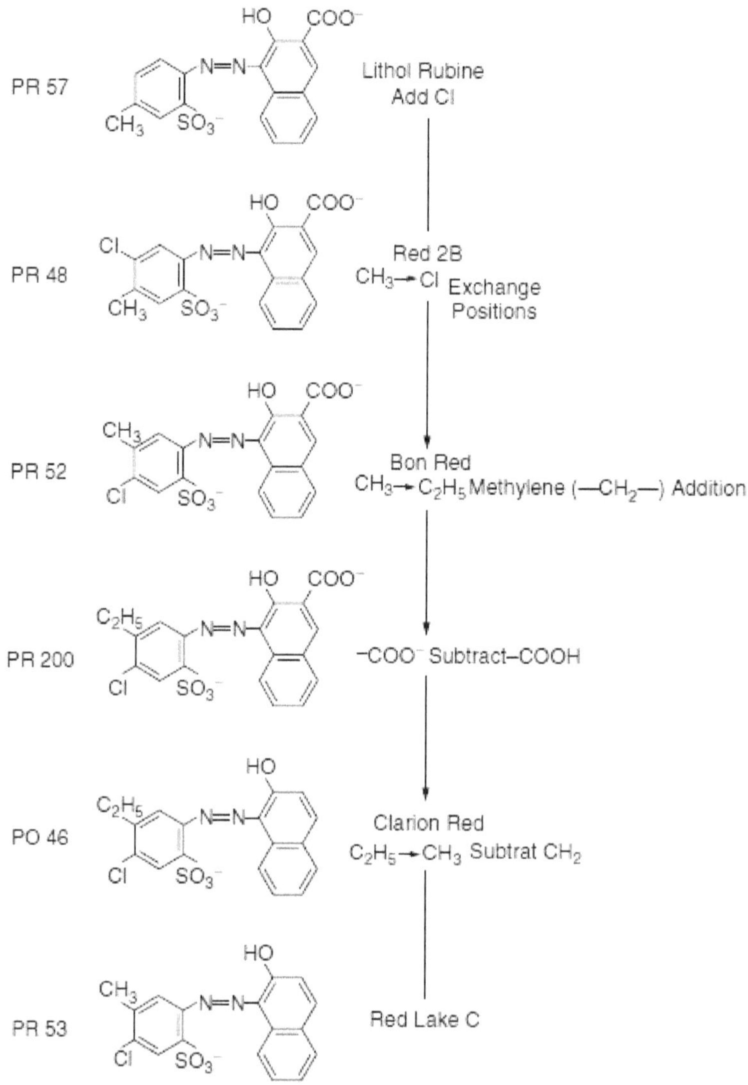

Figure 6 — Metallized azo reds

The diazonium salt is reacted quickly with the second half of the pigment, which is called the coupler.

The coupling reaction takes place rapidly in the cold to yield the sodium salt of the pigment. This sodium salt is all but useless as a pigment for the coatings industry because of its marked tendency to bleed even in the weakest of solvent systems. The pigment is, therefore, metallized to confer improved properties on the product. The pigment suspension is then filtered and washed to remove any residual inorganic derived from the reaction.

Figure .6 illustrates the structures of the different metallized azo reds that are readily available.

The Lithol Reds are primarily Barium Lithol (PR 49:1) and Calcium Lithol (PR 49:2). Although limited in their application in the coating industry, these pigments do find some use at masstone levels — that is, the pigment is not tinted with a white tint base — where their fastness properties are acceptable.

The pigments are bright reds with high tint strength and good dispersion characteristics. The barium salt is lighter and yellower in hue compared to the calcium salt. Permanent Red 2B is the generic name that encompasses Barium Red 2B (PR 48:1), Calcium Red 2B (PR 48:2), and Manganese Red 2B (PR 48:4). The major use of the calcium and barium salts is in baked industrial enamels, which are not required to be fast to outdoor exposure. Use in alkaline systems is severely restricted because of the poor alkaline fastness of these salts.

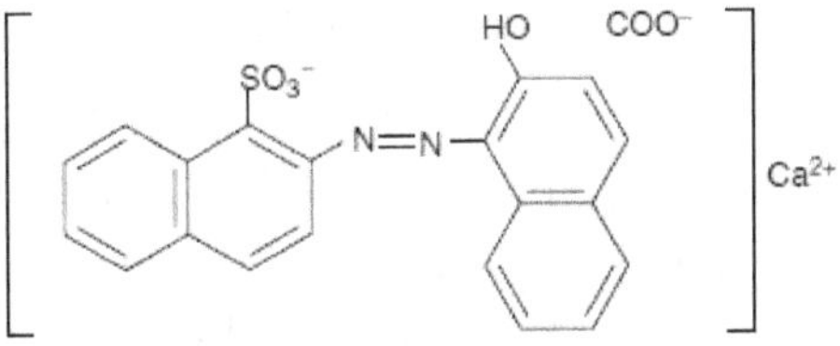

Figure .7: BON maroon

The barium salt is characterized by a clean, yellow hue and, although slightly poorer than the bluer calcium salt in light fastness and tinting strength, it does possess a slight advantage in bake stability.

Manganese Red 2B has a sufficiently improved light fastness to be used in implement finishes. The manganese salt is slightly bluer, dirtier, and less intense when compared to the calcium salt.

Rubine Red also known as Calcium Lithol Rubine (PR 57:1) is a clean, blue-shade red pigment exhibiting the high tinting ability typical of the azo reds of this class. Its major use in coatings is in interior systems that call for an inexpensive red with good solvent and heat resistance. Again, to maintain maximum fastness properties, use of the pigment at near to masstone levels is recommended.

BON red , used as calcium BON Red (PR 52:1) or Manganese BON Red (PR 52:2), is characterized by outstanding cleanliness, brightness, and purity of color. The manganese salt offers a very blue-shade red with improved light fastness compared to the calcium salt. As such, this salt is suitable for exterior coatings applications.

BON Maroon, (PR 63:1) is illustrated in Figure .7; the manganese salt of BON Maroon is of considerably more importance than either the calcium or barium salts. Its light fastness is such that the pigment can be used at masstone levels for implement finishes.

Nonmetallized Azo Reds

As implied by their classification, the nonmetallized azo reds do not contain a precipitating metal cation and, as such, offer increased stability against hydrolysis in strongly acidic or alkaline environments.

Synthesis of this class of pigment follows the previously described classical method of diazotization of a primary aromatic amine followed by coupling of the resultant diazonium salt. No anionic groups capable of accepting a metal cation are present in the molecule; thus metallization is not a factor in their synthesis. Typical nonmetallized reds are Toluidine Red (PR 3) and the wide range of Napthol Reds as represented by Pigment Reds 17, 22, and 23.

Toluidine Red is used in full shade in such coatings applications as farm implements, lawn and garden equipment, and bulletin paints, where a bright, economical red of adequate light fastness is required. Because of the pigment's poor durability in tint shades, it is rarely used at any level other than a full shade.

The individual properties of the Napthol Reds depend on the specific composition of the product as well as the conditioning steps used during pigment manufacture. As a class, they are a group of pigments that exhibit good tinctorial properties combined with moderate fastness to heat, light, and solvents.

Unlike the metallized azo reds, the Napthol Reds are extremely resistant to acid, alkali, and soap. These properties lead to their use in latex emulsion systems and masonry paint. In terms of performance and economic characteristics, the Napthols form a link between the toluidine reds at the lower end of the scale and the perylene and quinacridone reds at the higher end.

High Performance Reds

Pigments for the exacting standards of today's automotive coatings are required to show satisfactory durability to outdoor exposure in such states as Arizona and Florida for 2 and possibly 5 years before being approved for use in automotive finishes. Similar requirements are placed on pigments chosen for use in automotive repair systems and marine coatings.

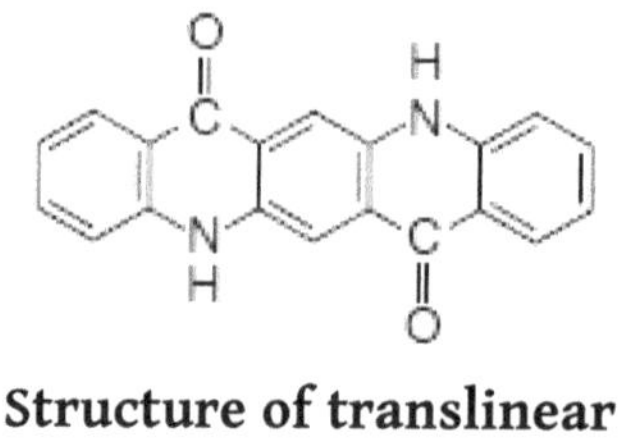

Figure .8 — Structure of translinear quinacridone

High performance reds fall into four basic classes: quinacridone reds and violets, reds based on vat dyestuffs known to include the perylene reds, reds derived from the benzimidazolone diazonium salts, and the disazo condensation reds.

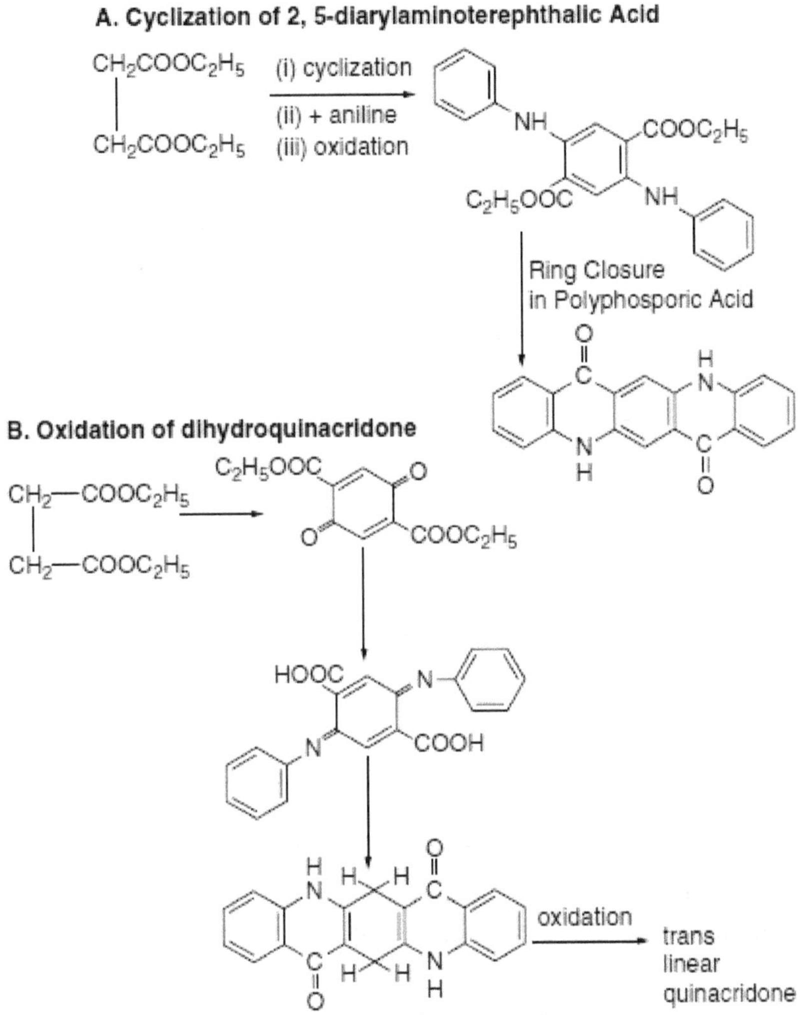

Figure .9 — Routes to quinacridone

Quinacridone Reds

Quinacridones may be described as heterocyclic pigments in that their structure comprises a fused ring structure in which the ring atoms are dissimilar. In the case of quinacridones, the ring atoms are carbon and nitrogen (Figure .8). Addition of differing auxochromic groups such as methyl (—CH 3) and chlorine (—Cl) gives Pigment Red 122 and Pigment Red 202, respectively. The two most commercialized routes in the synthesis of quinacridone (PV 19) involve either the oxidation of dihydroquinacridone or the cyclization of 2, 5-diarylaminoter-ephthalic acid as outlined in Figure .9. Subsequent conditioning leads to the desired crystal modification. Use of 2, 5-diarylamino-terephthalic acid at the cyclization stage yields the unsubstituted trans linear quinacridone. Use of 2,5-ditoluidinoterephthalic acid yields the 2,9-dimethyl quinacridone, Pigment Red 122.

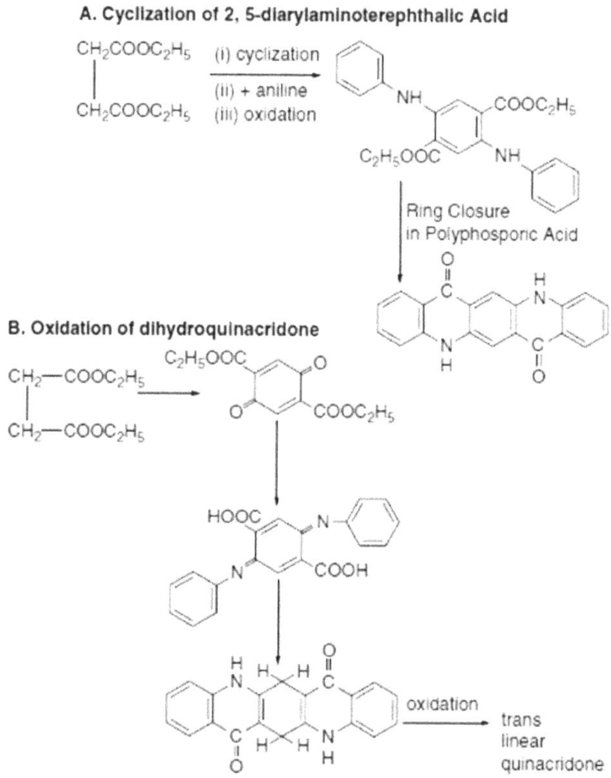

Figure .9: Routes to quinacridone

As a group of high performance pigments, quinacridones find their primary uses in automotivefinishes, both metallic and solid shades, industrial finishes, and exterior finishes.

Table .5 — Types of Quinacridone

Colour Index Name	Hue	Comments
PO 48	Gold	Quinacridone quinone
PO 49	Deep Gold	Quinacridone quinone
PR 122	Magenta-yellow	2,9-Dimethyl quinacridone
PR 192	Red-yellow	Unsymmetrical monomethyl quinacridone
PR 202	Magenta-blue	2,9-Dichloroquinacridone
PR 206	Maroon	Mixed solid solution
PR 207	Scarlet	4,11-Dichloroquinacridone
PR 209	Yellow-shade red	3,10-Dichloroquinacridone
PV 19	Violet-blue	β-Quinacridone
	Red-yellow	γ-Quinacridone
PV 42	Maroon	Mixed solid solution

The pigments combine excellent tinctorial properties with outstanding durability, solvent fastness,lightfastness, heatfastness, and chemical resistance. Table .5 lists the shades currently available.

The pigments combine excellent tinctorial properties with outstanding durability, solvent fastness, light fastness, heat fastness, and chemical resistance. Table .5 lists the shades currently available.

Vat Reds

The red pigments, based on anthraquinine, include such structures as Anthraquinone Red (PR 177), Perinone Red (PR 194), Brominated Pyranthone Red (PR 216), and Pyranthone Red (PR 226), as illustrated in Figure .10.

These anthraquinone-derived pigments are manufactured via a series of complex reactions to include such processing as sulfonation, nitration, halogenation, condensation, and substitution.

Reds

Perylene reds provide pure, transparent shades and novel styling effects when used in metallic finishes. These pigments offer improved flow characteristics when used in highly pigmented coatings formulations such as those required for high solids base coat/clear cost systems, as well as high transparency and good bleed resistance.

The perylenes possess high color strength, good thermal stability, excellent light- and weatherfastness and, with the exception of Pigment Red 224, excellent chemical resistance. Perylenes may also be described as vat pigments and in fact are the only class of vat pigments to be specifically developed as pigments rather than as dyestuffs. Almost all the perylene pigments have a structure as shown by the generic formula illustrated in Figure .11; that is; they are based on N,N - substituted perylene-3,4,9,10-tetracarboxylic diimide. A notable exception is Pigment Red 224 (Figure .12), which is actually derived from the perylene tetracarboxylic dianhydride.

Acenaphthene is oxidized to 1, 8-naphthalic anhydride followed by ammonation to yield the naphthalimide.

The napthalimide is condensed in a fused caustic medium to yield the perylene-3, 4, 9, 10-tetracarboxylic acid diimide. The diimide can then be conditioned to convert the crude into pigment and achieve Pigment Violet 26, or methylated to give Pigment Red 179.

The diimide may be hydrolyzed to produce the dianhydride, Pigment Red 224, or condensed with various aromatic or aliphatic amines to give such pigments as those featured in Figure .13. As with many of the pigments already described, the perylenes have to be conditioned to obtain the compounds in a pigmentary form.

Thioindigo Reds

The thioindiogoid chromophore serves as a nucleus for a wide range of red to violet pigments. The thioindigo reds include Pigment Reds 36, 87, 88, 181, and 198. These pigments are noted for their brightness of shade and all-around good fastness properties. These properties have resulted in the use of the thioindigoes in quality coatings; Pigment Red 88 is the prime pigment, followed by Pigment Red 198.

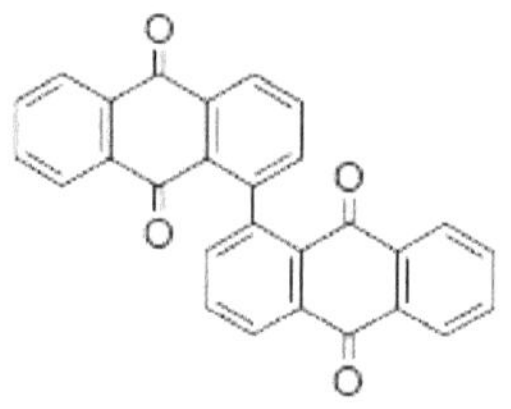

Anthraquinone Red (PR 177)

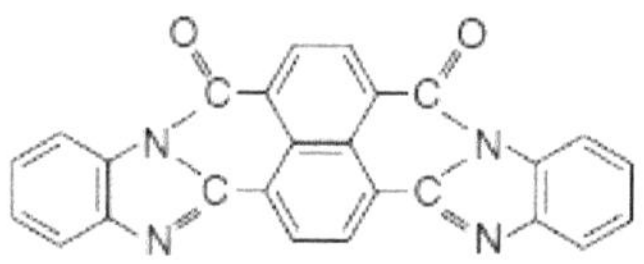

Perinone Red (PR 194)

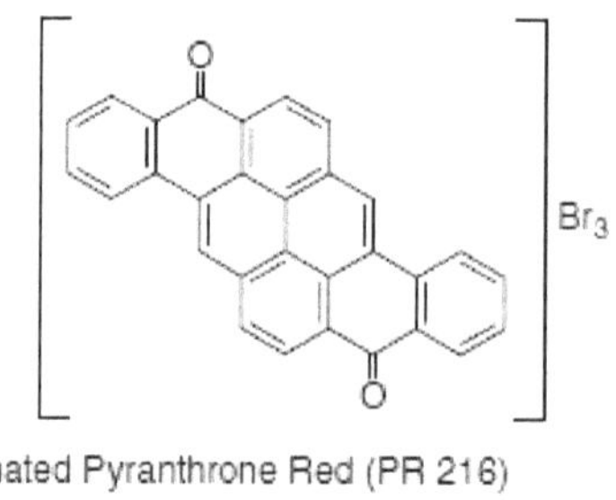

Brominated Pyranthrone Red (PR 216)

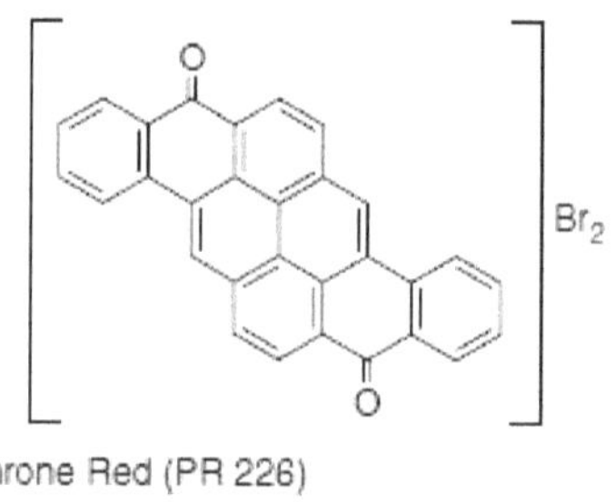

Pyranthrone Red (PR 226)

Figure .10 — Structure of translinear quinacridone.

Benzimidazolone-Based

The benzimidazolone pigments feature Pigment Reds 171, 175, 176, 185, and 208. These pigments are azo-based compounds that contain the benzimidazolone structure as part of their coupling component.

Although used primarily for the coloring of plastics because of their outstanding thermal stability, this class of pigments does find application in the coatings industry. Benzimidazolones show excellent fastness to light, good weatherability, and excellent fastness to over spraying at elevated temperatures. The benzimidazolone pigments are prepared using the diazotization and coupling techniques described earlier.

Generally, after the coupling process, the pigment is treated to obtain a uniform, controlled crystal growth and particle size distribution.

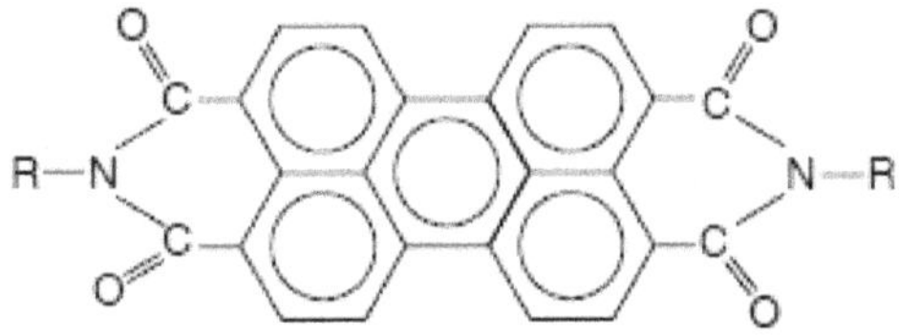

R	Pigment	Shade
R = $-C_6H_3OC_2H_5$	PR 123	Vermillion
R = $-C_6H_3(CH_3)_2$	PR 149	Scarlet
R = $-CH_3$	PR 179	Maroon
R = $-C_6H_5OCH_3$	PR 190	Red
R = $-H$	PBr 26	Bordeaux
R = $-C_6H_5Cl$	PR 189	Yellow Shade Red
R = 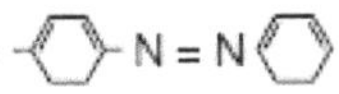	PR 178	Red

Table .11 — Perylene structures.

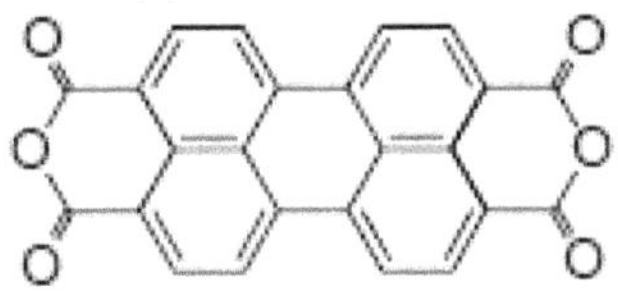

Figure .12 — Pigment Red 224.

Disazo Condensation Reds

These pigments feature such properties as high tinctorial strength and fastness to solvents and heat. Disazo condensation reds have found considerable use as replacement pigments for lead containing pigments.

Their outstanding fastness properties have resulted in their use in high quality industrial finishes.

Figure .14 illustrates three typical structures of the disazo condensation reds. Colour Index names for these three pigments have not been assigned. The synthesis sequence generally is similar for each of the disazo condensation pigments. The azo components are initially coupled to yield monoazo dyestuff carboxylic acids, which are converted to acid chlorides before final conversion to the disazo by condensation with the arylide component to yield the pigment in question.

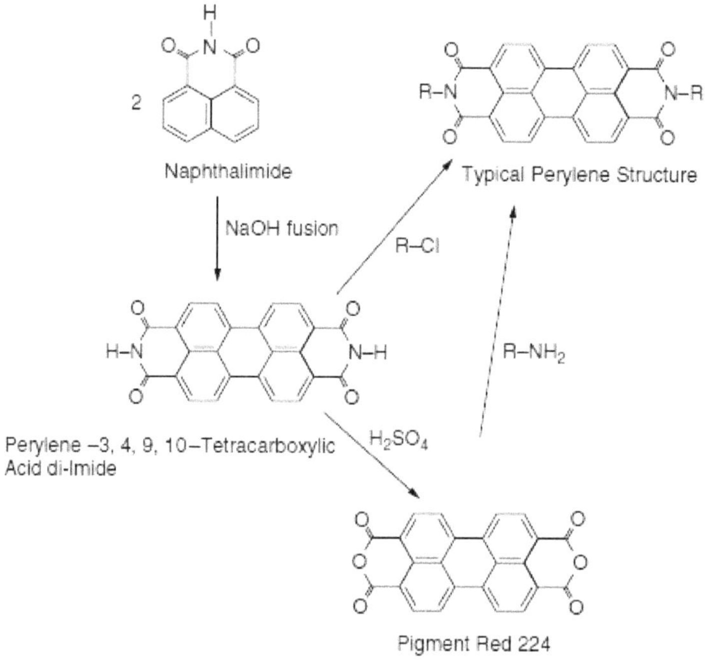

Figure .13 — Perylene synthesis.

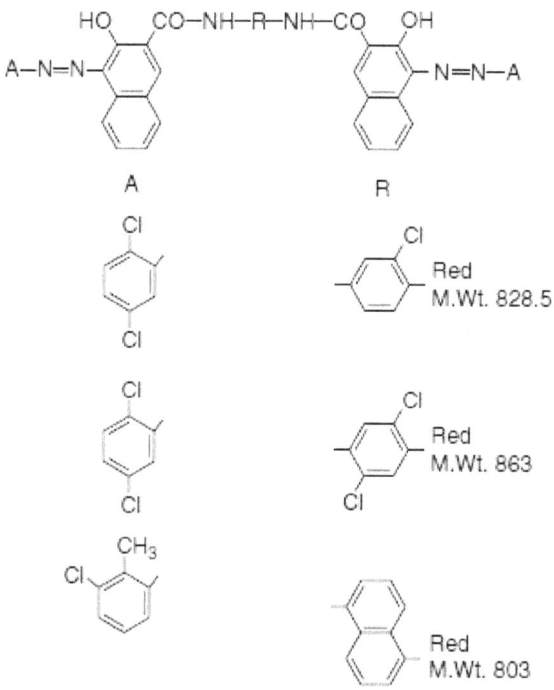

Figure .14 — Structures of the disazo condensation reds.

Miscellaneous High Performance Reds

In recent years a number of novel high performance reds have been commercialized by such companies as Sandoz and BASF. New from Sandoz are Pigment Reds 242, 214, and 257.

Pigment Red 242 is shown in Figure .15. The product is a yellow-shade red with a clean bright shade and very good all-around fastness properties. It is recommended for lead-free coatings formulations for the production of high quality finishes and bright red shades.

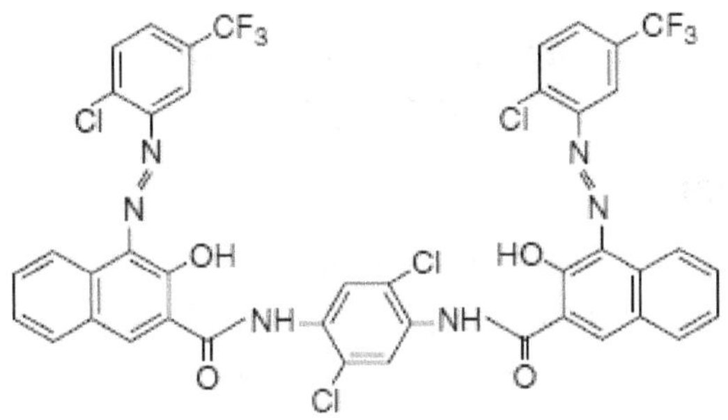

Figure .15 — Structure of Pigment Red 242.

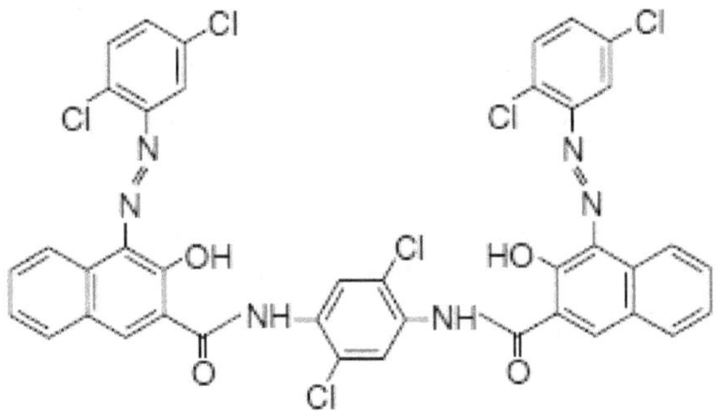

Figure .16 — Structure of Pigment Red 214.

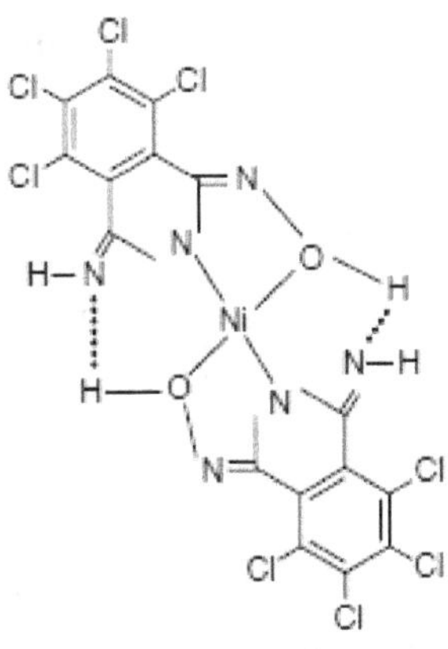

Figure .17 — Structure of Pigment Red 257.

Pigment Red 214 (Figure .16) is a bluish red with properties similar to those of Pigment Red 242.

Pigment Red 257 (Figure .17) is a nickel complex pigment with a red violet masstone and a magenta undertone; its fastness properties are similar to those of the quinacridone pigments.

Organic Yellows

Yellow pigments can be subdivided into four broad classifications based on their chemical constitution.

These classifications are comprised of monoarylide yellows, diarylide yellows, benzimidazolone yellows, and heterocyclic yellows.

Monoarylide Yellows

Monoarylide yellows are all azo pigments; their manufacture is based on the diazotization procedure, followed by coupling. The structures of the major monoarylide yellows are represented in Figure .18.

Pigment Yellow 1 is often referred to as Hansa Yellow G for historical reasons. This pigment is a bright yellow that finds outlets in trade sales, emulsion, and masonry paints. The pigment's major drawbacks are its poor bleed resistance, poor light fastness in tint shades, and markedly inferior bake.

Pigment Yellow 3, referred to as Hansa Yellow 10G, is a pigment that is considerably greener and cleaner than Pigment Yellow 1. The pigment finds a market in trade sales, water-based emulsions, and masonry paints. The pigment suffers from the same deficiencies of poor bleed resistance and poor tint light fastness exhibited by Pigment Yellow 1. Pigment Yellow 3 is, however, suitable for exterior use at high tint strength.

Pigment Yellow 65 is a monarylide yellow that finds use in trade sales, latex, and masonry paints. The pigment offers a light fastness in full shade of 7 and 6–7 in tint, a considerable improvement over Pigment Yellow 1.

Pigment Yellow 73 is a pigment close in shade to Pigment Yellow 1 which again finds use in trade sales, latex, and masonry paints. It is not, however, considered to be durable enough for exterior applications.

Pigment Yellow 73 plays an important role in interior, intermix systems because of its stability against recrystallization in the presence of glycols and wetting agents as used in aqueous systems.

Pigment Yellow 74 offers the user a pigment that is suitable for outdoor applications and considerably stronger and somewhat greener than Pigment Yellow 1. Again usage in trade sales, latex, and masonry paints is widespread.

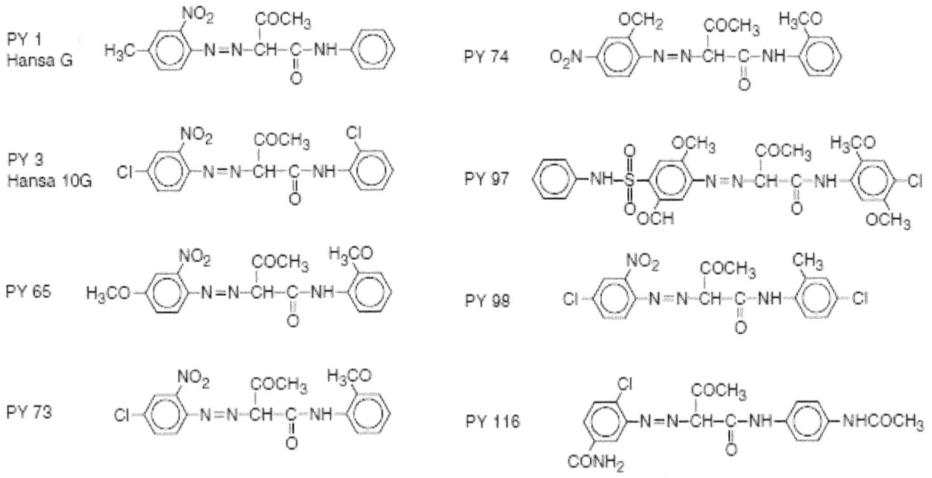

Figure E .18 Monoarylide yellow structures.

Pigment Yellow 97 is a newer pigment that shows the advantages associated with a smaller particle size, less agglomeration, and an improved particle size distribution when compared to the other monoarylide yellows. Pigment Yellow 97 gained significant use in trade sales applications primarily because it was available when demands for lead-free formulations were increasing. Additionally, the pigment finds use in high quality decorative paint.

Pigment Yellow 98 is a pigment that has met with only limited commercial success to date. The pigment is similar in shade to Pigment Yellow 3 but is considerably stronger and more heat stable; it finds use in trade sales, masonry, and latex-based paints.

Pigment Yellow 116 is a pigment similar in shade to light chrome yellow (PY34). Pigment Yellow 116 exhibits improved chemical resistance compared to the other monoarylide yellows, showing improved fastness to solvent, heat, and light. Its major area of use is currently in the formulation of lead-free coatings.

In summary, the monoarylide yellows constitute the most important class of organic yellows as consumed by the coatings industry. Because of the absence of any lake forming groups (e.g., —COOH or —SO3H) in their structure, they possess excellent alkali fastness, which enables them to be used in all major aqueous paint formulations. Additionally, when compared to their inorganic lead containing counterparts, the chrome yellows, the monoarylides offer considerably higher tinting strength, bright yellow hue of high chrome, and less tendency to darken on exposure to atmospheric pollutants.

Diarylide Yellows

Figure .19 illustrates the structure of the diarylide yellows, and Table .6 presents a summary of the properties of this class of yellows.

Basically, an inspection of the structures indicates that this class of yellows has a backbone structure that hinges on 3, 3-dichlorobenzidine with properties that differ depending on the nature of the coupling component. Properties that are common features of this group of yellows are low cost, reasonable heat stability, and moderate chemical resistance.

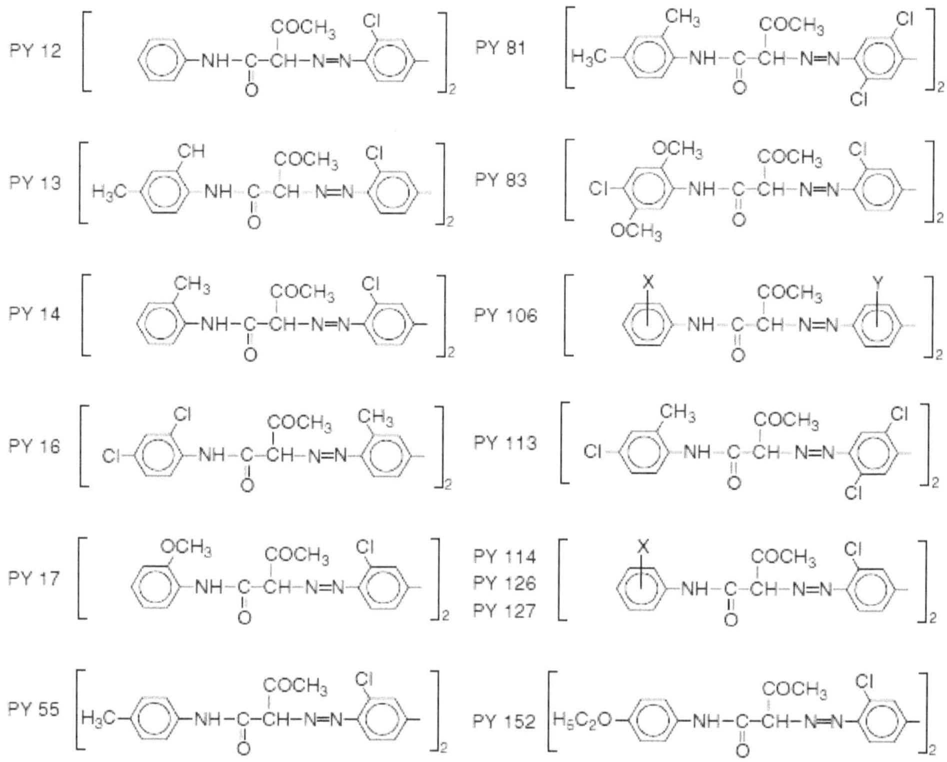

Figure .19: Diarylide yellow structures.

The major market for the diarylide yellows is the printing ink industry. The diarylide yellows are approximately twice as strong as the monoarylide yellows; additionally they offer improved bleed resistance and fastness to heat. However, none of the diarylide yellows has durability adequate for the pigment to be considered in exterior coatings applications; thus the diarylide yellows should not be used in any outdoor situations.

Benzimidazolone Yellows

Figure .20 illustrates the structure of the organic yellows that fall into the classification of benzimidazolone yellows because each is an azo pigment derived from 5-acetoacetyl-aminobenzimidazolone.

The exceptional fastness to heat and the excellent weather fastness of this class of pigments are attributed to the structural presence of the benzimidazolone group.

Used initially for coloring plastics, these pigments are finding increased use in coatings systems, where their excellent weather fastness, heat stability, and fastness to overstripping are required (e.g., when formulating high quality industrial finishes). Table .7 gives a summary of the properties of the benzimidazolone yellows.

Heterocyclic Yellows

This class of pigments contains an assortment of yellows that all contain a heterocyclic molecule in their structure as presented in Figure .21. In spite of the apparent degree of complexity in the synthesis of such structures, these new high performance pigments continue to be introduced into the marketplace to fulfill the exacting demands of the consuming industries. Compounds such as Isoindoline Yellow (PY 139) and Quinophthalone Yellow (PY 138) are particular examples of such complex, novel pigments.

Table .6 Summaries of Diarylide Yellow Properties

Colour Index Name	Common Namea	Comments
PY 12	AAA Yellow	Poor lightfastness; poor bleed resistance; major use in printing inks
PY 13	MX Yellow	Redder shade than PY 12; improved heat stability and solvent fastness; major use in printing inks
PY 14	OT Yellow (274–1744)	Green shade; some use in interior finishes; poor tint lighfastness
PY 16	Yellow NCG	Bright green shade; improved heat and solvent fastness; used in full shade for coatings
PY 17	OA Yellow (275–0562)	Green shade; some use in interior finishes; poor lightfastness
PY 55	PT Yellow	Red shade, some use in interior finishes; poor lightfastness; isomer of PY 14
PY 81	Yellow H10G	Bright, green shade; same shade but much stronger than PY 3
PY 83	Yellow HT (275–0570) (275–0050)	Very red shade; improved transparency, heat stability, and lightfastness over PY 12; some use in interior finishes
PY 106	Yellow GGR	Green shade; poor tint lightfastness; major use in packaging inks
PY 113	Yellow H10G	Very green shade; more transparent than PY 12 and offering better heat and solvent fastness; some interior finish use
PY 114	Yellow G3R	Red shade; improved solvent and lightfastness over PY 12; major use in oil- based inks

Colour Index Name	Common Namea	Comments
PY 126	Yellow DGR	Similar shade to PY 12 but offering improved heat and solvent fastness; major use in printing inks
PY 127	Yellow GRL	Bright, red shade; poor lightfastness; major use in offset inks
PY 152	Yellow YR	Very red, opaque product; poor lightfastness; some use in interior finishes as a lead chrome replacement

All these yellows offer the user additional high performance pigments that find applications in high quality coatings, where the end user can justify the economics of purchasing. Table .8 summarizes the properties of these heterocyclic pigments.

P.Y. 120

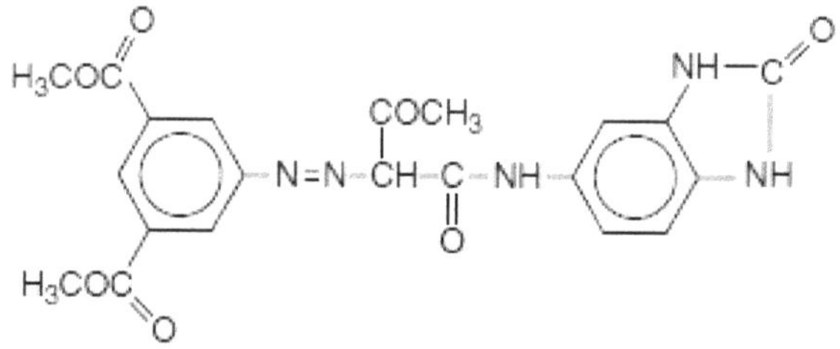

P.Y. 151

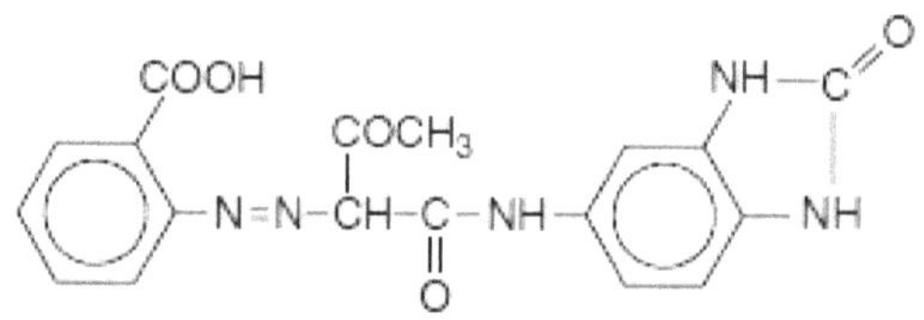

P.Y. 154

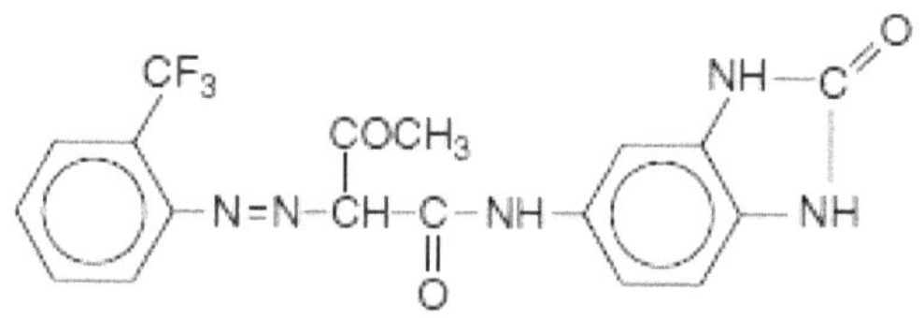

P.Y. 156

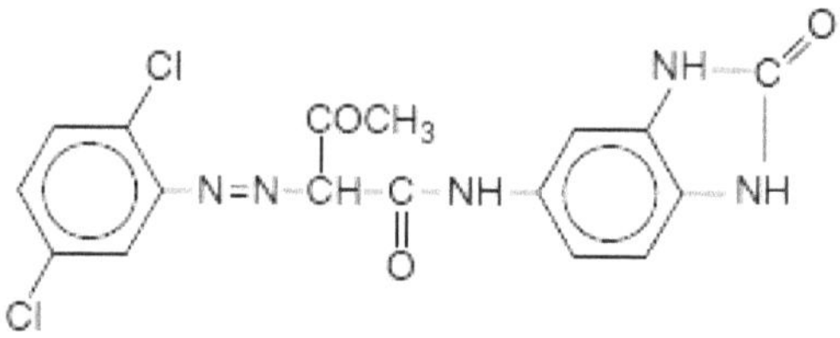

P.Y. 175

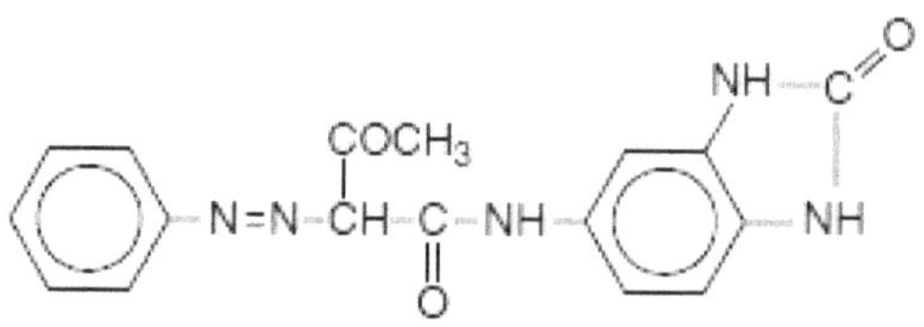

Figure .20 — Benzimidazolone yellow structures.

Table .7 — Summary of Benzimidazole Yellow Properties

Colour Index Name	Common Name	Comments
PY 120	Yellow H2G	Medium shade; good solvent fastness; excellent lightfastness; used in industrial finishes
PY 151	Yellow H4G	Greener shade; good solvent fastness; excellent lightfastness; industrial and refinish applications
PY 154	Yellow H3G	Green shade but redder than PY 151; good solvent fastness; excellent lightfastness; industrial and automotive refinish applications
PY 156	Yellow HLR	Redder shade; transparent; good exterior durability in full shade and tint; all exterior coatings and refinish applications
PY 175	Yellow H6G	Very green shade; good solvent fastness; excellent lightfastness; all exterior coatings and refinish applications

Table .8 — Summaries of the Properties of the Heterocyclic Yellows

Colour Index Name	Common Name/Description	Comments
PY 60	Arylide Yellow	Very red shade; moderate light and solvent fastness; trade sales, latex and masonry paints; good acid and alkali fastness
PY 101	Methine Yellow	Bright yellow; highly transparent and exceptionally brilliant; industrial finishes and specialty coatings; only moderate bleed and alkali fastness
PY 109	Tetrachloroisoindoline	Green shade; excellent brightness strength and durability; automotive finishes
PY 117	Greenish yellow complex of an azomethine	Excellent chemical, light, and heat fastness; specialty finishes
PY 129	Azomethine Yellow	Very green shade; excellent overall fastness; industrial and specialty coatings
PY 138	Green shade quinophthalone	Clean hue and excellent overall fastness properties; high quality industrial and automotive finishes
PY 139	Red-shade isoindoline	Similar in masstone to medium chrome (PY 34); excellent light and solvent fastness; industrial and automotive coatings
PY 150	Pyrimidine Yellow	Very green shade; good heat and lightfastness; industrial coatings

PY 153	Red-shade Nickel Dioxine	Excellent fastness properties; specialty coatings and baking enamels; poor acid resistance
PY 155	Azo Condensation Yellow	Green shade; excellent overall fastness properties in full shade; industrial and specialty coatings
PY 173	Isoindolone Yellow	Very green shade; excellent fastness properties; industrial and specialty finishes
PY 182	Triazinyl Yellow	Medium shade; excellent fastness properties at masstone levels; industrial finishes

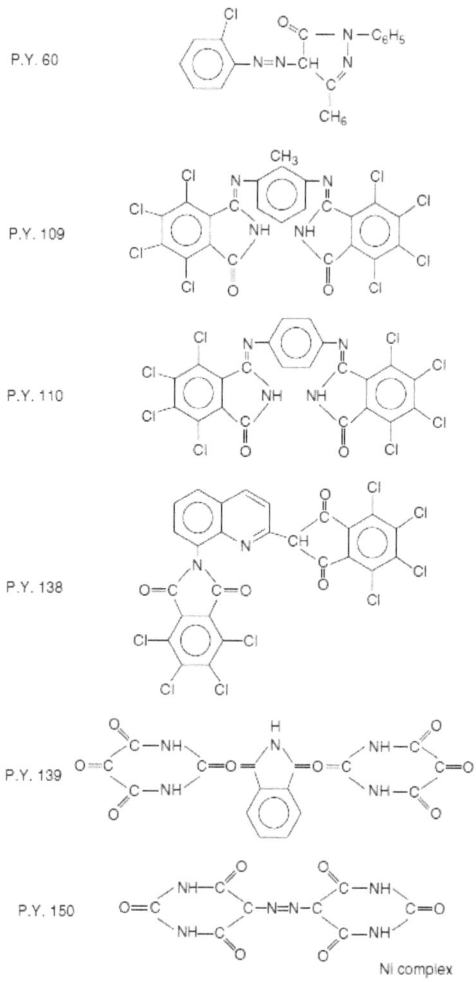

Figure .21 — Structure of heterocyclic yellows.

05 High Temperature Pigments

INTRODUCTION

High temperature pigments can be defined as chemical substances that impart color to a substrate or binder and retain their color and finish at elevated temperatures.

There are many everyday applications where consumers require aesthetically pleasing finishes, in the latest fashion colors, that last. There are many diverse, high performance applications that require careful pigment selection to ensure that the coloration is long-lasting; rarely will a consumer be aware of the technological considerations that apply when designing such products.

Chemically, high temperature pigments are inorganic compounds. Although many chemical classes potentially fall into this category, an important family of pigments is termed complex inorganic color pigments (CICPs), otherwise known as mixed metal oxides (MMOs) or complex inorganic pigments (CIPs). These pigments are heat stable to temperatures exceeding 1832°F (1000°C), suitable for the majority of applications. High performance pigments also include cadmium pigments, able to withstand temperatures of up to 752°F (400°C), and bismuth vanadate pigments, with heat stability of up to 392°F (200°C). These pigments exhibit excellent color properties .

There are two distinct classes of CICPs, similar in chemistry but differentiated by end market and by particle size; pigment-grade for plastics, surface coatings, building materials, and glass applications, and ceramic-grade for ceramic applications.

It is interesting to note that some colors are more heat stable than others. For example, black is a strong absorber of infrared (IR) radiation and therefore retains heat. Hence, black pigments and dark colors require better heat stability than lighter colors or pastel shades.

Recent technology advances include pigments that provide color and functionality by building in the ability to reflect IR rays away from the substrate or binder, and hence, lowering the heat buildup and prolonging the lifetime of the product. These products are promoted by Energy Star for their benefit to the environment in terms of energy-saving initiatives.

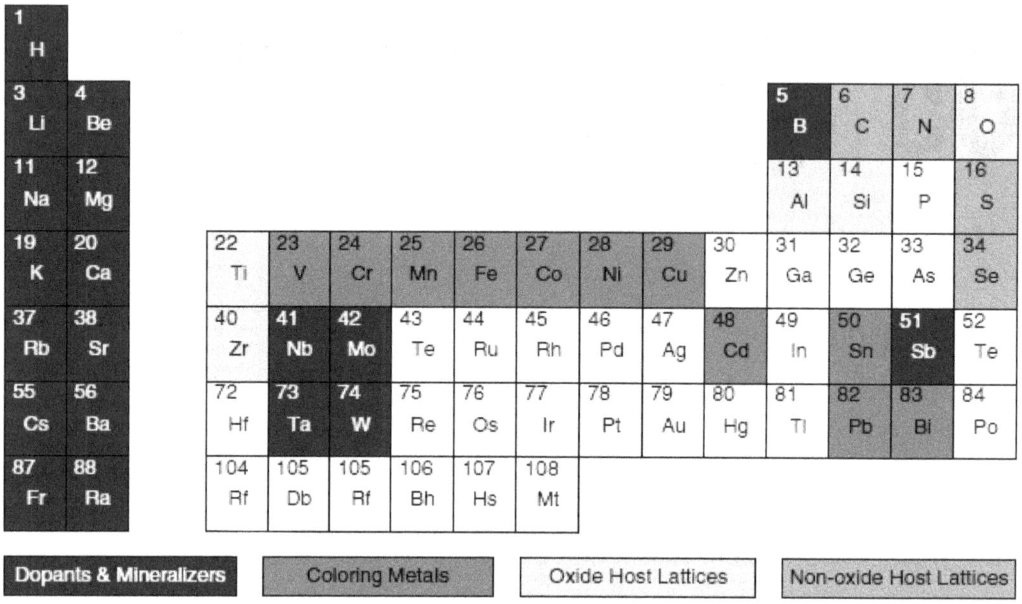

Figure .1 — Periodic table

THE TECHNOLOGY

Color Mechanism

Complex inorganic color pigments are synthetic, highly crystalline metal oxides, differentiated from simple metal oxides, as they contain two or more metals. The color of the compound is determined by the combination and location of the transition metal ions in the oxidic host lattice, otherwise known as chromaphores. The transition metals are responsible for generating electronic d–d transitions, which are energetically weak, and therefore, the colors produced are not as bright and intense as other pigment types, for example, organic pigments.

Transition metal ions can easily substitute for metal ions in the host lattice if they have equal electronic charge. Often, combining several metals with different valence states produces a charge-balanced pair, demonstrating the complexity of the formulations used. Although transition metals are the color-forming centers, the hue can be shifted with dopants, that is, noncolored ions that can either occupy interstitial sites in the lattice, distorting the lattice shape, or can substitute into sites the lattice, producing a charge imbalance. These structures may be inherently more unstable and may produce less durable pigments.

Chemical Structure

Complex inorganic color pigments are dominated by two crystal structures, rutile and spinel, although there are many others. Rutile pigments are largely titania-based, doped with metals such as nickel, chromium, or antimony at low levels.

Production Methods

Complex inorganic color pigments are produced commercially by solid state reaction, by heating intimate mixtures of metal oxides, salts, or hydrates at high temperatures between 1112°F (600°C) to 2552°F

Table .1 — Crystal Structure Types

Crystal Stucture	Typical Formula	Typical Pigment Type	Coloring Metal	Colour
Rutile	MO_2	$(Ni,Sb,Ti)O_2$	Ni	Yellow
Spinel	M_3O_4	$(Co,Zn)Al_2O_4$	Co	Blur
Zircon	M_2O_4	$(V,Zr)SiO_4$	V	Torquoise
Hematite	M_2O_3	$(Fe,Cr)2O_3$	Fe/Cr	Black/Brown
Cassiterite	M_2O_3	$(Co,Zn)SiO_3$	Co	Blur

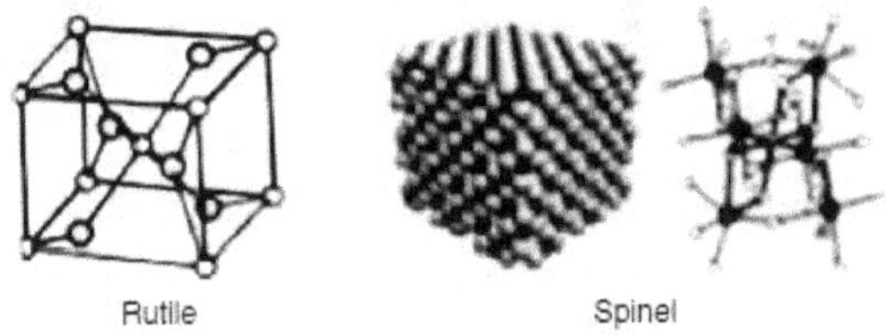

Figure .2 — Crystal structure types. (1400°C).

Intimate raw material mixtures are produced by wet milling techniques to ensure that all components are finely divided. The surfaces of grain boundaries are often coated to achieve the best reaction. Mineralizes can be used to aid the rate of reaction to ensure completion with minimum heat work; these are frequently used when producing zircon pigments.

Figure .3 — Saggar firing.

Figute .4 — Milling

The calcination process can be a batch process, loading the pigment manually into refractory saggars and firing in intermittent kilns, or it can be a continuous process using state-of-the-art rotary firing techniques.

The pigments have developed their full color during firing but are refined by milling to the required particle size, dependent on the application. Pigments for surface coatings, plastic, and glass applications are usually designed with a fine particle size and a narrow particle size distribution, for maximum tinting strength; conversely, pigments for ceramic applications tend to be coarser, with a wider size distribution, for maximum masstone color strength.

Table 2 — Common Rutile Pigments

Colour Index Reference No.	Typical Formula	Color	Main Commercially-Available Grades
Yellow 53	$(Ni,Sb)TiO_2$	Yellow	BASF K1011 and Ferro PK9415
Yellow 216	$(Sn,Zn)Ti_xO_y$	Yellow/Orange	Rockwood 34H1001, 34H1002, 34H1003
Brown 24	$(Cr,Sb)TiO_2$	Buff	BASF K2011 and Ferro PK6070
Brown 164	(Mn,Sb) TiO2	Dark brown	BASF K2711

PIGMENT TYPES

Rutile Pigments

Rutile pigments consist primarily of titania, 70 to 90 wt%. They represent the largest class of pigments, with Brown 24 being the most significant, followed by

Yellow 53. As with all CICPs, the coloring metal ions are trapped tightly within the host lattice and therefore are not extracted in use. The extractable metal content of all pigments is carefully controlled, making them suitable for food contact applications.

The color shade can be altered by substituting antimony with niobium, tungsten, or barium; however, this increases the unit cost of the formulation, and therefore, these pigments are not regularly encountered. Yellow 216 is the most recent innovation, introduced to the market in 2003. The range of yellow to orange colors is more intense than with the conventional titanates, with superior tinctorial strength. They contain a new Sn-Zn color center not reported in the industry before, surprisingly intense for two noncolored metal ions and probably caused by bandgap transitions, although this is still being investigated.

This new chemistry comes at a time when increasing environmental concerns surround certain heavy metals; it offers an attractive alternative. Manganese browns are gaining popularity, particularly in North America, as the best iron-free browns available for PVC claddings and vinyl sidings.

Spinel Pigments

Spinel pigments are largely aluminates or chromates. These formulations are modified by a number of dopant metals, each supplier striving to achieve unique shades. Generally speaking, titanates and aluminates offer brighter, cleaner shades than chromates.

Table .3 — Common Spinel Pigments

Colour Index Reference No.	Typical Formula	Colour	Main Commercially Available Grades
Green 50	$(Co,Zn)Ti_2O_4$	Green	Ferro PK4045, Rockwood 34K48, Shepherd 223
Green 26	$CoCr_2O_4$	Green	Rockwood 34K6, Ferro PK4047, Shepherd 179
Green 50	$(Co,Zn,Cr)Ti_2O_4$	Turquoise	Ferro PK5500
Blue 28	$COAl_2O_4$	Blue	Rockwood 34L2000, Ferro PK5091, Shepherd 3
Blue 36	$(Co,Cr)\ Al_2O_4$	Blue-green	Rockwood 34L2001, Ferro PK5095, Shepherd 9
Yellow 119	$ZnFe_2O_4$	Tan	Rockwood Mapico 10A
Brown 29	Fe,Cr_2O_4	Brown	BASF K2975
Black 28	$(Cu, Mn)\ Cr_2O_4$	Black	Rockwood 34E30, Ferro PK3085, Shepherd 1

Table .4 — Common Zircon Pigments

Colour Index Reference No.	Typical Formula	Colour	Main Commercially Available Grades
Yellow 159	(Pr,Zr)SiO4	Yellow	JM CS2300, JM CT1300
Blue 71	(V,Zr)SiO4	Turquoise	JM CS2500, JM CT1500
Red	(Fe,Zr)SiO4	Coral	JM 14M929, JM CT1100
Red	(Cd,Zr)SiO4	Red	JM CS2803, JM CT1805
Orange	(Cd,Zr)SiO2	Red	JM CS2202, JM CT1202

Zircon Stains

Zircon glaze stains can be divided into two groups: doped and encapsulated (or included) colors. Praesodymium, vanadium, and iron dope the zircon lattice by substituting into zirconium sites, and form the basis of the triaxial colors, so called as they represent three corners of a cyan, magenta, yellow color set and therefore offer extremes of ceramic color. They are blended together to form a wide color palette for high temperature applications. Cadmium zircon colors, however, are made by encapsulating discrete cadmium selenide pigment particles by zircon, producing a two-phase product. Encapsulation of the coloring compound is optimized by ensuring that the pigment phase is formed before the zircon phase, in a competitive reaction.

PIGMENT PROPERTIES

Complex inorganic color pigments are called high performance pigments because they are extremely durable in use. Further, these pigments show:

- Outstanding color retention after exterior exposure, in the presence of UV light, moisture, and heat
- Excellent heat stability
- Very good opacity to visual and ultraviolet (UV) light
- Good chemical resistance against attack from solvents, acids, and alkalis
- Chemically inert, combining well with other inorganic and organic pigment types
- Nonmigratory in polymers and plastics, showing no color bleed
- Highly refractive, with a tendency to abrade softer materials

Complex inorganic color pigments can actually enhance the overall lifetime of the coating or colored object. The pigment acts functionally by absorbing or reflecting UV radiation away from the binder.

Most organic binders will degrade photocatalytically in such conditions, particularly in the presence of titania, which catalyses free-radical formation and accelerates the aging process.

These pigments are often used in conjunction with high color strength pigments to impart opacity and durability to a formulation, reducing the cost of the organic component of a formulation.

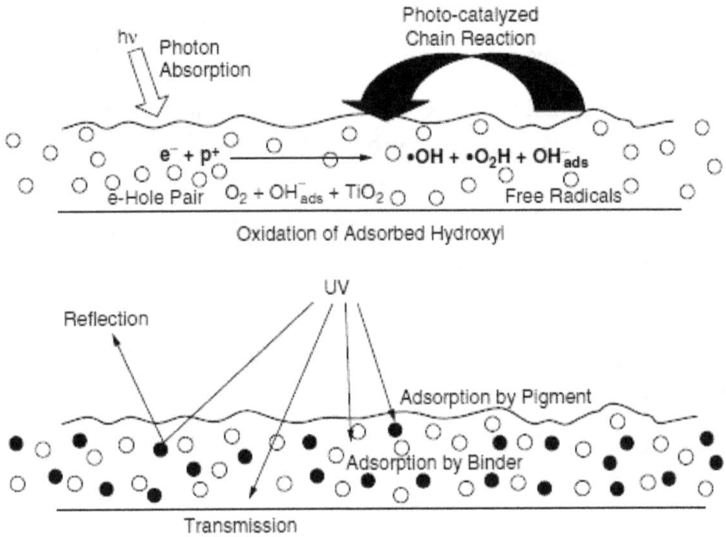

Figure .5 — Mechanism of photocatalytic degradation.

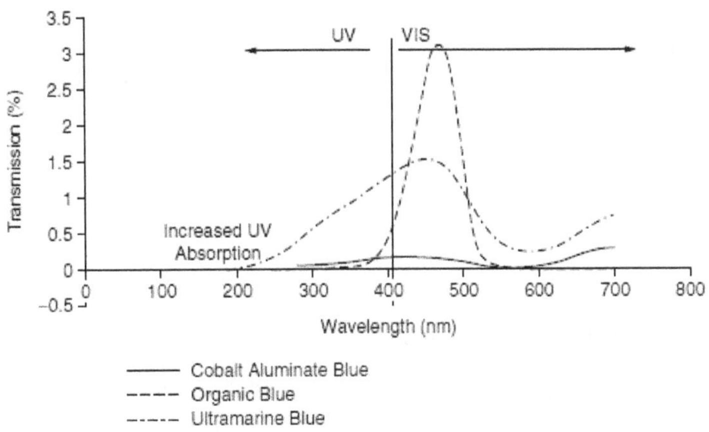

Figutr .6 — UV absorption of blue PVDF coil coatings.

TYPICAL APPLICATIONS

Complex inorganic color pigments are the pigments of choice where colored objects are subjected to high temperatures, UV light, or harsh chemical environments either during the manufacturing process or in use.

Surface Coatings

The highest performance surface coatings applications are usually for exterior applications such as coil coatings (used for building claddings), roofing, general construction parts, exterior paints (for example, masonry), and industrial and architectural paints.

Coil coatings usually adopt an acid catalyzed curing process that can be reactive to some pigment types. Coil-coated parts are bound by lengthy manufacturer guarantees of up to 25 years; hence, the highest performance pigments are required.

Exterior masonry paints can be based on permeable silicate paint technology to allow the underlying concrete or mortar to "breathe." These paints are highly alkaline and are not compatible with organic or ultramarine pigments. With the increasing use of universal color tinters, formulated to be equally compatible with solvent-, water-, and silicate-based paints, CICPs are growing in popularity because of their wide compatibility.

Heat-resistant paints are used for barbeques, wood-burning stoves, car exhausts, and engine parts.

These are largely silicone based, pigmented with copper chromite spinel black (CI Pigment Black 28).

Ceramics

A wide range of colors is available for applying to a range of tableware substrates, to suit all designs applied to elaborate dinner sets, hotelware, giftware, decorative ornaments, and collectables:

- Porcelain
- Bone and vitreous china
- Earthenware
- Stoneware

The same color palette is also used in the structural business to deliver high temperature fired color to sanitary-ware and floor and roof tiles.

Color is supplied as a glaze stain, where color pigment is supplied in powder form, or as a decorative color, supplied as a powder blend or liquid ink and containing a glass frit for glaze formation. It can also be applied as a predesigned decal transfer, ready for direct application.

Glaze Stains

The color palette typically consists of the following:

- Zircons (Praesodymium Yellow, Iron Coral, Vanadium Blue, Cadmium Red)
- Chrome Tin Pink
- Cobalt Aluminate or Silicate Blue
- Iron, Nickel, Chromium Browns and Blacks

Supplied as standard range color powders, they are blended to the desired custom colors and combined with transparent or opaque unleaded glazes for firing onto tableware and sanitary-ware.

Decorative Color

Decorative colors for tableware, giftware, and collectables are selected by firing technology:

- On-glaze color: twice-fired color applied and fired directly onto the substrate before the top glaze is applied and fired
- In-glaze and underglaze color: once-fired color applied onto or under the top glaze and fired simultaneously

Table .5 : Traditional Glaze Compositions

Glaze Effect	Al_2O_3 (%)	B_2O_3 (%)	CaO (%)	K_2O (%)	Na_2O (%)	MgO (%)	SiO_2 (%)	TiO_2 (%)	ZrO_2 (%)
Transparent	12.52	2.50	9.80	3.60	2.80	1.00	67.71	0.07	-
Opaqua	11.35	2.26	8.86	3.22	2.50	0.92	64.7	0.09	6.10

Decorative colors are supplied as powder blends, custom powder blends color-matched to customized colors, or as liquid inks. They are available in 25 color sets or four-color CMYK sets for four-color printing techniques.

Decal Transfer Designs

CICPs are supplied as prepared custom and standard color designs for tableware applications.

Building Materials

CICPs are widely used to color cement and concrete and mortar and grouting for outdoor pavings, swimming pools, and roof tiles. The alkalinity of the cementitious materials used demands inorganic pigments to ensure long-lasting coloration.

Glass

Glass enamel producers traditionally opt for CICPs largely in automotive and decorative applications.

Heated rear windscreens contain black copper chromite, because a pleasing jet black appearance is retained after firing at elevated temperatures, without interference with the glass frit or the silver conductive track. Decorative glass applications use a wide range of colored pigments, for example, the Pepsi red and blue trademark decorations for drinking bottles.

Plastics

Rigid PVC extruded or molded parts are used for window profiles, gutterings, and joints. These build up heat during the summer months and will readily degrade unless steps are taken to prevent it. Stabilizers are added to PVC to absorb UV light; careful selection of the pigment system used can also add value to the finished goods.

Silicone rubbers are frequently colored from compounded color concentrate, pigmented with CICPs.

Copper chromite spinel black is a common colorant for production of nonconductive electronic housings and spark plug casings. The pigments used in such applications must be carefully controlled for extractable metals, because silicone polymers often contain residual monomers that can initiate premature polymerization.

Engineering polymers, used in automotive interiors and many domestic appliances, require high processing temperatures and, therefore, require CICPs for coloration.

06 Dispersion

The next stage is to get the pigment into the paint. The pigment is usually supplied as a powder, in which the granules are actually aggregates of the fine particles produced by the pigment manufacturer. These particles must be dispersed or separated from one another and evenly distributed throughout the paint as a colloidal suspension. For this suspension to have the maximum stability in organic solvents, the surface of each particle should be completely wetted with the varnish: there should be no intervening layers of air or adsorbed water.

Wetting with solvent alone is not enough and pigment dispersions in solvent have poor stability. Each particle in the pigment suspension must be stabilized by polymer chains, anchored to its surface by intermolecular attractions, yet extending out into the varnish because of their attractions for the solvent molecules. When two such particles approach one another, they do not adhere. Contact between them involves the intermingling of polymer chains from the two particles. Locally, in the region between the particles, the concentration of resin molecules is higher than at other points in the varnish. This upsets the equilibrium of the system, so solvent molecules diffuse into this region, dilute the concentration and restore equilibrium.

The osmotic pressure involved is sufficient to separate the particles. A resin suitable for dispersion usually contains polar groups (which provide the attraction for pigment surface molecules) and is completely soluble in the solvent mixture of the dispersion. Sometimes, surfactants are used to bridge the particle-resin interface and assist wetting. In water, ionic surfactants (e.g. soaps) can provide the pigment surface with an electrical charge. The particles, being of like charge, repel one another and the dispersion is stable.

Dispersion is usually carried out in a mill, a machine in which the aggregates are subjected to the forces of shear and (sometimes) attrition. When shear is the dispersing action, the aggregates are squeezed between two surfaces moving in opposite directions, or in the same direction but at different speeds. It is just like making cocoa, where the powder has to be dispersed in a little milk to form a paste. Dispersion is carried out by a shearing motion between spoon and cup. In paint mills, the principle is the same, but the power and degree of dispersion are much greater.

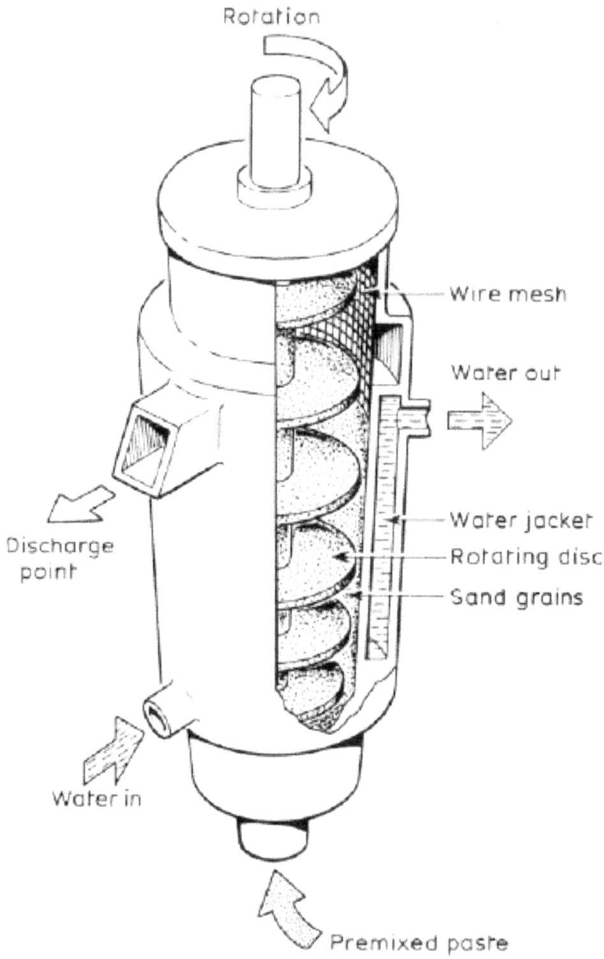

Figure .1 — A sand grinder.

During the manufacture of the pigment, the particles are reduced to their ultimate size by crushing in liquid slurry. Where attrition is part of a dispersion process, the conditions are much milder and there is no attempt to fracture individual particles. In the viscous medium of the varnish, aggregates-not particles-are broken.

Several types of mills are used. The high speed disperser is used for easily dispersed pigments and consists of a horizontal disc with a serrated edge, which rotates at high speed about a vertical axis. The ball mill is a cylinder revolving about its axis, the axis being horizontal and the cylinder partly filled with steel or steatite (porcelain) balls, or pebbles. The speed of rotation is such that the balls continually rise with the motion and then cascade down again, crushing and shearing the pigment. In the sand grinder (Fig. 1) or the bead mill, the axis of rotation is vertical, the grinding medium is coarse sand or glass beads and the charge is induced to rotate at higher speeds by revolving discs in a stationary container. The ball mill produces a batch of pigment dispersion; the sand grinder gives a continuous output of dispersed pigment.

Roller mills, which are rapidly disappearing from the paint industry, consist of a number of horizontal steel rolls placed side-by-side and moving in opposite directions, often at different speeds, with very small clearances

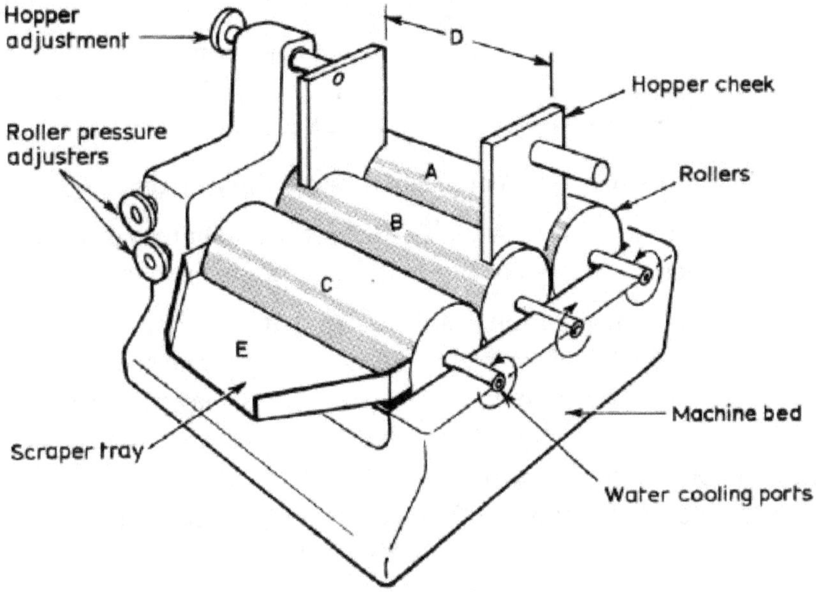

Figure 2 — A triple roll mill.

in between. In these gaps, the pigment/varnish mixture is sheared. The triple roll is shown in Fig. 2. A paste is fed in at D, is lightly sheared between rollers A and B and more severely between B and C. A scraper blade E removes the dispersion. All rollers are independently driven and run at different speeds to increase shear. A fourth type of mill is the heavy duty or 'pug' mixer, in which roughly S-shaped blades (Fig. 3) revolve in opposite directions and at different speeds in adjacent troughs (Fig. 4). A stiff paste is required. In modern versions of this type of mixer, the blades are mounted vertically and dip into the dispersion vessel (a vertical cylinder) like the blades of a kitchen dough mixer.

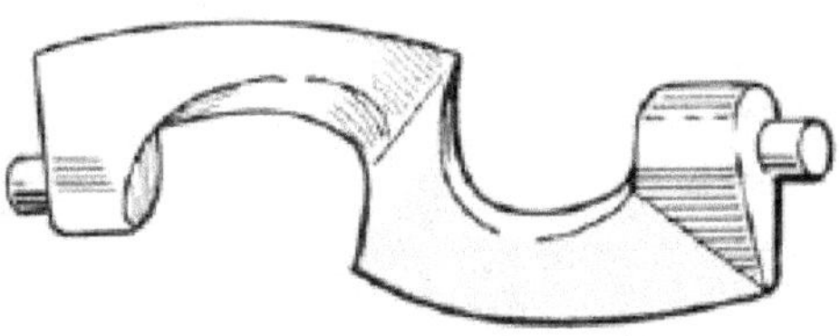

Fig. 3 — Blade for a heavy duty mixer.

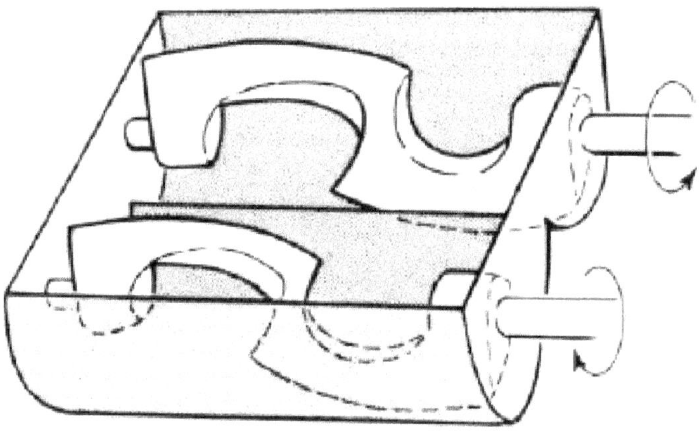

Figure 4 — Troughs of a heavy duty mixer.

Several alternative mills are available, which the reader may discover elsewhere. These few examples are quoted to illustrate the principles involved.

It is obvious from the mention of stiff pastes that the whole paint is not charged into the mill In fact, the paint maker aims to put in the maximum amount of pigment and the minimum amount of varnish to get the largest possible paint yield from his mill. This mixture forms the grinding or first stage. When the dispersion is complete (after a period varying from 10 minutes to 48 hours according to the materials and machinery involved), the consistency is reduced with further resin solution or solvent, so that the mill can be emptied as cleanly as possible. This is the 'let-down' or second stage and may take up to two hours. The third or final stage (carried out in a mixing tank) consists of the completion of the formula by addition of the remaining ingredients. A break-down of a possible ball mill formula looks like this:

	wt %	
Pigment	10·0	
Resin	1·0	Stage I (grinding). Then add:
Solvent	3·0	. . . .
Resin	1·0	Stage II (let down). Empty mill – then add:
Solvent	3·0	. . . .
Resin	29.0	
Solvent	51·5	Stage III (completion of formula)
Additives	1·5	
	100·0	

The exact composition of Stage I is found by experiment, to give the minimum grinding time and the most stable and complete dispersion. Stages II and III also require care, as hasty additions in an incorrect order can cause the pigment tore-aggregate (flocculate).

The amount of pigment in the formula is that required for the appropriate colour, hiding power, gloss, consistency and durability. As a rough guide, the amount might vary from one third of the binder weight to an equal weight (for a glossy pastel shade).

COLOUR MATCHING

The colour of paint is usually matched to some colour pattern. It must, at any rate, be matched to the colour of the previous batch of the same paint.

If a new shade is to be matched in paint, the procedure for matching is as follows:

1. The main hue and the undertones are observed; also whether the colour is 'clean' or 'dirty.
2. From experience and by application of the principles of subtractive mixing, the pigments likely to give the colour are chosen. The procedure is described under Pigment Selection.
3. Each one of these pigments is dispersed separately and each dispersion is converted to paint. The colour matcher now has several paints, each containing only one pigment. These are known as colour solutions.
4. The colour solutions are blended, the choice of solutions and proportions being varied, until finally the correct blend to give the colour is obt ained. The proportions of pigment in this blend are recorded.
5. The final paint is often made by blending colour solutions, but frequently a more stable dispersion is obtained if all the pigments are dispersed simultaneously in the same mill. This is called co-grinding the pigments. If the colour is then slightly off shade, it is adjusted by additions of colour solutions or concentrated pigment dispersions.

In larger paint companies, where spectrophotometers or colorimeters and a computer are available, these stages can often be speeded up considerably. The new shade can be measured by the colorimeter and expressed precisely as a group of numerical values. This information can be passed to the computer, which has been previously programmed with the values for numerous pigments and with the 'rules' governing the mixing of these pigments to give the desired colour. The computer will then print out a choice of pigmentations (type and proportions) for achieving the colour.

The formulator makes a selection and will usually find that the colour he produces to the computer's instructions will be close to the desired colour and that the recipe will require only minor modifications to give the exact shade.

Most colours, even dark ones, require substantial proportions of white pigment. Clean colours require organic pigments and should be matched with the minimum number of pigments. The subtractive effect increases with the number of pigments, reducing the reflected light intensity and hence 'dirtying' the colour. If darkening or 'dirtying' of a colour is required, it can be done with black pigment. The use of black pigment can markedly alter the shade. Thus brown is a mixture of yellow (or red) and black.

07 Solvents

Before we start, it should be explained that the word 'solvent' is used loosely in the paint industry to denote any liquid that may be used in paint: it tells you nothing about the liquid's ability to dissolve the polymers in the paint.

In this chapter, we will have to use the word in the paint sense and in the scientific sense. It is hoped that the meaning intended will be clear from the context and, to minimize the confusion, the word 'liquid' will be used as much as possible, where no description of solvent power is intended.

CHEMICAL TYPES

The following liquids are commonly used in paints to carry the pigment and binder:

1. Water
2. Aliphatic hydrocarbon mixtures: chiefly paraffins
3. Terpenes
4. Aromatichydrocarbons
5. Alcohols
6. Esters
7. Ketones
8. Ethers and ether-alcohols
9. Nitroparaffins
10. Chloroparaffins

1. WATER is the main ingredient of the continuous phase of most emulsion paints. It is also used alone, or blended with alcohols or ether-alcohols, to dissolve water-soluble resins (e.g in water-reducible paints) or dyestuffs (e.g. in water stains).

Any type of resin can be made water-soluble. The usual procedure is to incorporate sufficient carboxyl groups into the polymer to give the resin a high acid value of 50 or more. These groups are then neutralized with a volatile base, such as ammonia or an amine, whereupon the resin becomes a polymeric salt, soluble in water or water/ether-alcohol mixtures. For example, drying oils can be made acidic if they are heatbodied in the presence of maleic anhydride, reaction taking place between the double bonds in both materials. If the carboxyl groups are not fully neutralized, an emulsion can be produced.

The polymeric part of the salt is negatively charged and can be attracted to and discharged at the anode in an electrolytic cell. If the anode is a metallic object, say a car body, and then the discharged resin (or paint, if it is associated with pigment particles) will deposit on that object and coat it. This is called anodic electro-deposition. If the paint film deposited is not porous then, as the film thickness increases, so does its electrical resistance (or insulating power) and thus the deposition rate slackens and deposition eventually stops. This has the effect of automatically controlling film thickness and of encouraging deposition at points which are more thinly coated.

From this arise two of the main advantages of electro deposition: it can be a fully automated process and suitably formulated paints (with good 'throwing power') can be made to coat all parts of an article of any shape, however awkward, with a uniform paint coating.

Instead of acidic carboxyl groups, resins can be made to include basic, e.g. amine, groups, and these can be neutralized with relatively volatile acids, e.g. acetic or lactic acids, to produce resin salts soluble in water or water-solvent mixtures. Again, incomplete neutralization produces an emulsion. In these systems, the resin is positively charged and can be discharged at a cathode. This is the basis of cathodic electro-deposition.

Even if the polymeric salts are capable of anodic or cathodic electro-deposition, waterborne paints made from them do not have to be applied in this way. The solution or emulsion paints produced can be applied by virtually any technique. There will be an increasing use of this type of paint as legislation, economics and world petroleum shortages encourage the use of decreasing amounts of volatile organic solvents and paints of low flammability.

An alternative method of obtaining water-solubility, without salt or ion formation, is to incorporate short lengths of naturally water-soluble polymer into the main polymer structure. Often polyethylene glycol is used. This is a polyether of molecular weight around 1000, which is made from ethylene oxide and ethylene glycol. 5-20 per cent by weight is incorporated into the resin, usually by esterification. The resin thus modified can be emulsified in water by simple stirring, or can be dissolved in water/ether-alcohol mixtures. Whereas ammonia will evaporate from the paint film eventually, polyethylene glycol will not, so the film remains permanently sensitive to water.

The chief virtues of water are its availability, cheapness, lack of smell, non-toxicity and non-flammability. However, it is not an ideal paint liquid, because of its limited miscibility with other liquids and because the film formers designed to be dissolved or dispersed in it, usually remains permanently sensitive to it. In fact, its abundance in nature makes it any paint film's worst enemy.

2. ALIPHATIC HYDROCARBONS are usually supplied as mixtures, because of the difficulty of separating the individual compounds. A boiling range for the mixture is usually quoted, e.g. S.B.P. (special blend of petroleum) spirit No. 3, 98-122°C. Many mixtures also contain a percentage of aromatic hydrocarbons, e.g. white spirit, 155-195 oc, about 15 per cent aromatic.0

3. TERPENES commonly used are turpentine, dipentene and pine oil. Turpentine varies with grade, but is principally a-pinene; dipentene is mainly limonene; while pine oils are mixtures, chiefly of terpene alcohols. Turpentine was once the main solvent for house paints, but has now been replaced by white spirit. The terpenes can be used as anti-skinning agents.

4. to 5. The AROMATIC HYDROCARBONS, ALCOHOLS, ESTERS and KETONES supplied to the paint industry are fairly pure named compounds. Some aromatic mixtures are sold cheaply under proprietary names.

6. ETHERS are not often used, but ether-alcohols-which contain both the ether (-C-0-c-) and alcohol (-c-o-H) groups- are very common, e.g.

$CH_3(CH_2)_3{\cdot}O{\cdot}CH_2{\cdot}CH_2OH$ ethylene glycol monobutyl ether(2-butoxy ethanol)	$CH_3{\cdot}O{\cdot}CH_2{\cdot}CHOH{\cdot}CH_3$ propylene glycol monomethyl ether (1-methoxy propan-2-ol)
$HO{\cdot}CH_2{\cdot}CH_2{\cdot}O{\cdot}CH_2{\cdot}CH_2{\cdot}OH$ diethylene glycol (DEG)	$CH_3{\cdot}O{\cdot}CH_2{\cdot}CH_2{\cdot}O{\cdot}CH_2{\cdot}CH_2{\cdot}OH$ DEG monomethyl ether

7. and 8. NITRO - AND CHLOROPARAFFINS are uncommon solvents in paints, the latter because they are somewhat toxic, the former because of cost and some recent evidence of toxicity.

Solvent properties

The most important properties of a liquid for paints are:

(A) Solvency, i.e. whether it is a solvent or non-solvent for a given film former. This depends on the film-former and is not an independent property of the liquid.

(B) Viscosity or consistency.

(C) Boiling point and evaporation rate.

(D) Flash point.

(E) Chemical nature.

(F) Toxicity and smell.

(G) Cost.

SOLVENTS

Let us consider these properties in more detail, referring to Table 2, which lists some common solvents together with figures for properties (a)-(d).

Table .2: Solvent properties

Solvent	Formula	Solvency		Viscosity at 20 °C (centipoises)	Boiling point (°C)	Flash point (closed cup) (°C)
		H-bonding group	Solubility parameter			
Water	H_2O	III	23·4	1·002	100	None
Aliphatic hydrocarbons						
cyclo-hexane	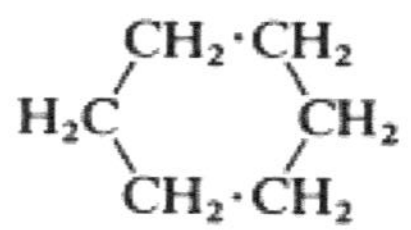	I	8·2	0·89*	81	3
White spirit		I			155–195	33[m]
Odourless white spirit		I	6·9		180–207	55
Terpenes						
Dipentene	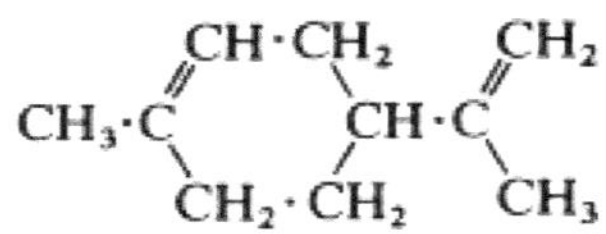	I	8·5	0·975*	175–190	32[m]
Turpentine	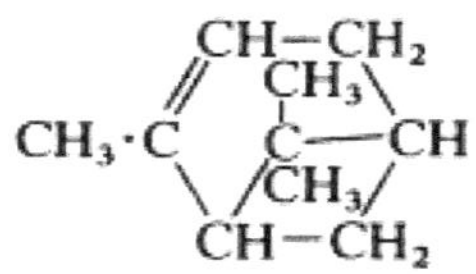	I	8·1	1·26*	150–170	33
Pine oil		I	8·6	6–26*	195–220	74–88[op]

Aromatic hydrocarbons						
Toluene	$C_6H_5 \cdot CH_3$	I	8·9	0·55*	111	4
Xylene	$C_6H_4 \cdot (CH_3)_2$	I	8·8	0·586	138–144	27
Styrene	$C_6H_5 \cdot CH{=}CH_2$	I	9·3	0·77*	146	31
Vinyl toluene	$CH_3 \cdot C_6H_4 \cdot CH{=}CH_2$	I			164–170	32
Alcohols						
Methanol	CH_3OH	III	14·5	0·547	65	12–14
Ethanol	C_2H_5OH	III	12·7	1·200	78	14
n-Propanol	$CH_3 \cdot (CH_2)_2 \cdot OH$	III	11·9	2·25	97	15
iso-Propanol	$(CH_3)_2 \cdot CH \cdot OH$	III	11·5	2·15*	82	12
n-Butanol	$CH_3 \cdot (CH_2)_3 \cdot OH$	III	11·4	2·948	118	35
iso-Butanol	$(CH_3)_2 \cdot CH \cdot CH_2 \cdot OH$	III		3·9*	108	25
sec-Butanol	$CH_3 \cdot CH_2 \cdot CH(CH_3) \cdot OH$	III	10·8	3·15	100	24
cyclo-Hexanol	$CH_2 \cdot CH_2$ $H_2C \quad CH \cdot OH$ $CH_2 \cdot CH_2$	III	11·4	81†	162	68
Ethylene glycol	$HO \cdot CH_2 \cdot CH_2 \cdot OH$	III	14·2	17*	198	111
Glycerol	$CH_2OH \cdot CHOH \cdot CH_2OH$	III	16·5	494§	290	160

* at 25 °C † at 38 °C § at 26.5 °C ‡ at 15 °C m = minimum op = open cup

Table 2 Solvent properties (continued)

Solvent	Formula	Solvency: H-bonding group	Solvency: Solubility parameter	Viscosity at 20 °C (centipoises)	Boiling point (°C)	Flash point (closed cup) (°C)
Esters						
Methyl acetate	$CH_3 \cdot CO \cdot O \cdot CH_3$	II	9·6	0·38*	57	−9
Ethyl acetate	$CH_3 \cdot CO \cdot O \cdot C_2H_5$	II	9·1	0·455	77	−4
Butyl acetate	$CH_3 \cdot CO \cdot O \cdot C_4H_9$	II	8·5	0·671*	127	23
Methoxypropyl acetate	$CH_3 \cdot CO \cdot O \cdot CH(CH_3) \cdot CH_2 \cdot O \cdot CH_3$	II	9·2	1·2	140–150	46
Ketones						
Acetone	$CH_3 \cdot CO \cdot CH_3$	II	10·0	0·316*	56	−17
Methyl ethyl ketone	$CH_3 \cdot CO \cdot C_2H_5$	II	9·3	0·423‡	80	−4
Methyl *iso*-butyl ketone	$CH_3 \cdot CO \cdot CH_2 \cdot CH(CH_3)_2$	II	8·4	0·546*	116	16
cyclo-Hexanone	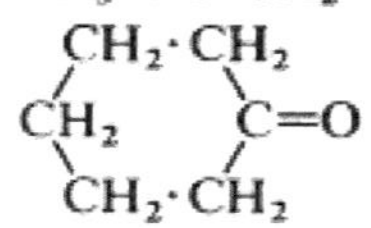	II	9·9	1·94	157	47
Methyl *cyclo*-hexanone	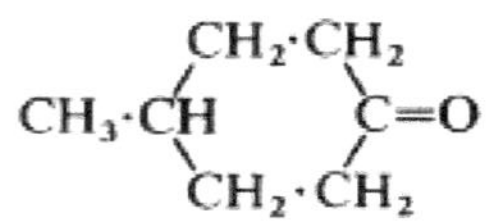 and isomers	II	9·3	1·75	165–175	47

Ethers and ether-alcohols						
Diethyl ether	$C_2H_5 \cdot O \cdot C_2H_5$	II	7·4	0·233	35	−40
1-Methoxy propan-2-ol	$CH_3 \cdot O \cdot CH_2 \cdot CHOH \cdot CH_3$	II	10.2	1.65*	120	38^{op}
1-Ethoxy propan-2-ol	$CH_3 \cdot CH_2 \cdot O \cdot CH_2 \cdot CHOH \cdot CH_3$	II	9.0	1.68*	132	43^{op}
2-Butoxy ethanol	$C_4H_9 \cdot O \cdot CH_2 \cdot CH_2 \cdot OH$	II	8·9	3·318*	171	61
Diethylene glycol (DEG)	$HO \cdot CH_2 \cdot CH_2 \cdot O \cdot CH_2 \cdot CH_2 \cdot OH$	III	9·1	30*	245–250	124
DEG monomethyl ether	$CH_3 \cdot O \cdot CH_2 \cdot CH_2 \cdot O \cdot CH_2 \cdot CH_2 \cdot OH$	II	9·6	3.53*	194	93^{op}
Nitroparaffins						
Nitromethane	$CH_3 \cdot NO_2$	I	12·7	0·62	101	35
Nitroethane	$C_2H_5 \cdot NO_2$	I	11·1	0·62	114	28
1-Nitropropane	$CH_3 \cdot CH_2 \cdot CH_2 \cdot NO_2$	I	10·7	0·81	132	34^{op}
Chlorinated paraffins						
Methylene chloride	CH_2Cl_2	I	9·7	0·425*	41	None
Ethylene dichloride	$CH_2Cl \cdot CH_2Cl$	I	9·8	0·838	84	13
1,1,1-Trichloroethane	$CCl_3 \cdot CH_3$	I	8·6	0·83	72–88	None

* at 25 °C † at 38 °C § at 26.5 °C ‡ at 15 °C m = minimum op = open cup

(A) Solvency

Polymers dissolve by the mechanism described for inorganic solids . A solvent for a polymer is a liquid, the molecules of which are strongly attracted by the polymer molecules. The molecules of a non-solvent are only weakly attracted. Simple inorganic and organic solids have fixed solublities because, above a certain concentration of molecules in the solution, the strong attractions operate often at short range to overcome the energy of movement of the free solid molecules, so that they group together, take up the crystal pattern and crystallize. The concentration of dissolved solid drops below the limiting figure and crystallization ceases. Amorphous polymers, as we have seen, cannot crystallize. Since the forces of attraction between polymer and solvent molecules are as strong as those between the polymer molecules, dissolving the polymer is like mixing a very viscous liquid with a very fluid one. There is no solubility limit; a true solvent is miscible with the polymer in all proportions.

It is not easy to look at the chemical composition of a polymer and predict solvents for it. A general guide is that like dissolves like, but this is limited, sometimes wrong, and gives no indication of what will happen if a liquid mixture is used. There is still no complete explanation of polymer solvency, but this property was rationalized for the paint formulator in 1955 by the American paint chemist, Burrell. He suggested that for every liquid two factors, or parameters, govern the solvency. The first is the hydrogen-bonding capacity of the liquid. He classified paint solvents roughly quantitatively into three groups:

1. Weakly H-bonded liquids (hydrocarbons, chloro- and nitro-paraffins).
2. Moderately H-bonded liquids (ketones, esters, ethers and etheralcohols).
3. Strongly H-bonded liquids (alcohols and water).

For each solvent, a second parameter is calculated from the latent heat of evaporation, by use of the equations of thermodynamics, and the numerical value of this parameter is a measure of the attractive force between the liquid molecules. This parameter is called the solubility parameter. A polymer has a solubility parameter range for each group of solvents. Polymer parameters may be calculated, but are usually obtained by experiment. A selection of solvents, uniformly covering the whole spectrum of parameters, is assembled. A typical selection is shown in Table 3.

Attempts are made to dissolve the resin at a technically useful 'solids' level (e.g. 20 per cent for nitrocellulose, 50 per cent for an alkyd) in chosen solvents. If the polymer dissolves in any two solvents in a group, it will dissolve in all the solvents of that group with parameters between those of the chosen pair. Thus the object is to establish the extremes of the range in each group, giving three parameter ranges for the polymer. Table 4 gives parameter ranges for some resins.

Group I	Group II	Group III
6.9 Low odour mineral spirits	7.4 Diethyl ether	9.5 2-ethyl hexanol
7.4 *n*-heptane	8.0 Methylamyl acetate	10.3 *n*-octanol
8.2 cyclo-hexane	8.5 Butyl acetate	10.9 *n*-amyl alcohol
8.5 Dipentene	8.9 DEG monobutyl ether	11.4 *n*-butanol
8.9 Toluene	9.3 Dibutyl phthalate	11.9 *n*-propanol
9.3 Trichloroethylene	9.8 2-butoxyethanol	12.7 ethanol
9.5 Tetralin	10.4 cyclo-pentanone	14.5 methanol
10.0 Nitrobenzene	10.7 DEG monomethyl ether	
10.7 1-nitropropane	12.1 2,3-butylene carbonate	
11.1 Nitroethane	13.3 Propylene carbonate	
11.9 Acetonitrile	14.7 Ethylene carbonate	
12.7 Nitromethane		

An untried liquid will dissolve the polymer, if its own parameter falls within the polymer's appropriate parameter range. Roughly speaking, the parameter of a mixture of liquids is the average parameter. For example, a polyester resin would not dissolve in n-butanol (11·4, group III) or xylene (8·8, group I). However, a 114 butanol/xylene mixture dissolved the resin. The mixture's parameter is roughly calculated thus:

butanol contribution	$1 \times 11{\cdot}4 = 11{\cdot}4$
xylene contribution	$4 \times 8{\cdot}8 = 35{\cdot}2$
mixture's parameter	$= 46{\cdot}6 \div 5 = 9{\cdot}3$

The H -bonding capacity of the mixture corresponds approximately to group II. Note that ethyl acetate (9·1, group II) is a good polyester resin solvent. Where many liquids are involved, the approximation may be too great, but the effect of adding a new liquid to paint may best be assessed by deciding what effect it will have on the H-bonding characteristics of the liquid mixture and whether it will raise or lower the average parameter.

The ideas of Burrell have been extended by others to make their use more quantitatively precise. Unfortunately, these extensions almost invariably make the concept of solubility parameters more difficult to use in practice. The simple initial concept of Burrell outlined above is still a very useful one for the paint chemist and technologist.

(B) Viscosity

1. MEASUREMENT. The viscosity of a simple liquid has been explained. It is the outward evidence of the internal resistance to flow and is measured in units called 'poises'. Viscosity can be measured by any method involving either the flow of the liquid, or the movement of some object in the liquid.

In the first category come the simple flow cups, perhaps the most widely used consistency control devices in the paint industry. In these, a given volume of liquid is timed as it falls through a hole of definite dimensions.

Table .4 — Solubility parameter ranges of some polymers

Polymer Chemical type	Chemical type	Prameter range in		
		Group I	Group II	Group III
N/C, RS 25 cps (dry)	Nitrocellulose	11·1–12·7	7·8–14·7	14·5
N/C, SS1/2 sec (dry)	Nitrocellulose	11·1–12·7	7·8–14·7	12·7–14·5
CAB1/2 sec	Cellulose acetate butyrate	11·1–12·7	8·5–14·7	12·7–14·5
E/C, N-22	Ethyl cellulose	8·1–11·2	7·4–11·0	9·5–14·5
PMM	Polymethyl methacrylate	8·9–12·7	8·5–13·3	0
Acryloid B-72	Acrylic copolymer	10·6–12·7	8·5–14·7	0
Vinylite AYAA	Polyvinyl acetate	8·9–12·7	7·4–11·0	0
45% O.L. linseed glycerol phthalate	Alkyd	7·0–11·9	8·4–14·7	9·5–11·9
30% O.L. soya glycerol phthalate	Alkyd	8·5–12·7	0	0
Beetle 227-8 (dry)	Urea formaldehyde	0	7·4–11·4	9·5–11·4
Uformite MX-61 (dry)	Melamine formaldehyde	8·5–11·1	8·9–13·3	9·5–11·4
Epikote 1001	Epoxy resin	10·6–11·1	8·4–13·3	0
Epikote 1004 and 1007	Epoxy resin	0	8·4–9·9	0
Epikote 1009	Epoxy resin	0	7·8–9·9	0
Epikote 1004 DHC ester	Epoxy ester	8·5–11·1	8·4–14·7	0
Propylene glycol maleate phthalate	Unsaturated polyester	9·2–12·7		0
Note: Some of the figures in this table do not appear in Table 3. This is because the choice of solvents in Table 3 is merely as example. other solvents can be, and often are, chosen for parameter range measurements.				

A viscosity in poises cannot be obtained, but the flow time is a measure of consistency. The consistency for a spraying paint, for example, can be specified as a given flow time in a particular flow cup at an appropriate temperature (usually 250c. If the hole is a long capillary, the answer can be converted to poises, but this sort of apparatus is more practical for pure liquids than for paint. An answer in poises is obtained easily if the viscometer contains a paddle or disc, mechanically driven to rotate the liquid in an enclosed space. The force tending to twist the stationary part of the apparatus in contact with the liquid is usually measured and the scale is calibrated in poises. Comparisons between measurements made in different types of instrument can then be made. Another advantage is that either the rate of shear (proportional to the speed of rotation), or the stress (proportional to the driving force) can be varied, depending on the instrument.

The second category of instrument may involve a ball falling through a liquid, or rolling in a tube containing the liquid. The time taken for movement over a given distance is measured, but conversion of the answer to poises is easily done. Another type, more suitable for very viscous liquids, involves the rise of a bubble in a tube containing the liquid. The time of rise in seconds can be converted to poises. Also for viscous liquids, is a method in which the vibration of a rod is damped by the liquid, the effect being related to the viscosity? Results can be obtained in poises.

2. FACTS AND THEORY. There are three essential facts about the viscosity of a polymer solution. Let us consider them in tum:

1. Early in this chapter it was stated that polymers do not have a solubility limit because they cannot crystallize. Nevertheless there are forces of attraction operating between the polymer molecules in solution.

Though these forces may be weak compared with those operating in, say, a sodium chloride solution, they operate over a very long length of molecule and there are frequent encounters with other long molecules, enabling those forces to come into play. In addition, there is the possibility of simple mechanical tangling, as with pieces of string. Since both factors have been quoted as causes of increasing viscosity, it is not surprising that even comparatively low concentrations of polymer can cause considerable 'thickening' of simple liquids. As the 'solids' of the solution rise higher, the encounters and entanglements between molecules become more frequent and so the viscosity raises, e.g. solution viscosities for RS 1/z sec nitrocellulose: 12%-1 poise, 20%-10 poises, 30%-100 poises. Eventually the solution becomes so viscous, that it cannot be used in paints unless further liquid is added. At higher 'solids' still, the solution almost ceases to flow and might be mistaken for a solid. So, although there is no limit at which the solution becomes saturated, there is a limit at which the solution becomes too viscous to use. No precise figure can be quoted for this, since it depends upon the use.

2. If we have two solutions of the same polymer, e.g. polymethyl methacrylate, in the same solvent at the same 'solids', the more viscous solution will contain polymer molecules of higher molecular weight. Since we have the same total weight of polymer in both solutions, there are fewer polymer molecules present in the high molecular weight polymer solution, but they are longer. The increased opportunity for entanglement reinforced by chemical attraction outweighs the reduced number of molecules. All the attractions between polymer molecules in a useful solution are continuously forming and breaking with the movement of the molecules. If they were permanent, the solution would not flow, since the polymer

molecules would form a semi-rigid, reinforcing network in the liquid. However, to double the molecular weight and halve the number of molecules, involves making a number of these temporary associations into permanent chemical bonds. The network is not made rigid, but it is a good deal less flexible, less amenable to being deformed in flow, than before. So high molecular weight polymers give more viscous solutions, e.g. 10% solutions of PV Ac of molecular weight 15 000-0·025 poises; 73 000-0·12 poises; 160 000-0·88 poises.

3. If we take a given sample of polymer of fixed molecular weight and dissolve it at the same 'solids' in a variety of 'true solvents', the viscosities of the solutions will be proportional to the viscosities of the original solvents, e.g. 12% polystyrene in methyl ethyl ketone (0·004 poises)-0·4 poises; in ethyl benzene (0·007 poises)-1·6 poises; in a-dichlorobenzene (0·013 poises) -3·3 poises. This is important, because we can reduce paint viscosity without lowering 'solids' or polymer molecular weight, simply by changing to a less viscous solvent, if a suitable one is available.

'True solvents' dissolve the polymer at all concentrations. Some liquids give solutions at certain concentrations, but further dilution precipitates the polymer. In these quasi-solutions, polymer molecules collect in clusters, raising the effective molecular weight of the polymer and giving an abnormally high viscosity. Solvents in the middle 80 per cent of the polymer's parameter range are almost certainly true solvents.

It is worth noting here that if true solvents in a solution are partly replaced by non-solvents, the viscosity can rise. This happens when the average solubility parameter of the mixture moves to the extreme of the range for the polymer and clustering of polymer molecules occurs. This is the stage prior to precipitation.

Finally, a word about the viscosity of emulsions. If a polymer is fully emulsified in a liquid and no part of it is in solution, then the molecular entanglements and associations do not occur. Consequently the emulsion viscosity must be independent of the molecular weight of the dispersed polymer. At low solids, the viscosity of the emulsion is the viscosity of the continuous phase liquid. The viscosity of the emulsion rises slightly as the solids 'rise', because of polymer particle collisions and weak inter-particle attractions, but there is no substantial increase in viscosity, until the concentration of particles becomes so high that the particles can scarcely move at all. A typical solids-viscosity curve is shown in Fig. 28. Thus the paint formulator can use a high molecular weight polymer at high concentrations, only by using it in emulsion form.

(C) Boiling point and evaporation rate

We have already seen that the flow of paint on vertical surfaces can be controlled by solvent evaporation. To get the right solvent balance for doing this satisfactorily, it is necessary to know the relative evaporation rates of the solvents available.

In particular we wish to know the evaporation rates of thin films of solvent at normal drying temperatures. These figures are available for many solvents at room temperature (20-25 oq. One system quotes relative evaporation rates, with butyl acetate, as reference solvent, rated arbitrarily as 100. /so-propanol at 200 evaporates twice as quickly. Another system quotes relative evaporation times. Diethyl ether is reference solvent, with the rating of 1·0. n-butanol at 33 takes 33 times as long to evaporate.

In practice, however, we are not concerned with the evaporation of thin films of single liquids under carefully controlled conditions of temperature, humidity and air movement. We are concerned with the evaporation of mixtures of liquids in the presence of polymers under varying atmospheric conditions. Evaporation rates of solvents in mixtures cannot be predicted from the individual evaporation rates: attractive forces between molecules can delay evaporation and these forces will vary from mixture to mixture.

Attraction for the polymer molecules will also delay evaporation, so solvents for the dissolved polymer will evaporate more slowly than evaporation rates suggest, while diluents will be unaffected.r At best, the published evaporation rates are a rough guide on which to base solvent selection and they are only suitable for paints that lose most of their solvent at room temperature. It is necessary to follow with actual trial of the proposed liquid mixture in the paint. This may need to be repeated several times, with logical replacement of solvents by those evaporating more slowly, to increase flow, or by others evaporating more rapidly, if excess flow (sagging) is occurring.

Because evaporation rates are no more than a guide, many paint chemists feel justified in using the boiling point (BP) of a pure solvent, or the boiling range (BR) of ~n impure solvent or mixture. The boiling point is a fixed property of a liquid and does not vary with the method of measurement, provided that the atmospheric pressure is 760 mm. It is usually true that liquids with low boiling points evaporate more rapidly at room temperature than those with high boiling points. But the boiling point tells you when a liquid will boil, not how fast it will evaporate at some lower temperature.

When a pair of liquids have boiling points within 30 oc of one another, it is difficult to say which will evaporate more rapidly at room temperature. Alcohols, in particular, are slower than their boiling points suggest, because they are strongly hydrogen-bonded.

In spite of this, boiling points are a sufficiently good guide for most paint purposes. Solvents are graded roughly into three groups: 'low boilers' (BP below 100°C), 'medium boilers' (BP 100-150°C) and 'high boilers' (BP above 150°C). Low boilers are used in spraying paints, because they evaporate between gun and surface and give the necessary rise in 'solids' and viscosity. High boilers are used to give flow and can be the sole solvents where the paint must be kept fluid for fairly long periods, e.g. where brushed paint must 'marry in' when two areas overlap. Medium boilers can be used in all types of paint to give flow at first, followed by fairly quick set-up, but in spraying paints they necessitate greater care by the user, since the nearer the spray-gun gets to the work, the wetter the applied paint will be: the solvent has less time in which to evaporate.

Most paints contain as much diluent as possible, since diluents are usually aliphatic hydrocarbons, which are much cheaper than the true solvents. The limit is decided for the formulator by the fact that the polymer must be kept in solution in the can and at all stages of application and drying. For this reason, the evaporation rates of diluents and solvents have to be balanced carefully, to make sure that there is always enough true solvent present and- particularly- to ensure that the last molecules to evaporate are solvent molecules. If there is too much diluent at any stage, the polymer precipitates and the film appear milky if it is unpigmented or low in gloss if it is pigmented.

(D) Flash Point

Flash points give an indication of flammability or fire risk. The flash point is the lowest temperature at which enough vapour is given off to form a mixture of air and vapour immediately above the liquid, which can be ignited by a spark or flame under specified conditions. Most countries have regulations concerning the storage, trapsport and use of products containing the more highly flammable solvents. Highly flammable materials are usually defined as those having a flash point below a certain figure and they may also be required to support combustion in a combustion test. The latter clause is introduced so that paints which are safer to use, but have a measurable- and perhaps low-flash point, are not penalized. For example, a waterborne paint may contain a small proportion of low boiling alcohol. This could be detected in a flash point test on the paint, yet the paint might not support combustion. The combination of a flash point below 320c and combustibility is used in the Highly Flammable Liquids Regulations.

Since regulations may vary so much from country to country and, even within one country, may vary according to use or method of transport, it is usually necessary to know accurately the flash point of any paint. The flash point of a mixture of liquids can be lower than the flash point of any of its constituents, so measurements should always be made on the actual sample.

These are usually carried out in 'closed' or 'open' cups of varying design and it is important to state which cup was used. Closed cups give lower flash points.

(E) Chemical nature

It is easy to forget that solvents are chemicals and can react with other paint ingredients. For stability in the can, this is undesirable..

(F) Toxicity and smell

Some liquids, e.g. benzene, have a cumulative poisonous effect and others can be harmful above certain concentrations in the air. The smell of a liquid may be enough to prevent its use for some purposes. Smell is largely a matter of customer's opinion, but toxicity information is obtainable from the sources quoted in the Appendix.

(G) Cost

Hydrocarbons (particularly aliphatic) are the cheapest solvents; some esters, less common ketones and nitroparaffins are the most expensive.

08 Paint Additives

These are those remarkable materials which, when added to a paint in amounts that can be as little as 0·001 per cent and seldom more than 5 per cent, have a profound influence on the physical and chemical properties of the paint.

ADDITIVES AFFECTING VISCOSITY

In the last chapter we dealt with the viscosities of liquids and polymer solutions. Emphasis was laid on the use of instruments which measure the viscosity in poises. Even better are the types which allow either the stress or the rate of shear to be varied.

With pure liquids, such variations are unnecessary, because the viscosity coefficient is a constant. As the stress increases, the rate of shear does so at the same rate and vice versa. This is also true for polymer solutions at paint viscosities. If, however, the 'foreign' material in the liquid is also insoluble in it, the equation may be true no longer. Since most paints contain undissolved materials, it becomes important in many instances to quote not only the viscosity in poises and the temperature of measurement, but also either the rate of shear or the stress at which the measurement was made.

Why does this deviation from constant viscosity occur? An ideal (pigment) particle dispersion, consists of a completely uniform distribution of isolated individual particles throughout the liquid. In the same chapter, we saw how chemical groups at pigment surfaces can exert strong attractions over other suitable bodies in the vicinity.

Our ideally distributed particles will of course move, since they are affected by gravity and by collisons with fast-moving liquid molecules, or slower but heavier polymer molecules. During the course of these movements they are bound to encounter other particles. Should suitable parts of the surfaces of two particles come into contact, inter-attractions may be strong enough to prevent the particles from separating by their own motion. If this pairing up process continues, quite large groups or clusters may form. A cage-like network or structure may even extend throughout the liquid (see Fig. 1).

Such an arrangement might give the appearance, in the can, of being very viscous, even jelly-like. But once the paint is stirred, the structure breaks up, the particles separate, the liquid 'gives' under the stirrer and the viscosity appears to drop. Should the stirring stop or slacken, the flocculation (as it is called) will begin again and the viscosity will rise. Thus the paint appears to have two viscosities: a high one when it is still and a low one when it is agitated or sheared. In fact, it has a wide range of viscosities, corresponding to all the rates of shear between zero and some value at which the viscosity becomes a minimum, or to all the stages of partial flocculation between complete flocculation and complete deflocculation.

If flocculation occurs slowly, the viscosity, measured at low rates of shear, increases with time during the rest period after an efficient shearing. When this happens the paint is said to be thixotropic. If there is no dependence on time or on the previous treatment of the paint and if the viscosity decreases as the rate of shear increases, then the paint is said to be pseudo-plastic. If there is a minimum stress required before any flow can occur at all, the viscosity behaviour is said to be plastic. All these types of behaviour are, of course, contrary to Newton's equation and are grouped together under the heading of non-Newtonian viscosity.

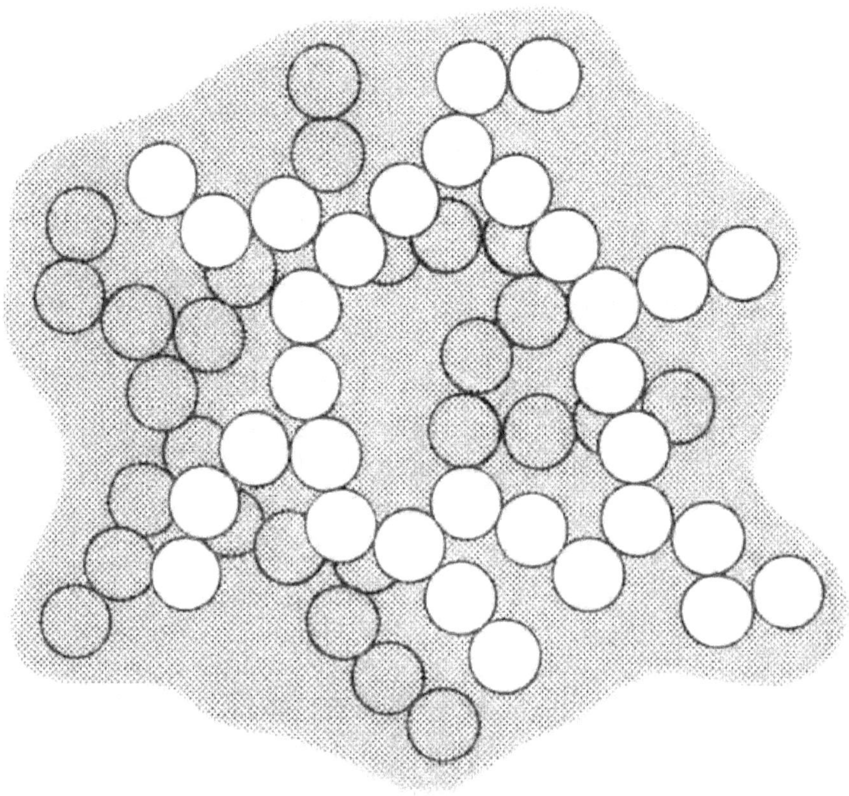

Figure .1 — A structure of flocculated particles

Most paints show non-Newtonian viscosity to some degree and, if it is marked, it can be most beneficial to the paint. Thus the viscous-even jellylike- paint in the can does not show hard pigment settlement, applies easily under the shearing action of, for example, a brush or spray-gun, and does not 'sag', because the viscosity rises as soon as the paint is still (or nearly so) on the object being coated. However, non-Newtonian viscosity should be introduced into paint with caution. While it is true that the paint will not 'sag', it may be extremely difficult to adjust it to obtain reasonable flow-out of application marks as well. Also, if the effect is obtained by pigment flocculation, the perfectly uniform pigment distribution, which gives the maximum light scattering and absorption, will be lost and hiding power will fall. Let us therefore consider ways and means of achieving non-Newtonian viscosity and discuss their relative merits.

(a) Pigment volume

If the level of pigment in the paint is high, the very bulk of pigment in the paint will cause non-Newtonian viscosity. The close packing of the particles makes some flocculation inevitable and the viscosity will be high simply because the particles impede each other's movements. A further phenomenon, dilatancy, can occur at very high pigment volumes. Here disturbing the arrangement of the particles can Rack them into an even tighter mass and, on stirring, the viscosity rises and the paint appears to set solid. When stirring ceases, fluidity apparently returns to the paint. Dilatancy, particularly of pigment settlement in cans of paint, is usually highly undesirable.

(b) Silica and silicates

Dispersion in the paint of a few per cent of very fine particle silica, Si02 (diameter about 0·015,um), produces a largely pseudo-plastic effect. The surface forces and very large surface area (e.g. 190-460m2 per gram or 1·3-3·3 acres per ounce!) are responsible.

Fine particle synthetic aluminium silicates (Al20 3·2Si02·2H20) are also used in paints containing organic solvents. Bentonite is used to impart non-Newtonian behaviour to water paints. It is a complex silicate of the montmorillonite type, containing aluminium, sodium, potassium, iron and magnesium. It has a plate-like particle and will swell to several times its dry volume by adsorption of water between silicate layers. The metal ions in montmorillonites can be exchanged for quaternary ammonium ions, ~N+, and when R is a long chain alkyl group, the character of the clay is changed considerably. Organic hydrocarbons are now adsorbed on to the platelet surfaces, because of their attraction for the alkyl groups. Thus these treated bentonites, given the trade name 'Bentones', will gel organic liquids or impart non-Newtonian viscosity to paints containing organic solvents. Attractive forces between the particles lead to structure throughout the paint (see Fig. 2).

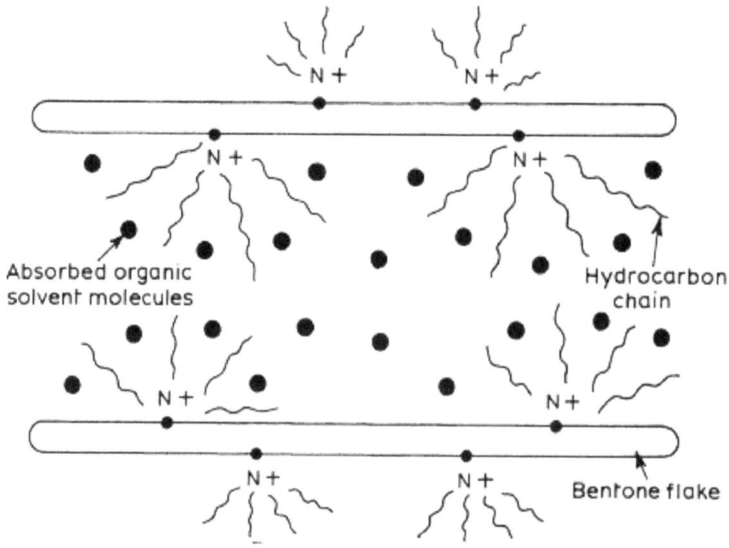

Figure 2 — The gelling mechanism of 'Ben tone'

(c) Resinous thickeners

There are a number of polymers, notably hydrogenated castor oil and its derivatives, polyamides and polyamide-oil or polyamide-alkyd reaction products, which can be used to impart non-Newtonian viscosity to paints, based on non-polar (mainly aliphatic hydrocarbon) solvents. The thickening mechanisms of polymeric additives are not fully established, but the resins have in common the following features: borderline solubility in the paints in which they are used and chemical structures involving lengthy soluble non-polar chains (e.g. fatty portions of fatty acids) and polar groups,

e.g. -OH, -CONH- and -COOH

The thickening mechanism is probably similar to that of the particles in section (b). A loose network forms, either between very finely divided colloidal particles of resin, via surface forces due to the attractions between polar groups, or between polar portions of large resin molecules. A colloidal particle might have a diameter of O·Ol, um (lOom) or less. If a single molecule of a polymer of density 1·0 was compactly coiled so as to form a sphere of diameter 10 nm, the molecular weight of the polymer would be 300 000. If a particle of the same size contained solvent molecules, the polymer molecular weight necessary to give that diameter could be much less than 300000. Depending on the molecular weight of the polymer in question, such colloidal particles in paint could be clusters of 2-100 molecules. The difference between small polymer particles and large molecules is not great.

Resinous thickeners are also used in aqueous paints. The mechanism of operation is similar, but the polarities of the functioning groups are reversed. Thus solubility is conferred by water-soluble polar chain lengths, such as polyethylene glycol or neutralized polyacrylic acid, and attractions form between the water-insoluble (or hydrophobic) portions of the polymer. The latter are usually concentrated as substantial chain segments at the ends of the chains (e.g. as in A-B-A block copolymers), or at intervals along the chains. In emulsion paints these associative thickeners not only form structures between their own molecules, but also anchor onto the hydrophobic portions of the latex polymer or pigment particles, which thus take part in the structure. High shear viscosity is controlled by thickener concentration, but low shear viscosity is related to the number and strength of the associations that are formed.

We should bear in mind the effect of the solvent environment on the thickeners described so far. In non-aqueous paints, polar groups on particle surfaces are even more accessible to small polar molecules (additives or solvents) than to polar groups on other particles or large molecules. Once these small molecules are in place, they prevent particle-particle or molecule-molecule associations by 'blocking' the active sites. Exactly the same is true (with polarities reversed) in aqueous systems.

Alternatively, the addition of polar solvents to hydrocarbon-based paints will improve the solubility of the thickener molecules and decrease the chance of particles, or of clusters of molecules forming because of borderline solubility. This will lessen or even remove the effectiveness of the thickener. Again the same principles apply to aqueous coatings: additions of water-miscible organic solvents (e.g. ether-alcohols) increase the solvency of the medium for the hydrophobic portions of the associative thickeners and lessen their tendency to associate.

The resinous thickeners above are thermoplastic and all ultimately soluble in some solvent blend or other. As we have seen, their low shear effects can be destroyed by solvent additions. If, however, the polymers are made as latexes or NADs by emulsion polymerization and a small amount of cross-linking is introduced via poly-functional monomer, then insoluble colloidal resin particles are produced. These are called micro gels.

In an alternative type of microgel, insolubility is obtained by strong hydrogen-bonding between regular, linear chains, leading to crystallinity, e.g. in polyureas made from hexamethylene diisocyanate and hexamethylene diamine.

Microgels thicken paints by forming flocculated network structures. They are not insensitive to solvent selection, but are much less sensitive and are effective in polar organic solvents. Indeed, some swelling of the polymer particles by solvent probably enhances their effect, perhaps by increasing particle diameter. They can be used in clear coatings without loss of clarity if they are designed to have the same refractive indexes as the film-formers.

(d) Metal chelates

Metal chelates are organo-metallic compounds in which organic chains are bonded to a central metal atom partly by covalent and partly by co-ordinate bonds, for example:

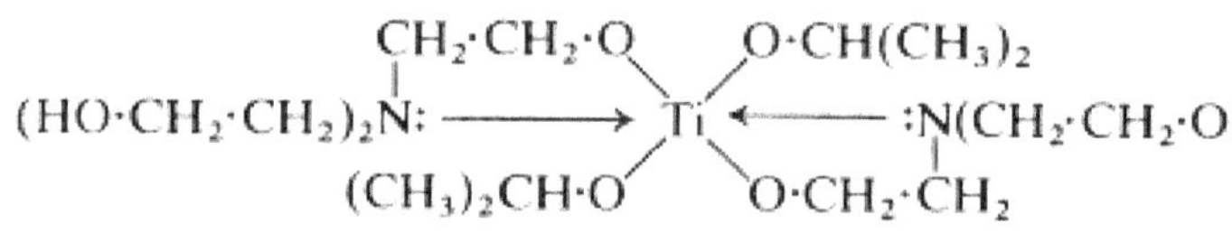

bis (triethanolamine) titanium di-*iso*propoxide

Some chelates of titanium and zirconium have proved especially useful for the thickening of colloid-stabilized aqueous latex paints.

They can induce considerable structure at low shear, leading to non-drip characteristics and good flow with control of runs and sags. The titanium chelates work best with latexes stabilized by cellulose ether colloids and thickening is thought to occur as the result of hydrogen bonding between hydroxyls in the chelate and in the colloid. An unacceptable irreversible thickening occurs with polyvinyl alcohol (PVA) colloid, due to interchange at the covalent bond between the PVA and the alkoxy group. High surfactant levels interfere with thickening.

The zirconium chelates work with latexes containing sodium carboxymethyl cellulose or cellulose ethers or polyvinyl alcohol, but in each case the pH of the paint must be carefully controlled.

Both types of chelate are effective at or below 1% and the structure formed is thixotropic. If stirred, the paint slowly recovers structure, but never quite to the original value and there is some loss of structure on each repetition of stirring. Methods of manufacture of the paints must be closely controlled.

Choice of Additives Affecting Viscosity

The additives that induce non-Newtonian viscosity in paints should be used as sparingly as possible. The paint formulator will find that it is extremely difficult to make several batches of paint to the same formula and the same viscosity and hence the same application characteristics. Some of the resinous thickeners are themselves difficult to reproduce exactly from batch to batch and the mineral thickeners are difficult to disperse reproducibly. As we have seen, all types are influenced by the ingredients of the paint itself and these too may vary slightly from batch to batch. Such variations may not matter normally, but they can play havoc when coupled with the variations in the additive.

The paint formulator must devise suitable tests, which will inform him of variation in viscosity characteristics and must know how to adjust the paint to give the required characteristics. Tests can include measurements of viscosity at several rates of shear. Points plotted on a viscosityrate of shear graph must fall within a given area. Alternatively, if the paint forms a gel in the can, the strength of the gel in the container can be measured by inserting a razor blade-shaped paddle on a spindle. The paddle is then slowly rotated by a motor. At first the paint and container turn with it against a spring restraining the magnetic platform on which the can stands, but eventually the gel breaks and, at this point, the twisting force or torque is noted and recorded as the gel strength.

The choice of resinous thickener type will depend largely on the solvent system involved. For non-polar organic solvents, linear polymer thickeners are suitable. For polar organic solvents, microgels can be used. For aqueous systems, linear (associative) thickeners, micro gels or (if colloids are present) metal chelates can be used. Mineral thickeners are available for all types of solvent.

Compatibility of the soluble resinous thickener with the main binder needs to be checked and excessive amounts of particulate thickeners can reduce gloss, especially in high gloss paints. Large differences in refractive index between binder and particulate thickeners can cause loss of clarity in clear coatings and should be avoided. The use of high pigment levels is seldom a chosen method of introducing non-Newtonian viscosity. Pigment level is usually dictated by gloss and only matt finishes and undercoats have really high pigment levels.

Additives Affecting Surface and Interfacial Tensions

Apart from the defects introduced by the application method, other troubles occur from time to time, often apparently without rhyme or reason.

'Popping'- the appearance of holes or craters in the film- is usually due to inadequate drying at room temperature before the paint is stoved, to the presence of too much low (or medium) boiling solvent, to reaction in the paint producing a gas, or to a faulty undercoat. It can be caused by contamination of the undercoat. 'Blushing', the whitening of the surface of a clear film or the loss of gloss of a pigmented one, is due to condensation and subsequent emulsification of water in the film. If the water evaporates after the film has set, it leaves fine air bubbles in the film. Refraction, reflection and diffraction lead to a milky appearance. These and other defects are put right by changes in drying conditions. One defect, however, usually requires additives to overcome it. This is known as 'cissing'.

'Cissing' is the appearance of small, saucer-like depressions in the film's surface. These are caused by particles or droplets of incompatible material, which either land on the film during drying, or are present in the paint itself.

The weight of the particle causes it to sink but, since there is no attraction between paint molecules and molecules in the particle's surface, the surface tension of the paint resists the particle's entry into the liquid film.

Immediately under the particle, the paint molecules have no alternative but to be in close proximity to the particle. In the zone surrounding the particle, the liquid 'skin' is depressed by the particle's weight, but attractions from within the paint film pull the surface molecules away from the particle. There is said to be a high interfacial tension between paint and particle. The net result, as Fig. 3 shows, is a 'ciss' mark.

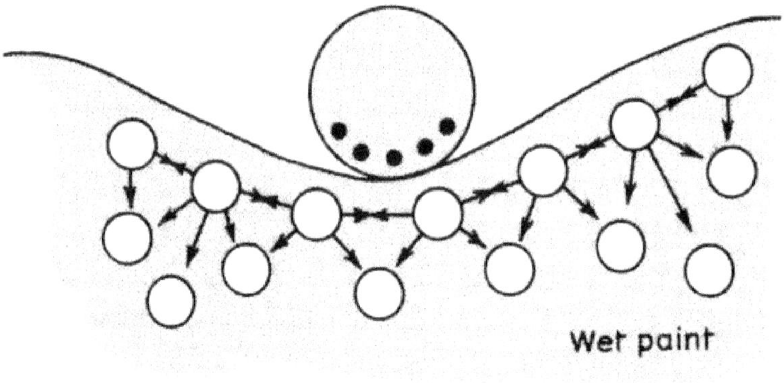

Figure 3 — A ciss mark

To prevent cissing, an additive that reduces interfacial tension is required in the paint. When interfacial tension falls, the particle is wetted by the finish and absorbed into the film. Surface-active agents reduce interfacial tension. Alternatively, an agent can be added that will reduce the liquid surface tension so much, that the interfacial tension also becomes low. Silicone oils do this effectively. Very little silicone oil is required, because it finds its way almost entirely to the surface. Silicones are semi-organic compounds of the general type where R may be a hydrogen atom, or some group containing carbon and hydrogen atoms.

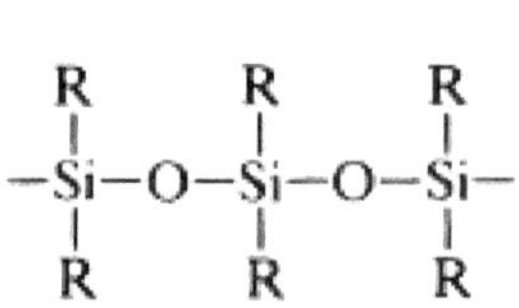

The higher molecular weight silicones are resins. Since the Si-0 and Si-C bonds are much more resistant to heat than C-C bonds, the resins may be used as film-formers in heat resistant finishes. If ·-R in the general formula is a methoxy (-0.CH3) or hydroxyl group, cross-linking is possible, or reaction with hydroxyl or carboxyl in other resins to produce, for example, silicone-modified polyester resins. These are used in Coil Coatings because of their high resistance to degradation by UV in sunlight. However, the pure resins are very expensive and are incompatible with most of the other paint polymers. They therefore find limited use in paint. As additives, silicone oils are widely used. The silicone oil must be compatible with the finish, or it will itself cause cissing.

Other materials which can prevent cissing are higher molecular weight linear polymers. The mechanism by which they act is obscure. All such additives are sometimes called ‘flow agents’.

SURFACTANTS

Surfactants are those chemicals whose molecules have two parts of widely differing polarity and solubility. The soaps, for example, have an ionizing salt ‘head’ to the molecule and a long non-polar hydrocarbon ‘tail’. A different type are the polyoxyethylene ethers of dodecyl alcohol, e.g. $C_{12}H_{25} \cdot 0 \cdot (CH_2 \cdot CH_2 \cdot 0kH$. These contain the non-polar dodecyl group ($C_{12}H_{25}$-), from dodecyl alcohol, at one end of the molecule and the successive polar ether linkages at the other. Whatever the formula or type, all surfactant molecules have in common these polar and non-polar portions.

Thus one end of the surfactant molecule is attracted to polar molecules (e.g. water) and polar surfaces, while the other prefers a non-polar environment. If we have the problem of two materials which will not wet or make chemical contact with one another, surfactants can bridge the gap. In the example already discussed under ‘cissing’, we will assume that tiny oily particles are dropping on to the surface of a particularly polar resin solution.

There is incompatibility and no wetting. If a suitable surfactant can be found with an aliphatic ‘tail’ attracted to the particle surface and a polar portion attracted to the resin molecules, chemical contact from particle to paint will be established, interfacial tension will drop and cissing will not occur. The surfactant provides a bridge across the gap between the two unlike molecules.

Wherever this ‘bridging’ is required, surfactants can be used. If pigment surfaces have poor attraction for binder molecules, surfactants can assist dispersion. When two liquids will not mix, surfactants will stabilize droplets of one liquid in the other, i.e. they will emulsify. Different surfactants are required for different systems and different applications: the nature and proportions of the two parts of the molecule will vary from use to use.

ADDITIVES AFFECTING GLOSS

A surface that is sufficiently smooth will be glossy. To reduce gloss we must roughen the surface and break up its smooth outline. In pigmented finishes, this is usually done by increasing the number of pigment particles present in the paint and hence at the surface of the paint. The protruding particles break up the smooth outline. In varnishes or clear wood finishes, we cannot do this and reduction in gloss (to a ‘satin’ sheen at least) is usually obtained with additives.

Either a few per cent of fine particle silica are used, or else an insoluble wax is dispersed in the finish and floats to the surface during drying. The silica, being an extender, is of course transparent when wetted by the finish. Its high oil absorption makes a small percentage as effective as a much larger quantity of ordinary pigment. However, in many cases the silica, like the wax, tends to float to the surface. It does so as large aggregates containing many particles. The air trapped in the unwetted spaces between the particles reduces the density of the aggregate, allowing it to rise to the surface. When wax is used, only very small quantities are needed and this method gives a very smooth feel to the finish. Polyethylene and polypropylene waxes are particularly efficient and, being relatively high in molecular weight, do not impair the heat resistance of the coating.

Additives Affecting Chemical Reaction

These are the items described as 'activators' (or 'catalysts'), 'accelerators','driers', 'inhibitors' (or 'retarder solutions'). Detailed examples will be given later; for the moment it can be said that they fall into two categories:

1. Those that initiate the drying reaction (e.g. 'activators')
2. Those that affect the rate of drying (the rest).

An activator may be one portion of two pack paint, one reactant of the resin-forming ingredients. In this instance the activator can hardly be called an additive. The true additive activator is a chemical which, when added as a minor ingredient, sparks off the chemical reaction in the paint. Such an additive is usually a chemical which decomposes to give free radicals which, in turn, initiate an addition polymerization. Driers and accelerators are true catalysts in the chemical sense, since they can speed up the chemical reaction responsible for drying, without being consumed in the process.

An inhibitor slows down the chemical reaction, usually by reacting with free radicals, thus preventing them from initiating addition polymerization. Inhibitors are therefore chemically changed in the process.

Retarder solutions are simply solutions of inhibitor. Retarder can also mean a lacquer thinner containing high boiling solvents. These retard drying simply by being slow to evaporate.

Anti-skinning agents (or anti-oxidants) are mild inhibitors of oxidative drying and are often included in paints drying by that mechanism, in order to improve stability in the can. They are usually fairly volatile and evaporate once the paint is applied. Moisture scavengers are included to inhibit reaction between paint ingredients-e.g. isocyanates or zinc or aluminium pigment-and water, which may have entered the paint during manufacture. Such a reaction might prematurely gel the coating in the container or produce gas under pressure. Calcium oxide has been recommended for zinc-rich coatings and other moisture scavengers.

Additives Affecting Living Micro-Organisms

Paints are susceptible to deterioration brought about by micro-organisms. These include particularly bacteria, yeasts and fungi. All these types of organism can cause deterioration of liquid paint in the can, but most reported incidents are caused by bacteria. Bacteria can cause gassing, viscosity reduction and colour drift in latex paints. They can enter the paint via infected intermediates and raw materials (including water) or unsterile equipment. The best defence against infection lies in high standards of plant hygiene with regular sterilization of equipment. Nevertheless, the inclusion of an in-can biocide at low concentration in latex and paint is a wise precaution. Commercial biocides for paint are often mixtures of complex organic chemicals, which together give protection against a wide range of bacterial types.

Once a paint film has been applied it is exposed to a wide variety of airborne spores from yeasts, fungi and algae. If they find nutrients in the paint or in dirt or debris adhering to it, plus moist conditions ideal for growth, then they will multiply forming unsightly colonies. Algae also require light for growth and so are found mainly on outdoor paint. As the mildew, moulds and algae multiply, film damage eventually results, leading to loss of protective properties. To prevent this fungicides and algicides are included in the paint formulation. Again a range of complex organic compounds are effective, as are various organometallic compounds (especially tin complexes). No one compound is completely effective against all species and widest protection is given by a mixture. Selection is based not only on biocidal effectiveness, but also on solubility, stability in the can and long life in the film.

Algae are also found among the large number of plant and animal organisms that cause fouling of ships' hulls below the water line. These organisms, which also include barnacles, tubeworms, mussels, polyzoa, hydroids, ascidians, sponges and sea anemones, settle on the hull and adhere strongly when the ship is stationary or moving very slowly. Once established, they colonize the surface, causing resistance to movement.

Reductions of speed by even half a knot lead to large extra costs, arising from lengthened journeys or higher fuel consumption. To prevent such growths, ships bottoms are coated with anti-fouling paints containing socalled anti-fouling additives. These are biocides which leach slowly from the film and are effective when released at only a few micrograms em – 2 day-1. In the paint, concentrations are usually much higher than normal additive levels. Many 'additives' are in fact pigments, including the traditional copper compounds, cuprous oxide and cuprous thiocyanate. They are used at high concentration, either in a slowly dissolving film former (e.g. rosin), or in an insoluble matrix (e.g. chlorinated rubber). Against enteromorpha seaweed (algae), organometallic compounds (often based on tin) are more effective and are used as non-pigmentary solid solutions in e.g. acrylic binders. Even more ingenious is the direct modification of the acrylic binder by copolymerization of a tin acrylate or methacrylate (e.g. a trialkyl tin acrylate). The surface of the polymer is gradually eroded away releasing the tin, so that the coating is kept smooth by a 'self-polishing' action.

09 Surface Coatings

THERMOSET POWDER COATINGS

Introduction

In recent years, awareness of environmental conservation and pollution prevention has risen steadily. Governmental regulation and true concern for the environment have motivated chemists to modify all types of coatings to reduce environmental impact. The concept of environmental "friendliness" has dramatically changed the way that coatings are formulated.

Powder coatings are arguably the most environmentally "friendly" coatings. They do not contain solvents to be released as hazardous air pollutants (HAPs). Powder coatings release very low amounts of volatile organic compounds (VOCs) during the baking cycle. They produce virtually no waste material. And, they contain very few hazardous chemicals. (Note: The few hazardous chemicals that have found their way into powder coatings are decreasing as they are replaced with safer materials.)

Powder Coatings Defined

Powder coatings can be described as "ground up dry paint." They may also be referred to as pulverized plastics. They have properties in common with both materials. The polymeric resins that are used in producing powder coatings are similar in nature to those used in both paints and plastics. All three materials are composed of combinations of resin, pigment, filler, and various additive materials. They may be thermoplastic or thermosetting.

The primary difference between the three types of compounds is the molecular weight range of the polymers used as binders. Plastics use the highest molecular weight resins, paint the next highest, and powder coatings the lowest. Paint, of course, also contains various solvents to dissolve or dilute the coating for easy application.

Thermoset Beginnings

Up to this point, all powder coatings were thermoplastic in nature. Thermoset powder coatings were just beginning to be explored. The first trials were made with existing resins or slightly modified polymers. They had low glass transition points (T g) and tended to sinter (cake) upon standing for any length of time.

In the early 1960s, two developments occurred almost simultaneously. The first development was the invention of storage stable thermoset powder coatings. They were epoxies based on diglycidyl ethers of Bisphenol A. The second invention was the electrostatic spray gun, which was capable of applying a powder coating under controlled conditions. Electrostatic application of powder coatings is performed by pumping a stream of coating particles in an airstream past a high voltage cascade generator. The powder acquires an electrostatic charge and flows through the generated electrical field to a grounded metal substrate. The substrate is then transported to an oven where the coating melts, flows out, and is cured.

PROCESSING AND EQUIPMENT

The process of producing a powder coating is somewhat more complex than that for other types of coatings. The equipment needs are more numerous. The production time is also somewhat longer. This section will describe each of the necessary pieces of equipment in the order that they are used during the process of making a powder coating.

Premixture

The first step in the manufacture of a powder coating is the premix stage. This is the most crucial phase of production. The various sizes of resin flake, pigment powder, etc., must be transformed into a homogenous blend of similar particle size before they enter the following stage — extrusion.

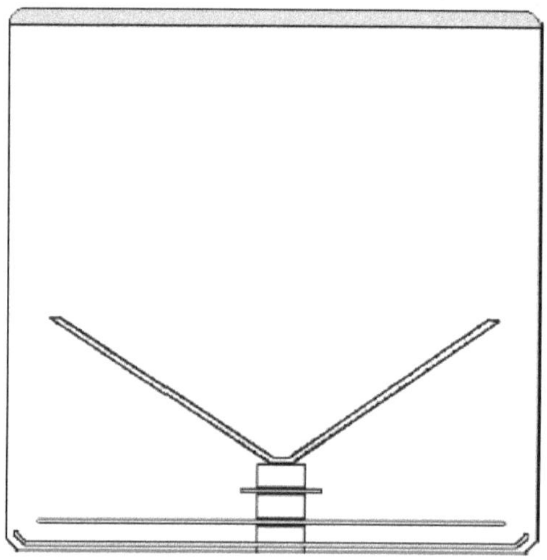

Figure .1 — Generalized schematic of premix equipment

Because the extruder is a simple melt mixer, and not a very efficient disperser, the premix stage must allow for the best possibility of good dispersion of all ingredients in the final product. Thus, premixture is performed in equipment with sufficient ability to break down the large resin flakes, thoroughly blend any liquid components, and disperse the smaller quantity constituents, such as additives.

Premixers

Most premix equipment consists of a mixing bowl with a locking cover. The equipment is usually fitted with a cooling jacket capable of being filled with running water or, in rare cases, cryogenic gasses. The mixing blades within the bowl may have various configurations. However, there are generally two or three blades placed at different heights and orientation to one another (Figure .1).

Extrusion

The second stage of powder coating manufacture is extrusion of the premix. The extruder is basically composed of a horizontally placed stainless steel tube or "barrel." Within the barrel are one or two "screws" that are slightly smaller than the barrel in diameter. The screws turn and move the premix forward through heated "mixing zones." The mixing zones have special attachments, called mixing paddles or blades. The paddles knead the melted premix and blend the various ingredients.

Contrary to what may be imagined, the extruder is not a very good disperser. The space between the screws and the barrel walls is relatively large and allows material to travel through without dispersing it thoroughly. Also, the coating mixture passes through the barrel fairly rapidly and does not have time to be dispersed efficiently.

Extruders

There are three basic types of extruder. The first type, the single screw extruder, is most often used in the plastics industry. It does not move material forward as rapidly as other types, and thermosetting coatings could react in the barrel and gel prematurely. Thermoplastic materials used in injection molding are commonly prepared in single screw extruders. However, there is one type of single screw extruder used in the thermosetting powder coatings industry. It has special attachments along the sides of the barrel that aid in the blending of the ingredients.

The second type of extruder is the counter-rotating twin screw extruder. It is composed of two screws placed side-by-side and turning in opposite directions (i.e., one turns clockwise and the other counterclockwise). This type is also usually used in the plastics industry.

The third kind of extruder, which is used in powder coating production, is the corotating twin screw extruder (see Figure 2). In this type of equipment, the two side-by-side screws turn in the same direction. This greatly increases throughput and decreases residence time. Reduced residence time is important in thermoset coatings, especially in the most reactive systems with short gel times.

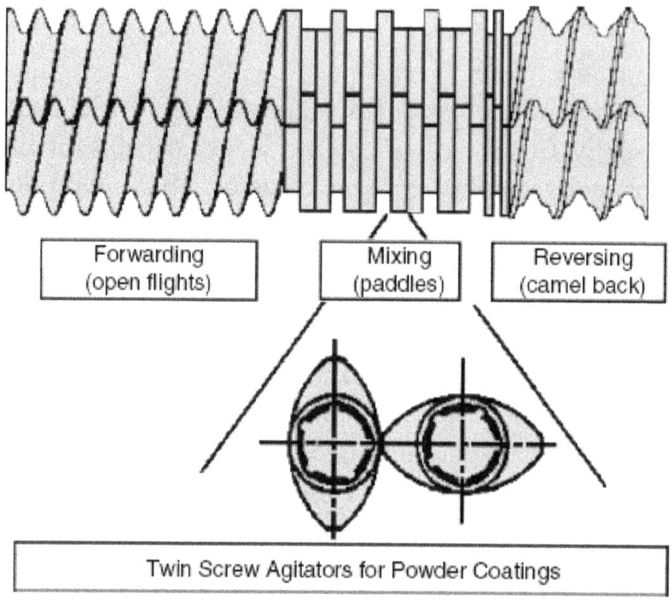

Figure 2 — Typical configuration for twin screw corotating extruder

Grinding

The grinding phase in the manufacture of a powder coating is important to its final performance. The average particle size and distribution are important in the application properties and final surface appearance of the coating.

Once a powder coating has been extruded, it passes through one or more stages to prepare it for grinding. The molten material is dropped onto chilled metal rollers, where it is pressed flat and cooled. The coating is then broken into flakes or granulated and fed to the grinder.

Grinders

There are three favored types of grinding equipment used in the powder coating industry. They are the impact or hammer mill, the air jet mill, and the air classifier mill. The impact or hammer mill grinds the coating materials by throwing them or hitting them against a stationary screen or stator. The particles are broken down until they are small enough to fit through the screen. The particles are then passed to various sieves or cyclonic separators for final particle size distribution adjustments.

The air jet mill is an extremely efficient grinder. Powder coatings are fed into a high volume and high velocity airstream. The particles circle the cylindrical grinding chamber striking other particles and breaking them into very fine powders. The fine material rises out of the chamber with the escaping air, through conduits, to separation devices that structure the particle size distribution.

The air classifier has elements of both of the previous types. There is an impact chamber that is supplemented with airstreams that rapidly break down the coating materials. This provides excellent control of particle size distribution, and the moving air tends to cool the powder, minimizing heat effects, such as impact fusion.

Sifting and Classifying

The final structuring of particle size distribution is accomplished by passing the ground materials through sieves or cyclonic separators. Sieves "scalp" the largest or smallest particles, which narrows the particle size distribution. By judicious use of sieves, the distribution and average particle size may be chosen for best application and surface properties.

Cyclonic separators or "classifiers" function by forcing powders in airstreams to circle conical chambers.

The heaviest (largest) particles will fall to the bottom of the chamber and will be collected for use or dumping. The smallest particles will flow out through the top with escaping air, will flow through conduits, and will be saved or discarded as the need arises. By adjusting the volume and velocity of the airstream, a particle size distribution can be varied.

APPLICATION EQUIPMENT

There are three primary methods of applying powder coatings. The first uses a fluidized bed. As described earlier, the fluidized bed transmits a diffuse stream of air upward through the bottom of a container holding the coating to be applied. The airstream is kept at low enough pressure so that the powder rises into the air but is not expelled from the container. As air escapes past the floating particles, the surface of the powder mass appears to bubble or "percolate." The term "fluidized" refers to this fluid-like motion. The metal part to be coated is heated above the melting point of the coating and is "dipped" into the fluidized bed. Powder touching the part melts and adheres to the metal. The part is then transported to a curing oven.

The second and third methods of application utilize electrostatic spray guns. There are two basic types of gun: corona-charged and "tribo"-charged. The corona-charged spray gun applies an electrostatic charge to the powder as the material flows through an electrical field generated at the gun tip of the gun. The tribo-charged gun has a polytetraflouroethylene (PTFE) barrel, and powder flowing through it acquires an electrical charge from friction. The charged powders are sprayed onto grounded metal parts, where they adhere through electromagnetic attraction. The parts are then transported to the curing oven.

Chemistry

The chemistry of powder coatings is not very different from that of liquid coatings. The main resin types used in thermosetting powder coatings are all familiar to the liquid coatings chemist. The epoxy, polyester, and acrylic polymers are well known in liquid coatings. However, the molecular weight and curing agents are somewhat different. From these three resin types, a number of different cross-linked systems are possible. The most common will be discussed here. Note that each subsection describing a cross-linking pair includes a common industry "nickname" in parentheses.

Epoxy Systems

The most common epoxy resins in use today are diglycidyl ethers of Bisphenol A made from reaction of epichlorohydrin with Bisphenol A. Diglycidyl ethers of Novolac resins also have significant commercial impact.

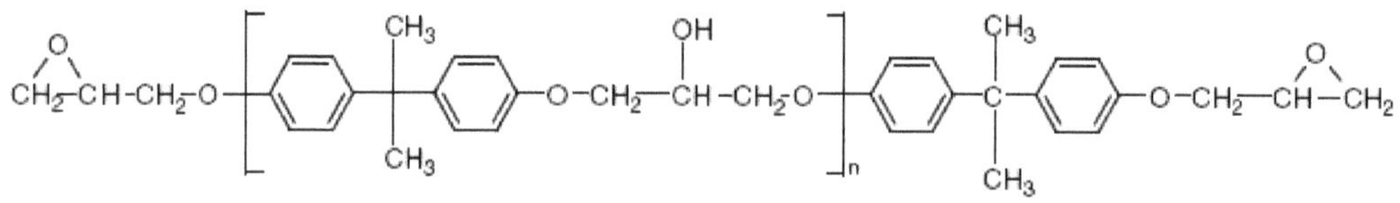

Diglyceride Ether of Bisphenol A Resin

The epoxy groups terminating each molecule react with acidic or basic curing agents. The three most common are the phenols, dicyandiamides (DICY), and carboxylic acids, including carboxy-terminated polyesters. Various acids, anhydrides, amines, and imidazoles are also used as cross-linkers with epoxy resins.

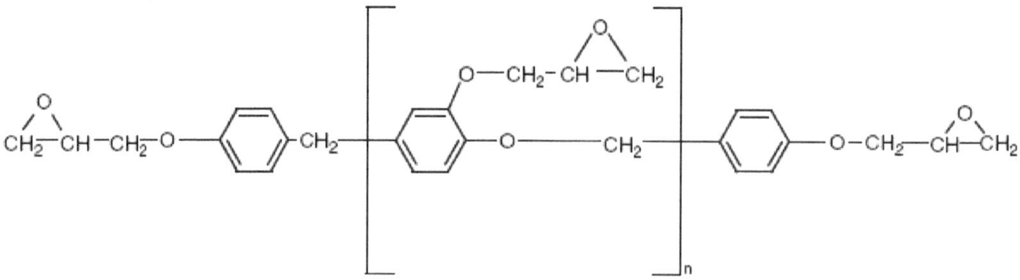

Diglycidyl Ether of Novolac Resin

Epoxy-Phenols (Phenolic)

The curing of epoxy resins with phenols results in the opening of the epoxide ring and the formation of a hydroxyl group, at either the primary or secondary position. The hydroxyl group is available for reaction in the cross-linking of the resin. The aromatic ring attaches to the unreacted carbon of the epoxide.

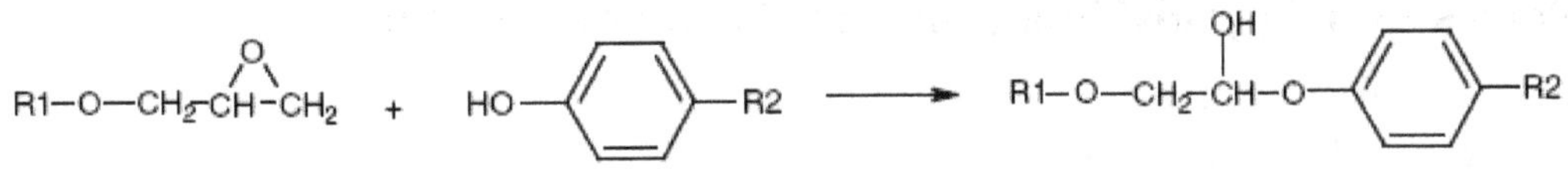

Epoxy Resin | Phenolic | Crosslink Intermediate

Epoxy-Dicyandiamide (DICY)

DICY cured epoxy coatings react in a similar manner to that of the previous type, where nitrogen-bearing groups replace the aromatic ring. All four functional groups will react with the epoxide, acting as a primary or secondary amine.

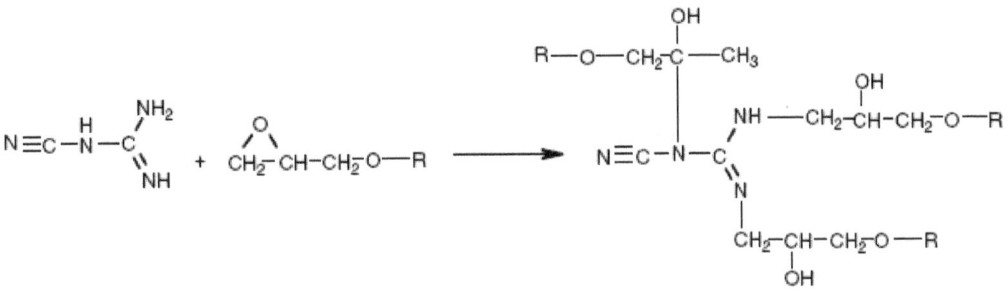

DICY | Epoxy Resin | Crosslink Intermediate

Epoxy–Polyester (Hybrid)

Epoxy resins react with carboxy-functional polyesters in the same way as carboxylic acids. The hydroxyl portion of the acid group reacts with the epoxide. The rest of the reaction follows as we have seen in the previous examples. Because both reactants are considered primary resins, the system is referred to as a "hybrid."

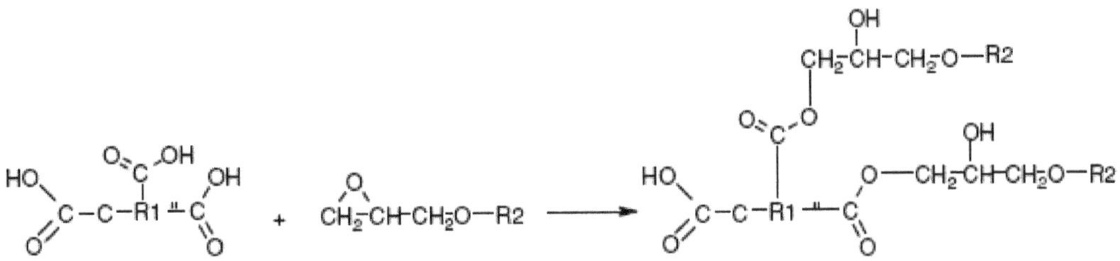

COOH-Polyester | Epoxy Resin | Crosslink Intermediate

Polyester Systems

There are two types of polyester resins used in thermoset powder coatings. They are carboxy- and hydroxyfunctional. There are several different curing agents used with each. We will examine the four most common systems.

Polyester-Triglycidyl Isocyanurate (TGIC or Polyester)

Triglycidylisocyanurate (TGIC) is a heterocyclic tri-epoxy curing agent. As such, it reacts with carboxyfunctional polyester resins similar to those used in hybrid coating systems. The major difference is in the trifunctionality. One mole of TGIC will react with three moles of polyester resin. The cross-link network is also more complex because of the extra reactive sites available.

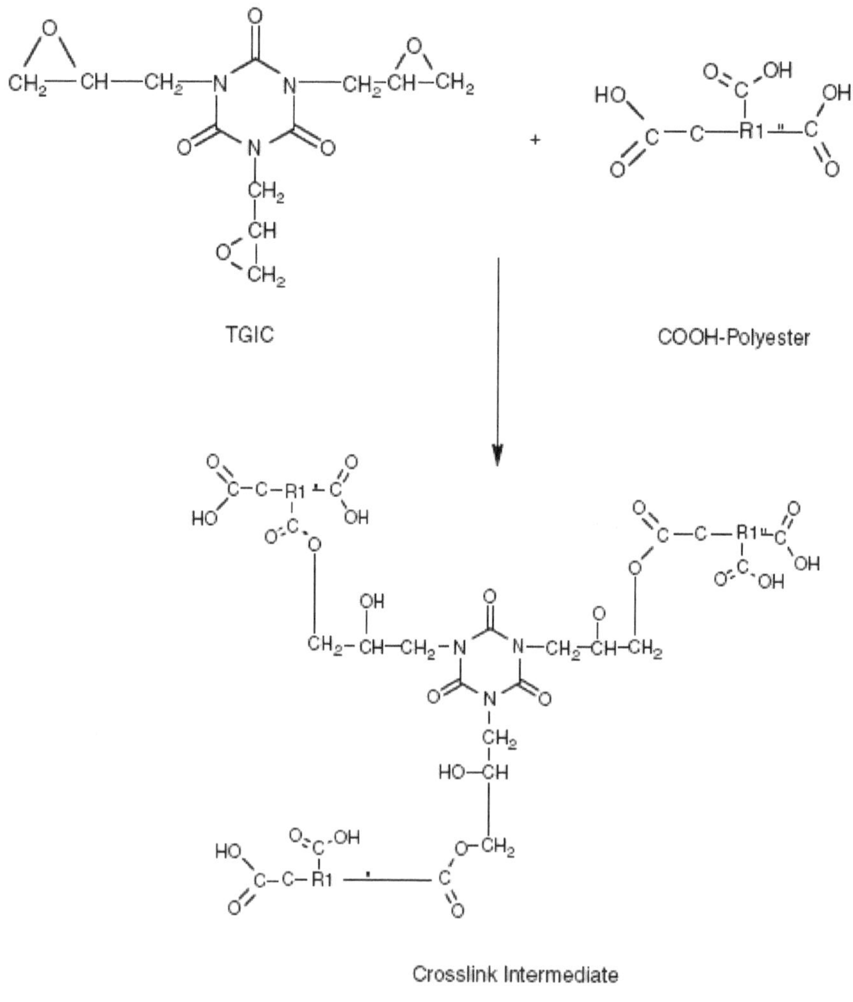

Polyester–Isocyanate (Polyurethane)

Polyester resins used to produce urethanes have hydroxyl functionality. They react at the carbon–nitrogen bonds in an isocyanate. The two isocyanates most used in powder coatings are isophorone diisocyanate (IPDI) and toluene diisocyanate (TDI). However, TDI melts at room temperature, and IPDI is a liquid. This would cause any powder coating in which they were used to have poor package stability. The coating would sinter (cake) into a solid very quickly. Furthermore, they would react with the polyester and gel during the heat of extrusion, again making them unusable.

Consequently, isocyanates are reacted with triol materials to give those higher melting points and better package stability when compounded in powder coatings. Then they are blocked, most commonly with (epsilon) å -caprolactam, to prevent them from reacting with the polyester until they unblock naturally at curing temperatures in the oven.

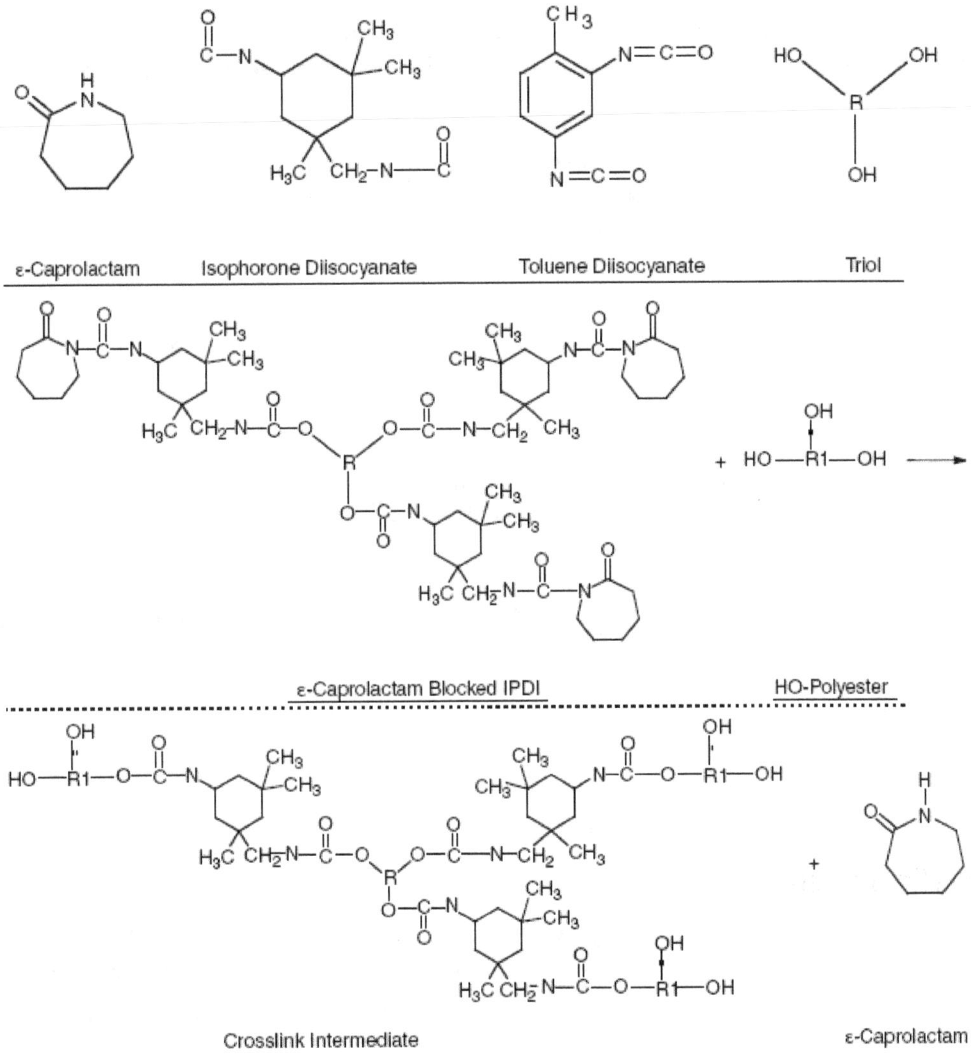

Polyester–Non-TGIC (n â -Hydroxyalkylamide, Tetramethoxymethyl Glycoluril)

As we discussed earlier, TGIC was found to have mutagenic properties. While some controversy still remains as to its hazardous nature, many coatings companies and regulatory bodies have taken a conservative stance and limited its use. There have been regulations enacted that require warnings to be placed on labels of coatings that contain TGIC. Coatings companies have also turned to curing agents that do not have these types of hazards associated with them.

The two types that are used most often are the â -hydroxyalkylamides and tetramethoxymethyl glycoluril. The hydroxy-amides are tetrafunctional, which makes them highly reactive at curing temperatures. Their major drawback is that they release water from the reaction that must escape the curing film. This requires further formulation to insure defect-free films. The glycoluril is also tetrafunctional and releases a VOC, methanol, as part of the reaction process.

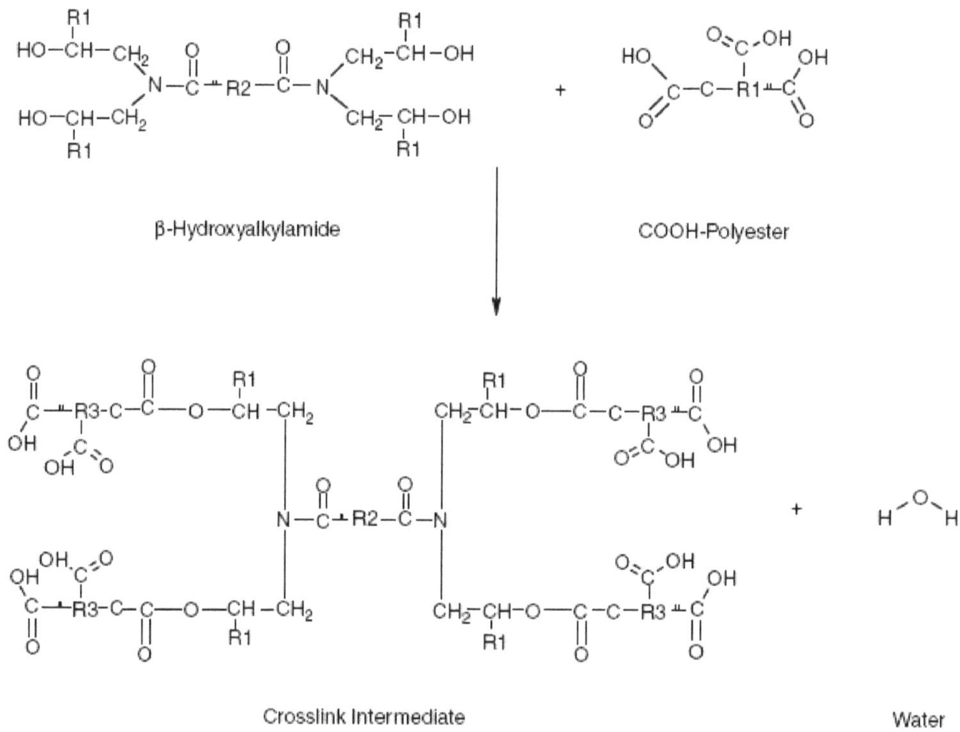

Acrylic Systems

There are two primary acrylic systems: those based on hydroxy-functional acrylic resins and those using epoxy- or glycidyl-functional polymers. Carboxy-functional materials have been produced. However, they have not made much progress into the industry.

Acrylic–Isocyanate (Acrylic–Urethane)

Acrylic–urethanes are formed in exactly the same way as their polyester counterparts. The acrylic resins are linear instead of aromatic. However, they use hydroxy-functionality and blocked-isocyanates to form the urethane bonds.

Acrylic–Diacid (Glycidyl–Acrylic)

Epoxy- (glycidyl-) functional acrylic resins can be compared to the hybrid systems discussed earlier. They are generally reacted with dicarboxylic acids or anhydrides. The most common cross-linker is 1,12-duodecanoic acid (1,12-dodecane dioic acid).

FORMULATION

Resin Systems

Formulation of thermoset powder coatings is much the same as that for liquid coatings used for similar purposes. A coating may be chosen for functional or decorative purposes. All of the coating types used in powder coatings can offer decorative options. Nonetheless, it is function that dictates system choice.

The resin chemistry must be chosen to suit service needs. Various pigments, fillers, and additive materials are then included to enhance decorative or functional requirements. There are two primary functions for any protective coating. They are chemical protection and exterior durability. As with many coating properties, these two tend to be in opposition. The best systems for chemical protection are usually the poorest for exterior durability.

Epoxy powder coating systems deliver the best chemical and corrosion resistance. However, they have the least effective exterior durability. The double bonds in the aromatic rings are easily broken by the ultraviolet (UV) light from the sun. Glossy finishes will go flat with as little as 6 months of exposure, with film degradation following soon thereafter.

Urethane systems are also very good for chemical resistance, and they have fairly good exterior durability, as well. Hybrids offer good chemical resistance, however, the polyester component makes them less effective. They are also poor in relation to exterior exposure due to the epoxy portion of the crosslink network.

Acrylic and TGIC powder coatings provide the best exterior durability. Some systems can survive for up to 20 years of exposure. They offer fair to good chemical resistance. Urethane systems seem to be the best compromise between chemical and exposure properties. As mentioned, they are very good in both areas.

Pigments and Fillers

Most pigments and fillers used in liquid coatings are suitable for use in powder coatings.

There are only a few special requirements for use. They must be sufficiently heat stable so they withstand the heat of extrusion and curing without degradation or color change. Normally, the heat of extrusion is 125°C or less for a minute or two. The heat of cure is usually 160 to 200°C for 10 to 20 min.

Second, they must be insoluble and nonreactive in the resin system. Blooming and color shift are the most common results from pigments that are partially soluble or reactive with the binder. Some epoxy system curing agents are especially susceptible to reaction with pigments.

ADDITIVES

Flow and Leveling

Flow and leveling agents are designed to minimize surface defects such as craters, pinholes, and orange peel. The mechanism of their function alters the surface tension and rheology of the coating. The likelihood of a smooth defect-free film is improved by reducing one (or both) of these properties in a coating.

Most flow and leveling agents are liquids. Many are blended with inert inorganic materials to offer them in a conveniently solid form. The chemistries of these are usually polyacrylates or polysiloxanes. A few new flow agents are available, however, in solid organic form.

Debubbling (Degassing)

The most common debubbling agent is benzoin (2-hydroxy-1,2-diphenyl ethanone). It is used to keep the surface of a curing film open long enough to allow for entrained air and evolved gasses to escape. Trapped air and gas bubbles are cause for premature failure of coating films, because they make the coating brittle. The one drawback to the use of benzoin is its tendency to cause yellowing in lighter colors. A number of new advances have entered the market in an attempt to match the efficiency of benzoin without the challenge of yellowing.

UV Inhibitors

Various UV light inhibitors are available to aid coating resistance to degradation by the sun's rays. The most common are hindered amines, phosphites, sulfates, and phenolics. Most will have some positive effect on any coating's UV resistance. However, each system will require testing to determine the best combination of inhibitors. Some systems, like epoxies and hybrids, will not develop any substantial UV resistance due to their aromatic nature.

Catalysts

Catalysts or accelerators are used to reduce the reaction time or curing temperature of the resin and cross-linker. They allow for faster production time by shortening the gel or "set" time of the thermosetting coating. Energy can be conserved, because full cure may be attained at lower oven temperatures. The most common catalysts are thiazoles (used in polyesters), phosphines and ammonium halides (used in epoxies), and thiocarbamates (used in urethanes).

End Uses

The following table details some of the applications currently using the chemistries discussed in this chapter.

The following table details some of the applications curently using the chemistries discussed in this chapter.	
Type	Typical Applications
Epoxy	Shelving, transformer cases, primers, bathroom fixtures, refrigerator racks, sweepers, sewing machines, power tools, room air conditioners, office furniture, instrument cases, garden tools, kitchen furniture, fire extinguishers, toys, refrigerator liners, dryer drums, microwave ovens, mixers and blenders, fertilizers spreaders, screening, oil filters, automobile springs, hospital equipment, bus seat frames, business machines, glass bottles.
Hybrids	Tool boxes, farm equipment, electrical control boxes, hot water heaters, hot water radiators, primer/surfacers, grain storage bins, transformer covers, 01.1 filters and air cleaners, air conditioner housings, fire extinguishers, toys, screening wire, power tools, shelving, office furniture.
Urethane	Fluorescent light fixtures, steel and aluminum wheels, patio furniture, playground equipment, fence fittings, chrome wheels and trim, garden tractors, range side panels and broiler, ornamental iron, air conditioner cabinets, restaurant furniture supports, transformer cases.
TGIC	Irrigation pipe and fixtures, outdoor furniture, air conditioning units, steel and aluminum, wheels, wire fencing, fence poles and fittings, farm equipment, aluminum extrusions, transformers.
Acrylic	Range side panels, refrigerator cabinets and doors, washing machine parts, dishwasher exterior, aluminum extrusions, microwave ovens, garden tractors, automotive trim coating.

PEELABLE MEDICAL COATINGS

Introduction

Prepackaged sterile medical devices and supplies became necessary in the late 1960s with the growth in prepaid health insurance programs. Insurers required that health care providers itemize the cost of all the supplies used during a procedure. This led to the rapid growth in the development of disposable or single-use devices. These were packaged in paper/plastic pouches, trays, or containers and then sterilized.

At the time of use, the package was torn open to expose the sterile device. When the package was torn, the device was showered with particulates and bacteria that caused a great deal of concern. At that time, the medical device industry was not regulated by the U.S. Food and Drug Administration (FDA).

In 1968, the E. I. DuPont Company introduced a polyethylene, paperlike material (Tyvek ®) that had most of the properties needed for packaging medical devices. Tyvek is a unique material that meets almost all of the critical requirements for medical packaging. It is a good bacterial filter — very porous, waterresistant, and puncture- and tear-resistant. Being made from polyethylene, it is stable during both ethylene oxide gas and radiation sterilization. It does not stand up well during steam sterilization, so only a few adhesives have been developed for this purpose.

With an acceptable packaging material available, peelable coatings were developed to seal Tyvek to plastic films or thermoformed trays to form pouches and trays that could be peeled open at the point of use without compromising the sterility of the device.

There are five basic types of adhesives used to seal Tyvek to plastic surfaces to form sterile packages.

These are cold seal, lacquer-based heat seal, water-based heat seal, hot-melt-based heat seal and lowdensity polyethylene. All of these adhesives are very difficult to formulate and apply so that they meet all of the requirements of the medical device manufacturers.

Some of the requirements these adhesives must meet are as follows:

- Must peel cleanly without generating particles
- Have peel strength over 1 lb per inch of width and less than 3 lb per inch of width regardless of the peel angle
- Be very stable both before and after sterilization (shelf life requirements can be as long as 10 years)
- Meet the U.S. Pharmacopoeia requirements for medical device plastics

Cold-Seal Coatings

Cold-seal coatings, also known as cohesive coatings, are made from natural latex and are sealed by pressure alone. They will not seal to other surfaces, only to themselves. This type of coating has been used for self-sealing envelopes, candy wrappers, and in a variety of industrial packaging. Peelability with cohesive coatings is achieved by coating a release layer between the adhesive and the substrate. This allows the adhesive to peel away from the substrate without leaving a residue or tearing fibers. The main advantage of cold seal is that it can be sealed at very high-line speeds, because only pressure is required to form the seal. The biggest disadvantage is that it is unstable during the coating process and is sensitive to ultraviolet light after coating.

Heat-Seal Coatings

The second type of coating, heat-seal adhesives, is more complex in its requirements than the cold-seal adhesives. The latter generally are applied only in the seal area, and seal only to themselves. The heatseal adhesives are normally coated over the entire surface of the web and have to seal to a wide range of plastics. Heat and pressure are required to form a seal.

Many different types of heat-seal coatings are used for medical packaging, but all must meet the same requirements. The most important requirement is that they be stable after sealing, so that a sterile package is assured for several years. Disposable devices are sterilized after packaging and must remain sterile until used. They must also have porosity of fewer than 200 sec Gurley to prevent the buildup of excessive internal pressure in the package during a normal gas sterilization cycle. The coating must be nontoxic and must meet the same requirements listed in the U.S. Pharmacopoeia for the plastics used in the packaged device.

Other desired attributes are good hot tack, clean color, enough strength to pass shipping tests, and good transfer of the adhesive onto the seal surface when the package is peeled open. The coatings used on heat-sealable Tyvek and paper come in many forms. They can be solvent-type lacquers applied in a grid pattern to maintain porosity, water-based dispersions dried very carefully so that they do not melt, or hot-melt adhesives applied in small dots.

The lacquer-type adhesives are applied by the gravure process at a rate of 3 to 6 pounds per 3000 square feet. The adhesive is applied in lines that cross at an angle, leaving an uncoated area between them resulting in good porosity. Also, the lines break up the peel forces to give better peelability.

The water-based dispersions are based on ethylene-vinyl acetate copolymer dispersions blended with waxes and tackifiers to form a blend of particles. This blend is coated on the web by air knife coating or by gravure. The coating must be dried at a low temperature to keep the coating from melting and forming a film. The porosity is achieved by having large particles in the blend and by mud cracking of the coating as it dries. The dried coating weight is 5 to 8 pounds per 3000 square feet.

Hot-melt coatings are a blend of an ethylene-vinyl acetate copolymer, terpene resins, and wax. These are melted together and applied by a gravure roller engraved with 50 to 70 cells per inch. The small dots of adhesive that result from this have a small uncoated area between them that makes the coating porous.

The coating is applied at a rate of 9 to 12 pounds per 3000 square feet. The optimum amount of coating will vary for each type of adhesive. The controlling factor is the melt viscosity of the adhesive at the sealing temperature. If the viscosity is low, a portion of the adhesive will be pushed into the substrate, reducing the amount of adhesive available to absorb stresses and to transfer during peeling. Good adhesion to a smooth surface with coated paper or Tyvek can be achieved with 5 to 12 pounds of adhesive per 3000 square feet. An accepted rule of thumb in the coating industry is that 15 pounds per 3000 square feet is equal to 0.001 in. of thickness.

While there are no regulations in this country requiring nontearing packages, the medical device industry places a high priority on this feature. Another desirable feature of the coating, which goes along with peelability, is the ability of the coating to transfer to the plastic surface to which it is sealed. The transfer pattern must be unbroken when the package is opened. This is interpreted to mean that the seal integrity has been maintained. A clear track through the seal area will cause the package to be rejected, because there is a path for bacteria to enter the package.

Formulating a coating that is nontoxic and strong, with good hot tack and good adhesion to a variety of plastic surfaces, and that is able to peel with controlled peel strength, is a difficult task. Several methods have been used to achieve peelability. The use of primer coats between the adhesive and the substrate is the one most often chosen; this creates a parting layer between the adhesive and the substrate. Release coating on the substrate surfaces are also used when paper is used instead of Tyvek.

Most people in the industry feel that the best method is to have an adhesive with a controlled cohesive strength. When the package is peeled open, the adhesive will split with a controlled force that is below the delamination strength of the substrate. By using this method, the surface fibers of the substrate are not raised or broken.

Cohesive failure of the adhesive can be achieved by formulating an adhesive that has two phases: one a strong adhesive to hold the package together and the other a weak friable material that breaks up the structure of the first phase. By varying the percentage of the two phases, the cohesive strength of the coating can be controlled within narrow limits.

The nature of peelable medical packaging materials is currently undergoing a change. There is a drive to reduce the cost of medical devices, causing a shift to paper and away from Tyvek. Also, environmental concerns are pressuring medical device companies to use some method other than ethylene oxide gas sterilization. The requirement for a peelable package, however, remains strong.

The medical device industry is now under the control of the FDA, which requires complete validation of processes and materials. Formulating and validating a new adhesive will generally take from 1 to 3 years. For this reason, packagers of medical devices are very reluctant to change suppliers. They do not have the engineering staff they had in the earlier years when the industry first started. To change an adhesive and develop the documentation required by the FDA is very expensive. Most companies have alternate suppliers they can use if they have problems.

CONDUCTIVE COATINGS

Introduction

In 1986, sales in the coatings industry exceeded $10 billion, and production approached a billion gallons. The breakdown of sales was $4.1 billion for architectural coatings, $3.5 billion for industrial coatings, and $2.4 billion for specialty coatings. Conductive coatings — a minuscule part of these trade sales — have been used both as industrial coatings and as specialty coatings. Regulations of the Federal Communications Commission (FCC), in Docket No. 20780, which regulates electromagnetic emissions from computing devices, have provided a strong impetus for the commercial development of conductive polymeric materials (including coatings and paints). Since October 1, 1983, it has been necessary for any computing device that generated signals or pulses in excess of 10 kHz to comply with the emission standards set forth in the docket. Although conductive polymeric coatings have made inroads in areas where metallic coatings previously were used, progress has been slow.

A product related to conductive coatings is metallized plastic. The most important commercial processes for metallizing plastics are electroless plating, metal spraying, sputtering, and vacuum metallizing.

The first commercial plating of plastics was recorded in 1905. Metallizing of plastics occurred during World War II, and large-scale production started in the early 1960s. All these processes are now multimillion- dollar industries. Large quantities of plastics are metallized each year, with automotive items making up more than 60% of the market on a plated area basis.

There are various reasons for metallizing plastics. In the automotive industry, metallized plastic combines the consumer appeal of metal with light weight. Electroless copper metallization is an indispensable part of the modern electronics industry. Printed circuit boards use electroless copper to coat nonconductive plastic surfaces to define the circuit patterns. Zinc arc and flame-spray techniques provide electromagnetic interference shielding on many plastics. The plastics that account for most of the substrates metallized are acrylonitrile-butadiene-styrene (ABS), polypropylene, polyphenylene oxide, epoxies, phenolics, polyimides, and polyesters. The commercial process for metallization of plastics merits separate discussion and is not further considered in this chapter.

Polymers (coatings) with conductivities greater than 1(Ù cm) –1 are defined as conductive polymers (also metallically conducting plastics, synmetals). Unfortunately, the literature is not clear-cut, and often, materials that are semiconductors with conductivities less than 1 (μ cm) –1 are also called "conducting."

Types of Conductive Coatings

We discuss metallic, filled polymeric, polymeric, and organometallic conductive coatings.

Metallic

Substrates are coated with conductive coatings mainly for functional and rarely for decorative purposes.

The most important commercial metallic coatings are nickel, copper, chromium, gold, palladium, platinum, silver, zinc, cadmium, iron, cobalt, tin, and lead.

Surface metallization of molded liquid-crystal polymer parts has been accomplished. To increase the adhesion of a metal coating to an organic substrate, the substrate is heated to 0.6 to 0.8 T cure of the substrate and held at that temperature while the metallic coating is deposited. This procedure provides for optimum intermixing between the metal atoms and the substrate, thus producing maximum adhesion.

A simple technique for plating ABS plastics consists of mixing a mixed catalyst solution for simultaneous sensitization and activation of the surface.

Adhesion between polyethylene and/or polypropylene and aluminum is improved if the plastic surface is pretreated with an oxygen ammonia, or sulfur dioxide low pressure plasma.

An electroless metal plating of plastics involves a plastic reducible by one metal salt followed by a redox reaction between the reduced plastic and the main group metal salt to produce the plated plastic.

A process for metallizing shaped articles of filled plastics with cuprous oxide has been described. The surface of the filled articles is subjected to a reducing agent such as a borohydride to convert the cuprous oxide to a conductive free-metal surface.

Filled Polymeric

Plastic coatings and paints can be made conductive by adding metallic fillers. Typical metal powders that have been used in coatings, paints, and lacquers are nickel, copper, gold, aluminum, iron, cobalt, palladium, zinc, and platinum, as well as various other alloys. A step beyond the use of powdered, more or less spherical, fillers is the incorporation of conductive fillers with high aspect ratio: length-to-diameter ratio of more than 100. This family of materials includes such fillers as metallized glass fibers, graphite fibers, metallized carbon fibers, quick-quenched aluminum flakes or fibrils, and fine diameter (6 to 8 μ m) steel fibers. The most common metallic conductor, copper, is not used, because it oxidizes within the plastic and impairs its physical properties.

Depending on particle form (sphere, flake, and fiber), size, and orientation, there is a certain critical concentration at which conductivity increases drastically (many orders of magnitude). To achieve significant conductivity with more or less spherical fillers (carbon black, metal powders), volume percent loadings above 40 are required. A striking difference in conductivity is observed when a high aspect ratio particle of the same material is used. Thus, with metal or metallized glass fibers of aspect ratio of 750 as fillers, metallic conductivities can be achieved at as low as 3 vol%.

At critical concentration, the filler can form a continuous phase through the matrix in the form of microscopic conductive channels (percolation). The use of high aspect ratio conductive fillers, when they are properly oriented, will always produce a superior conductivity at a lower volume percent concentra- tion than comparable spherical fillers. However, for these materials to have high efficiencies, they also must survive the preparation and application processes. For example, the initial high aspect ratio of a very brittle material may quickly reduce to powder in a high shear mixing environment, if proper techniques are not developed for incorporating such brittle materials.

Organic solvent-based coatings of acrylic or urethane resins containing nickel, silver, copper, or graphite powders are used in the electromagnetic and radio frequency interference applications. For cost reasons, the nickel-filled acrylic coatings are most often used. Lacquers based on silver give the highest conductivities and, therefore, the best screening attenuation, but they are the most expensive. Copper lacquers are cheaper, and the attenuation is similar to that of silver; the oxidation problem of copper lacquers has been solved. Lacquers containing graphite are the cheapest, but their performance is also the least effective.

An initiation layer for the electroless deposition of copper, nickel, Ni-P, Ni-B alloy coatings has been described. The layer was formed by applying polyester or its polymerization material solution containing such metal powders as nickel, copper, silver, aluminum, cobalt, palladium, zinc, or platinum. Large differences in the specific conductivity of filled conductive plastics coatings were demonstrated between materials filled with 15 vol% aluminum alloy fiber.

Techniques have been developed for generating and incorporating in situ metallic filler in organic coatings or paints. For example, the use of AgNO and erythorbic acid, formaldehyde, paraformaldehyde, or terephthalic aldehyde will result in the generation of metallic silver in a chosen organic coating or paint system.

Polymeric

Although the unfilled, molecularly conductive polymeric materials have attracted much attention recently, the successful developments in surface-metallized plastics make it unlikely that inherently conductive polymers will be a significant factor in the marketplace in the near future. Also, the commercially entrenched, filled conductive polymeric materials seem to become more efficient as time goes by. One projection of worldwide markets for filled and inherently conductive polymers is as follows:

	1982	1987	1992
Filled conductive, lb	1×10^6	1×10^7	1.5×10^7
Inherently conductive, ld	0	1×10^5	1.5×10^6

This projection, made in the mid-1900s, is clearly very optimistic about the prospects of inherently conductive polymers.

The major problems of molecularly conductive polymeric materials (coatings) are still their instability under environmental conditions and difficult or impossible processing from solution or melt. Consequently, they have to be synthesized and treated in their final form of application. In the undoped state, the basic polymers are either semiconductors or insulators. It is only after they are doped (oxidized or reduced) chemically or electrochemically that they become metallically conductive.

Doping can be done by means of electron-accepting (oxidizing) agents, such as AsF5, iodine, bromine, SbF5, NOPF6, Al (ClO4)3,or HClO4 ; or electron-donating (reducing) agents, such as lithium, sodium, potassium, or sodium naphthalide. Unfortunately, most of the doped materials generated are extremely reactive because of the carbonium ions and carbanions formed upon doping and thus are not stable under ambient conditions.

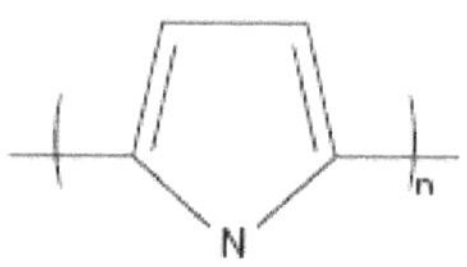

Another heterocyclic, conducting, environmentally stable polymer is polythiophene. Polythiophene and its derivatives can be synthesized and doped chemically or electrochemically to conductivities as high as 100 (Ù•· cm) –1.

Recently, some progress has been made in this area. The first environmentally stable, conducting polymer was polypyrrole. Coatings or films of it were prepared electrochemically from aqueous solutions of sodium n -alkylsulfonates, and disodium 1,10-decane disulfonate or toluenesulfonic acid. The products possessed conductivities as high as 230 (cm) –1. The films possessed good mechanical properties and environmental stability in excess of 5 years (thus far tested).

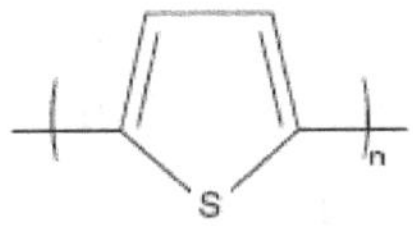

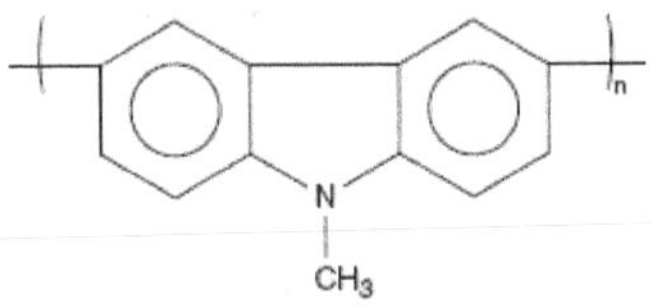

Environmentally stable polycarbazole was synthesized and doped to a conductivity of 5 (cm) –1.

A polymer that is attracting more attention recently is polyaniline. It can be prepared both chemically and electrochemically, and when doped with various acids in aqueous solutions, it achieves conductivities as high as 10 (cm) –1.

A recent and exciting development is the preparation of aqueous polypyrrole and polyaniline latex. Spherical, submicrometer polypyrrole latexes have been prepared using poly(4-vinyl pyridine)-based steric stabilizer. The use of such a stabilizer allowed the polypyrrole particles to be controllably and reversibly aggregated or stabilized, depending on the pH of the dispersion medium. The solid state conductivity of films can be as high as 2 (cm) -1.

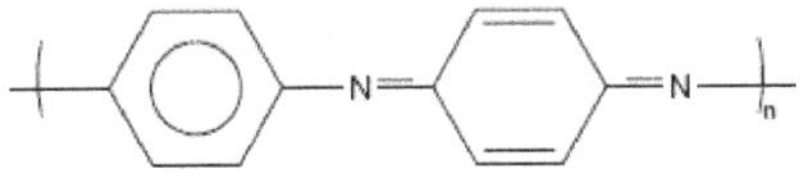

Except for the electrochemically prepared polypyrrole in a continuous sheet form, obtainable in developmental quantities from Badische Anilin- und Soda-Fabrik (BASF), Ludwigshafen, West Germany, none of the other materials have passed the research stage.

Organometallic

Although most of the current interest has been in unfilled molecularly conducting polymer systems, conductivity in metal containing polymers has also been a topic of interest. Almost all the molecularly conducting polymers are doped systems and, thus, may contain metals from the doping agent. Such systems, however, are air and moisture sensitive and are difficult to handle. In general, organometallic polymers do not have these instability problems, but their processibility is still a problem.

A new class of low-dimensional materials — for example, polymeric metallophthalo-cyanines with such metals as aluminum, chromium, tin, and gallium at the center of the ring — has been synthesized and made conducting [~1 (cm) –1] by halogen doping (partial oxidation).

Various transition metal ions have been coordinated with conjugated ligands to synthesize polymers with metal ions within the main chain.

Ligands such as tetrathiosquarate, tetrathionaphthalene, tetrathiafulvalene, and tetrathiooxalate have been used. The tetrathiooxalate complexed with nickel ions gave linear polynickel tetrathiooxalate oligomers with conductivities as high as 20 (cm) –1.

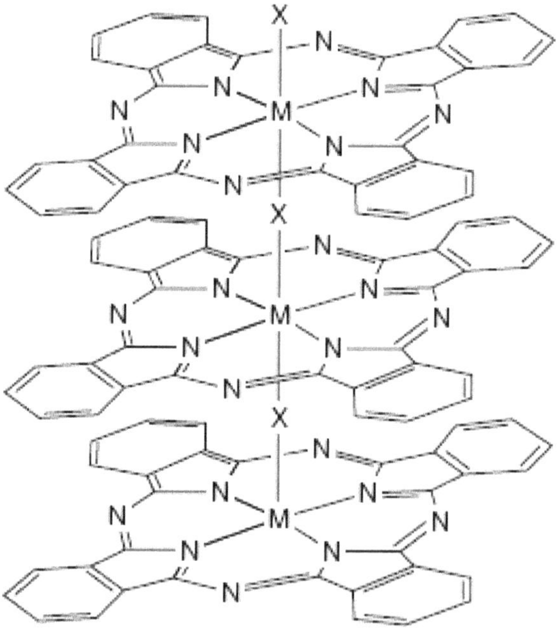

A completely different approach to depositing conducting organometallic coatings involves the use of a low-pressure plasma (LPP) environment. The LPP environment may be used to deposit a polymeric organometallic coating (or powder) from organometallic monomers, or it may be used to convert a deposited organometallic coating into a metallic one.

The metals introduced in the organometallic coatings by LPP/organometallic monomers were iron, tin, mercury, tantalum, lead bismuth, and metal coatings by the LPP posttreatment were gold, platinum, palladium, silver, and lead. The generation of metal surface coatings from certain organometallic coatings can be also achieved by thermal means (controlled pyrolysis).

The advantage of the LPP process is that it permits a metallic coating to be formed on a heat-sensitive substrate without the use of elevated temperatures. The process also permits formation of adhering gold and platinum coatings otherwise difficult to deposit on plastic substrates. None of the conducting organometallic coatings or their deposition processes has gone beyond the research stage. However, the conducting organometallic coatings effort is very new compared with the other types mentioned before.

COMMERCIALLY AVAILABLE CONDUCTIVE COATINGS

Whereas the metallized plastics effort is a multimillion-dollar industry, the commercial application of conductive polymer coatings as a paint, lacquer, ink, adhesive, or a solution of some kind forms a very small industry indeed. Below are summarized several typical products available commercially.

A proprietary aluminum containing paint, AG 9680, manufactured by A.I. Technology, Inc., Princeton, New Jersey, is claimed to approach the shielding effectiveness of 70 to 75 dB, effectiveness similar to that of pure silver.

A silver-filled silica matrix elastomer, Aremco-Shield 615, has been developed by Aremco Products, Inc., Ossining, New York. This material has been formulated into a conductive paint that can be applied by either brush or spray; it cures at room temperature and bonds to metals, glass, and plastics.

A silver lacquer (Eccocoat CC-2) and an elastomeric, silver-filled, conductive coating (Eccocoat CC- 40) have been developed by Emerson and Cuming, Canton, Massachusetts. The lacquers and the elastomer coatings may be applied by dipping, spraying, silk screening, roll coating, or brushing. In most cases, a simple spray coat is adequate to produce a highly conductive surface [up to 20 (Ù·cm) –1] using the air-dry method. Oven curing will give improved conductivity.

A rather extensive series of conductive coatings under the trade name Evershield has been developed by the E/M Corp. of West Lafayette, Indiana. The series of products consists of a graphite-filled acrylic resin system, EC-G-102, intended mainly for applications for electrostatic charge dissipation; a high performance nonoxidizing copper-filled acrylic resin system, EC-C-301, is suitable for spray gun application.

The coating has an attenuation performance of 50 to 70 dB at 10 to 1000 MHz. A popular nickelfilled acrylic resin system, EC-N-501, easily paintable and with superior adhesion characteristics for a wide variety of plastic substrates, is also available. It has an attenuation performance of 50 to 60 dB at 30 to 1000 MHz.

APPLICATIONS

Shielding from Electromagnetic Interference

The advent of the FCC Docket 20780, which regulates electromagnetic emissions from computing and communication devices used in industrial and residential locations, has really provided a stimulus for the industry to come up with cost-effective methods for limiting the level of electromagnetic interference (EMI). To meet the set standards, the manufacturers have adopted a variety of methods for controlling EMI. These methods have ranged from redesigned basic circuitry to incorporation of conductive shielding materials in the devices. The incorporation of conductive shielding may take different routes:

- Use of metal enclosures
- Metallic foil tapes
- Metal coatings on plastic enclosures
- Conductive paints on plastic enclosures
- Conductive plastic enclosures
- Flexible laminates with metal foil

The shielding effectiveness of a homogeneous medium, such as a conductive coating, is related to the propagation of the electromagnetic field through the coating. The shielding effectiveness is directly related to the electronic and magnetic properties of the coating; therefore, for best shielding effectiveness, materials with both high relative magnetic permeability and high electrical conductivity are necessary. Thus, it has been found that the various metals and alloys form the following "series" in decreasing order of effectiveness:

$$\text{Ag} > \text{Cu} > \text{Au} > \text{Al} > \text{Zn} > \text{brass} > \text{Ni} > \text{Sn} > \text{steel} > \text{stainless steel}$$

Currently, the most cost-effective and the most problem-free materials for shielding are claimed to be nickel-filled acrylic or polyurethane conductive paints.

The Stealth

One of the more glamorous applications of conducting coatings has been in the "stealth" technology. Until the Pentagon revealed the top-secret Stealth (Figure .1) fighter on November 10, 1988, little could be said about its materials technology. The so-called stealth materials can provide a minimal radar profile for military aircraft and naval vessels. This profile is achieved primarily by a combination of geometric design and materials properties. The active components consist of several classes of materials: carbon fiber composites, radar-absorbing coating, ferrite layers, and interference layers in the form of certain pigment-filled polymer coatings.

From other EMI work, it had been known for years that incorporating short, electrically conducting fibers (e.g., stainless steel) into thermoplastics in as low as 0.5 vol% fraction gave sufficiently high levels of dc conductivity for electromagnetic shielding purposes.

Figure.1 — Stealth fighter.

One of the radarabsorbing coatings has been iron fibril-filled elastomeric epoxy. Presumably, such coatings are most effective against short wavelength radars, and the longer the wavelength, the thicker the required coating.

Miscellaneous

A novel application of conducting coatings is to use them as ambient-responsive elements for remotely readable indicator devices.

Spacecraft charging problems have required efforts in the separation, shielding, and filtering to decouple discharges from susceptible circuits, and this decoupling involves the use of conductive surface coatings.

Conductive coatings are also used on waveguides, radio frequency reflectors, and radar dishes, and as radiofrequency shielding on capacitor plates.

Electrochemically generated, conducting polyheterocycle coatings are good candidates for electrochromic displays. Aluminum is being used on video disks and in large-area solar cells. Conductive paints are used to enhance the electromagnetic field attenuation of equipment rooms or entire buildings.

Conducting coatings are used to coat rusty areas and bolts to suppress background noise in high frequency transmitters and receivers on ships and aircraft. Conductive coatings are candidates for passivating layers against photocorrosion of photoelectrodes and modifiers of electrodes for electrochemical reactions.

New Developments

The field of laser direct-write metallization of organometallic films has been expanding rapidly as the potential for accomplishing one-step pattern definition and metallization is realized. Recently, ion implantation has been conducted on a variety of polymers with a variety of ion species. The result has been significant increase in electrical conductivity. Thus, for example, Kapton H film implanted with 1 MeV 15N2 + at several doses reached a conductivity of 1 (Ù·cm)–1. It was suggested that the ion implantation led to graphite particle formation, probably along the penetrating ion tracks.

Conducting silver lines have also been generated by focusing a laser beam on a polyimide film containing dispersed silver nitrate.

A novel chemical method was used to generate an Ag–Hg alloy on the surface of a polyacrylamide film. The technique involved holding the polyacrylamide aqueous solution with AgNO3 in a mercurysaturated atmosphere.

A recent symposium on laser and particle-beam chemical processing for microelectronics described laser direct writing of aluminum, copper, palladium, and carbon.

SILICONE RELEASE COATINGS

Introduction

Silicone release coatings are vitally important to the tag and label industry, which could not exist in its present form without reliable release agents. Silicones possess unique physical and chemical properties that make this class of substances ideal for the purpose of releasing pressure-sensitive adhesives. Silicone release agents worth \$130 to \$150 million were sold worldwide in 1988, contributing to products with a total value that exceeds \$3 billion.

The term "silicones" as commonly used refers to linear (two-dimensional) polydimethylsiloxanes, which may be structurally depicted as follows:

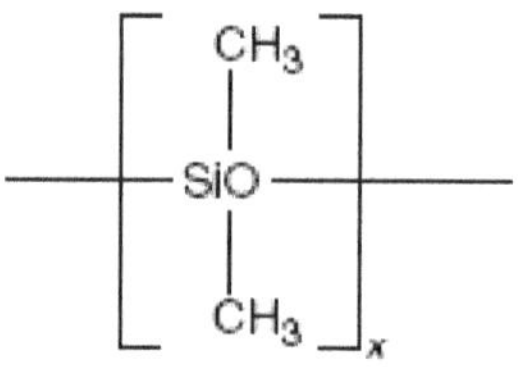

Where x is an integer greater than 1. Silicon is tetrafunctional, so an infinite number of silicone polymers may be devised with different organic groups replacing methyl, or with three-dimensional resin structures wherein silicon atoms are incorporated in the polymer structure via three or four —Si–O— linkages.

Since, however, the low surface tension, nonpolarity, chemical inertness, and low surface energy responsible for the outstanding release characteristics of silicone coatings all derive from the linear dimethylsilicone structure, this discussion focuses on linear polymers.

Silicone coatings that release pressure-sensitive adhesives have been in use for some 35 years. The chemistry and applications of silicone release coatings have undergone remarkable change during this time, with the pace of development accelerating in recent years. In the face of increasingly sophisticated and demanding requirements, silicones remain the only proven means of providing pressure-sensitive adhesive release for the tag and label industry.

The laminate structure normally used by the label industry is illustrated in Figure .1. The liner most often used is paper, usually a machine-calendered (i.e., supercalendered kraft), claycoated, or glassine paper designed to minimize penetration during coating and curing of silicone.

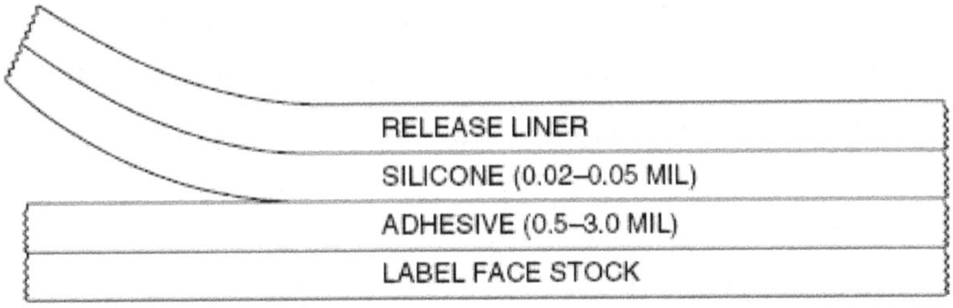

Figure .1 — Pressure-sensitive label stock laminate

Good holdout is essential for complete coverage at economically viable coat weights (<1.2 g/m2). Converters are also increasing their use of plastic films and film laminates to take advantage of plastics' dimensional stability and strength. Such substrates are often thermally sensitive, requiring suppliers to develop novel low temperature cure or radiation cure silicone systems to meet the new demands.

In use, the laminate construction herein depicted is split into two separate lamina: the release liner silicone lamina is peeled away from the label-adhesive lamina. Silicone coatings therefore function as a vehicle for the adhesive, permitting it to be applied far from where it is ultimately used. The low surface energy of a cross-linked polydimethylsiloxane surface (usually ~22 dynes/cm) compared with the surface energy of organic pressure-sensitive adhesives (typically >30 dynes/cm) prevents tight physical bonding of the adhesive to the silicone. That is, nonpolar silicone coatings are not compatible with highly polar adhesives, besides possessing minimal attractive forces toward other molecules. Chemically, the carbon– silicon and oxygen–silicon bonds of a dimethylsilicone molecule are unreactive, so that properly cured silicones are not prone to chemical interaction with pressure-sensitive adhesives with which they are in contact.

THERMAL-CURED SILICONE RELEASE AGENTS

Silicone release coatings widely used today can be considered to be thermosets: they are applied in liquid form (100% silicone fluids or dispersed in solvent or water) to a release liner, then irreversibly converted to cross-linked films via thermally accelerated chemical reactions between reactive silicone polymers and silicone cross-linkers. "Thermal cure" is a somewhat misleading term, because the chemical reactions that effect cross-linking into a cured state take place at ambient conditions (25°C), as well as at oven temperature (~150°C). Upon conversion to absolute (Kelvin) temperature scale (room temperature = 300 K; oven temperature = 425 K), it is clear that an oven need furnish only half again as much energy to a reactive chemical system as is already present in the coating pan environment to bring about cross-linking.

Mixing reactive silicone polymers, catalysts, and cross-linkers together therefore initiates the curing process; maintaining a complete coating mix at a coatable viscosity while still permitting rapid oven cure mandates the use of dispersing media, inhibitors, or other modifications, as described in this section.

Heat-accelerated silicone cross-linking (curing) reactions are of two types: condensation cure and addition cure. Condensation cure systems for paper release applications were the earliest development of this technology. Condensation cure utilizes silanol-chain-stopped dimethylsilicone polymers, polymethyl-hydrogensiloxane cross-liners, and condensation catalysts. The polymers are illustrated below:

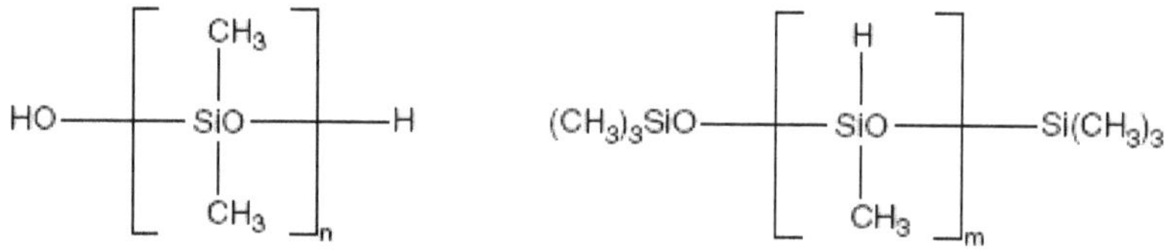

SILANOL POLYMER METHYLHYDROGEN SILICONE

Where n can vary from about 50 to more than 4000, while m is much less than n; m normally is 10 to 50. SiH is a very reactive chemical species that readily condenses with silanol (SiOH) groups, forming extremely stable siloxane bonds and liberating hydrogen in the process:

$$\equiv SiOH + \equiv SiH \rightarrow \equiv SiOSi\equiv + H_2$$

Many different catalysts accelerate or initiate this condensation reaction; metal soaps and driers such as dibutyltin acetate are the most efficient and economical, and are therefore in general use.

Condensation cure systems are applied as solutions in organic solvents (toluene or heptane, or mixtures thereof), or as oil-in-water emulsions, because in the absence of a dispersing medium, a catalyzed mixture of a silanol-stopped silicone plus polymethyl-hydrogen-siloxane cross-linker sets up to an insoluble cross-linked gel in a few minutes at room temperature. There is no known means of retarding the condensation reaction sufficiently at room temperature to permit solvent-free coating without rendering the composition uncurable at oven temperatures. Solvent (or water in the case of emulsions) therefore acts as a bath life extender through the dilution effect, while also permitting easy, convenient coating of the silicone material.

Although use of solvents or water mandates high oven temperature and solvent recovery, and entails fire or explosion risk, such materials are readily coated via simple techniques such as direct gravure, reverse roll, metering rod, and doctor blade. Coating out of a solvent vehicle also gives the silicone supplier wide latitude in silanol molecular weight; such dispersion products as General Electric SS-4191 consist of approximately 30 wt% solutions of high molecular weight silanol gums (MW in aromatic solvents).

Even at 70% solvent, these products as furnished have viscosities exceeding 10,000 cps, requiring further dilution to about 5 wt% silicone solids with more solvent to render them coatable. The cross-linker is normally packaged in the silanol solution; catalyst is added to the fully diluted bath at time of use.

Controlling silanol molecular weight is a proven means of controlling the release characteristics of the cured condensation-cross-linked coating. Long chains of polydimethyl-siloxane between cross-linking sites provide a rubbery, elastomeric coating; shorted intercross-link intervals lead to higher cross-link density and a harder, more resin-like coating. The rubbery coatings provide tight (high) release, which displays a marked dependence on delamination speed in comparison to the low (easy) release independent of stripping speed obtained from highly cross-linked silicone films. Accordingly, silicone suppliers offer several different molecular weight silanol-based dispersion products, permitting the end user to obtain a desired range of release. The relationship between silanol chain length and nominal release level is graphically shown in Figure .2.

Addition cure silicones resemble condensation cure silicones in some respects: both types of system rely on thermally accelerated cross-linking reactions between polymethyl-hydrogen siloxane cross-linker molecules and a separate reactive dimethylsiloxane polymer. Addition cure processes utilize catalyzed reaction of unsaturated organic groups attached to otherwise unreactive dimethylsilicones with SiH groups present on the cross-linker. Polymers in use are vinyl-functional silicones, the general structure of which may be represented as follows:

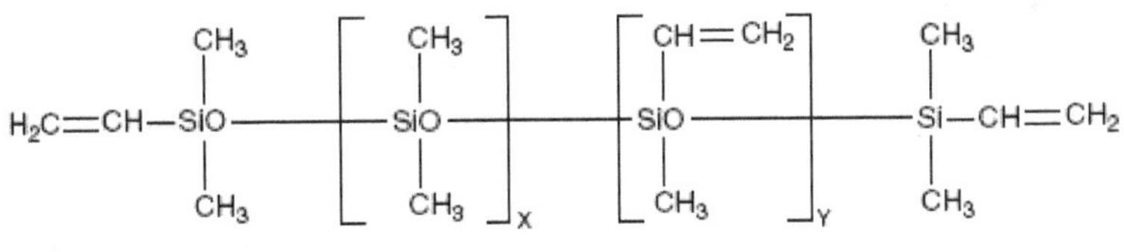

X + Y = 50 to >4000; Y ≥ 0

The curing reaction is an addition to the SiH group across the olefin double bond, also known as a hydrosilation process:

With each polymer containing multiple reactive sites, it is clear that a very complex, highly crosslinked thermoset structure is obtained upon cure. Hydrosilation can be triggered by heat, peroxides, azonitriles, and high energy ultraviolet or gamma radiation. These particular reactions are free radical in nature and are therefore not well suited for normal coating equipment, as radicals so generated are quenched by atmospheric oxygen. Hydrosilation catalyzed by silicone-soluble Group VIII precious metal compounds (certain complexes of platinum, rhodium, and the like) is the best means of curing solventless silicone coatings.

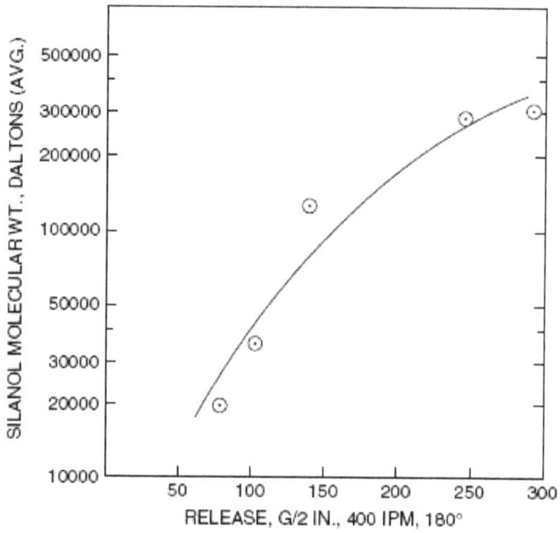

Figure .2 — Release of an acrylic pressure-sensitive adhesive (Monsanto Gelva 263) as a function of silanol molecular weight.

These catalysts effectively promote SiH + Si-vinyl cross-linking at concentrations in the part-per-million range, making their use economical despite their cost. The addition cure reaction is an example of homogeneous catalysis. The precise mechanism is not firmly established.

But it is thought to be a concerted series of events in which the catalyst metal atom acts as a site to which reactive Si—H and olefin groups simultaneously bond in the proper proximity and stereochemical configuration for addition to occur.

This description of addition cure silicone chemistry is still incomplete. Whatever the system chosen, useful pot life of catalyzed mixtures of vinyl silicone and methylhydrogen silicone polymers seldom exceeds a few minutes at room temperature, like the condensation cure silicone system described previously.

However, use of an important ingredient — the inhibitor — permits solvent-free coating of addition cure silicone release agents. Inhibitors function by tying up catalyst atoms at low temperature while still allowing rapid cure to occur at elevated temperature. An ideal 100% solids silicone coating composition combines instant cure at oven temperature with infinite pot life in the coating pan. Actual commercial systems depart from the ideal behavior, as shown in Figure .3.

A clue to inhibitor function is provided by the hydrosilation mechanism. During the cross-linking process, SiH and Si-vinyl groups must at some time be coordinated to the platinum metal center. An inhibitor molecule that prevents access to catalyst atoms at low temperature must not block coordination sites at oven temperature. Inhibitors are therefore often selected on the basis of colatility, or because complexes formed with platinum are unstable and break down at high temperature. Many of these proprietary additives are described in the patent literature, ranging from classical complexing or chelating agents to electron-deactivated olefins and acetylenes.

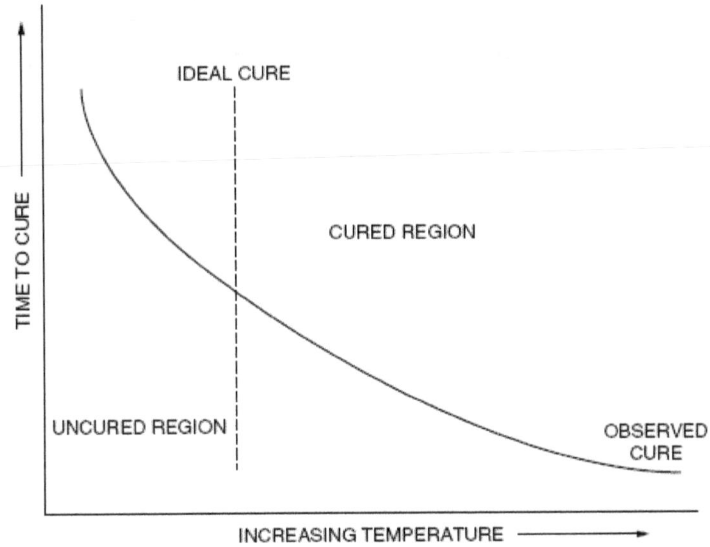

Figure .3 — Cure time as a function of temperature

Addition cure silicone release agents are available as solvent-free, low viscosity vinyl silicone fluids, as solvent-dispersed vinyl silicone gums (analogous to most silanol-based condensation cure products), and as emulsions. In each case, sufficient vinyl functionality is built into the linear silicone molecules to promote formation of highly cross-linked resinous cured coatings. These products therefore provide uniformly low (premium) release from most pressure-sensitive adhesives. Controlled release is not obtained by varying molecular weight of these vinyl silicone polymers (unlike the silanol case), and thus a different approach has been taken by silicone suppliers to combine the advantages of addition cure chemistry (particularly solventless packages) with a controllable range of release.

Studies of cured dimethylsilicone release coatings by electron spectroscopy for chemical analysis have confirmed that the surface is much more organic than would be predicted from the stoichiometry of the (CH 3) 2 SiO polymer unit.

An adhesive in a laminate construction is therefore largely in contact with unreactive, bulky, freely rotating methyl groups; more polar —Si—O—Si— polymer backbones concentrate beneath the coating surface. A depiction of this postulated structure is offered in Figure .4.

Since the surface orientation of Si—CH 3 groups governs the release characteristics of highly crosslinked nonelastomeric silicone coatings, it follows that breaking up this nonpolar, featureless "methyl landscape" by inclusion of materials that alter the polarity of the silicone should alter the release of the coating. This is, in fact, accomplished by adding vinyl-functional silicone resins to the basic linear vinyl silicone polymers. "Resin" is here defined as nonlinear silicone structures bearing high concentrations of functionally. Solventless high release additives are therefore mixtures of vinyl silicone resins with vinyl silicone fluids. Because these resins are normally friable solids when isolated, their blends with vinyl

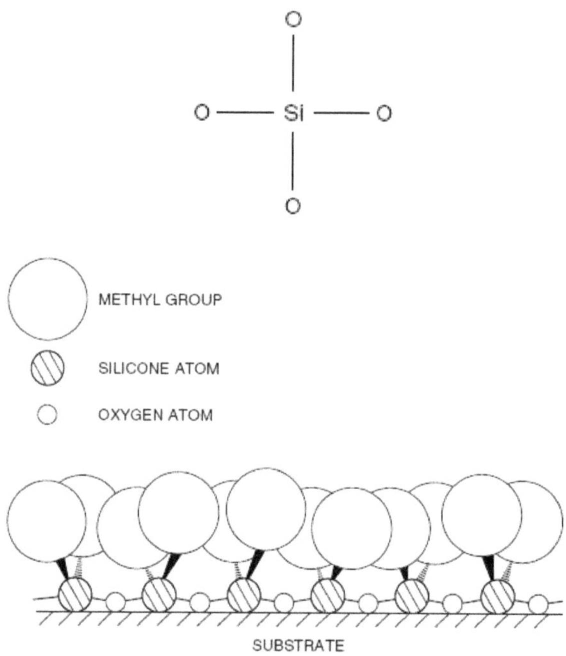

Figure .4 — Representation of silicone release coating

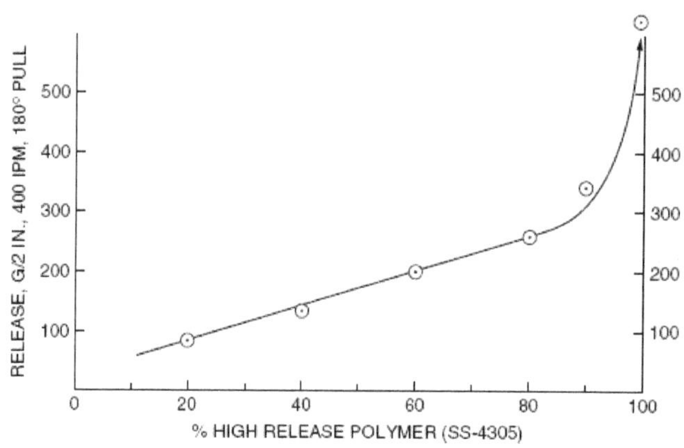

Figure .5

Solventless silicone (GE SS4300/SS4305 system) controlled release versus a pressure-sensitive adhesive based on styrene-butadiene rubber silicone fluids are much more viscous than the unblended fluids, which limits the amount of resin that can be present in solventless high release additives. Nonetheless, a wide range of controlled release can be obtained by use of suitable blends of high and low release solventless vinyl silicone products, as is illustrated in Figure .5.

Two solventless silicone product designs are available. The simplest system consists of three products:

- Low release polymer + catalyst + inhibitor
- High release polymer + catalyst + inhibitor
- Cross-linker

It has recently been discovered that more efficient use of expensive hydrosilation catalysts results from maintaining catalyst and inhibitor apart until time of use.

Multicomponent systems taking advantage of this consist of the following:

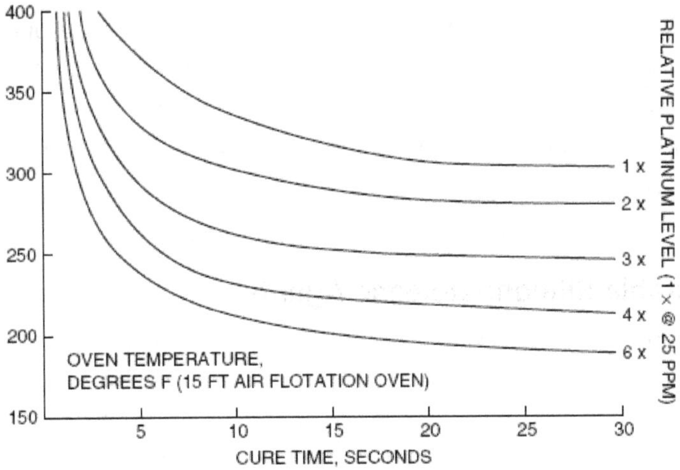

Figure .6 — Cure as a function of catalyst concentration

- Low release polymer
- Low release polymer + high catalyst concentration
- Inhibitor
- High relzase polymer
- Cross-linker

Separation of the solventless release package into its individual components permits converters to custom formulate their coating baths to optimize performance for any given application; determining the catalyst and inhibitor levels that provide best combinations of cure and bath life permits end users to derive the most benefit per silicone dollar. While in principle the hydrosilation reaction is zero order in catalyst, increasing the catalyst concentration speeds cure at a given temperature and lowers temperature required for cure at a given oven dwell time. This effect is illustrated in Figure .6.

Another approach to lowering siliconizing cost is the use of silicone-compatible olefin extenders. In principle, a nonvolatile unsaturated hydrocarbon miscible with addition-curable dimethylsilicone fluids should cure in the coating with little effect on release performance.

In practice, replacement of silicone with less expensive organic olefins (or acetylenes) has not been firmly established as a means of reducing coating cost without performance penalty.

More than 30 domestic solventless silicone coating sites are currently in operation; new release coating capacity expected on stream in the near future will add to this number. These machines use offset gravure heads with multiple differential roll speeds to provide continuous coverage at low deposition. This coating technique works best with coating materials of less than 2000 cps viscosity. Silicone suppliers therefore furnish solventless silicones in viscosity ranges of 200 to 2000 cps (lower viscosity results in too much penetration of paper substrates), limiting polymer molecular weights to, roughly, 5000 to 20,000 daltons.

Novel coating techniques have been reported that do not require gravure rolls and are claimed to work with higher viscosity materials.

Radiation-Curable Silicone Release Agents

Solventless silicone coating application extends to radiation-curable silicones, for there is no reason to furnish UV or electron beam (EB) processible materials in solvent carriers. Polydimethylsiloxanes are not cross-linked by UV or EB radiation at economically practical intensity or dose, respectively. Practical radiation cure of silicones requires silicone polymers incorporating radiation-sensitive organofunctional groups, as illustrated below:

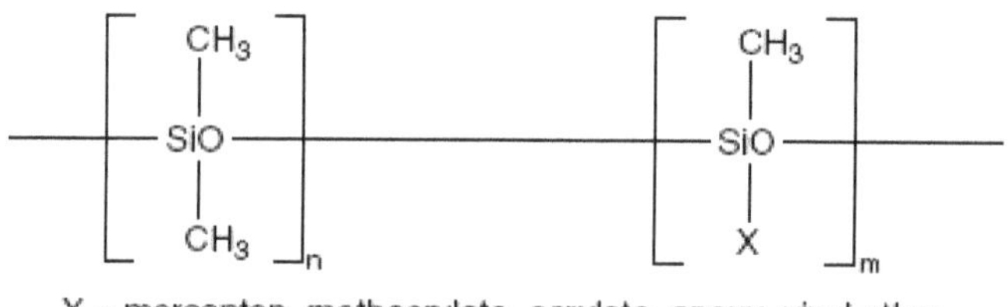

X = mercaptan, methacrylate, acrylate, epoxy, vinyl ether

In addition, UV cure normally requires high concentrations of photosensitizers or photoinitiators, such as benzophenone, benzoin ethers, and cationic-type "onium" salts. The presence of polar organofunctional moieties plus photocatalysts in radiation-cured release agents causes significant performance differences between radiation-cured and conventional thermally cured silicones. Nonetheless, industry demands for low (or "zero") temperature processing of silicone coatings to permit use of thermally sensitive films and to prevent demoisturization of papers have prompted considerable efforts by major silicone suppliers to develop, then improve, radiation-curable products. Radiation-curable silicones are now available from several sources.

Acrylated and methacrylated silicones were the focus of the earliest patented work in this area. Acrylated silicones specifically developed for release applications were introduced by Goldschmidt and are in commercial use at a small number of coating facilities equipped to perform EB cure. These materials have an important performance drawback inherent in free radical acrylate cross-linking chemistry: because cure is subject to severe inhibition by atmospheric oxygen, efficient inerting (<500 ppm O2) of EB or UV cure chambers with nitrogen is essential for fast, complete cure to occur. While nitrogen blanketing is not impossible, inerting adds complexity and cost to the coating operation.

Cure chemistry pioneered by W. R. Grace & Company overcomes oxygen inhibition problems that interfere with radiation cure of acrylates. Mercapto-olefin addition is initiated by UV light in the presence of suitable photosensitizers, or by EB radiation. The reaction is analogous to hydrosilation:

$$—RSH + CH_2{=}\,CHR' \text{®} —RSCH_2CH_2R'$$

The chemistry has been extended to release coatings by development of mercaptoalkyl-functional silicone polymers.

While mercaptovinyl silicone systems can be UV cured in ambient atmosphere, market acceptance of products based on this technology has been slowed by their objectional odor (skunk fragrance is derived from mercaptans) and by the tendency of unreacted mercaptan residues in the cured coatings to chemically react and bond (via addition) to free acrylate usually present in cross-linkable acrylic pressure-sensitive adhesives. The same issues affect hybrid acrylic silicone-mercaptosilicone systems developed to take advantage of the oxygen insensitivity of mercapto-olefin addition.

Certain "onium" (sulfonium and iodonium) salts are known to be capable of initiating photopolymerization of epoxides and vinyl ethers. Epoxy-functional silicones are readily prepared, so application of cationic UV cure to silicones was an obvious extension of the technology.

A major performance advantage inherent in epoxy silicone-iodonium salt photocurable systems results from the non-free-radical nature of this cross-linking. This particular cure mechanism is not subject to oxygen inhibition, making UV-curable epoxy silicone based release agents particularly well suited to wide web converting operations, as nitrogen blanketing is not needed.

The epoxy silicone UV cure system has been shown to combine exceptionally fast UV cure response with premium, stable release versus cross-linkable acrylic, styrene-butadiene rubber, and hot-melt adhesives.

As with other radiation-curable silicone release systems, however, controlled release additives capable of providing a broad, predictable range of release for the UV epoxy silicone coatings have remained elusive. Another problem associated with these cationic cure silicone materials is substratedependent performance. Excellent cure, anchorage, and release are obtained when corona-treated films and film laminates (such as polyethylene, polyethylene kraft, polypropylene, polystyrene, and certain polyesters) are coated, but slower cure, higher release, and inconsistent anchorage often occur when porous paper substrates are utilized. In many respects, this difficulty is analogous to problems encountered when solventless silicones were introduced and converters were first attempting to coat 100% solids silicones on paper release liners. More highly densified (or glassine) papers, perhaps treated with acidic coatings or sizing agents, ought to prove compatible with the UV-curable epoxy silicone release agents.

THE FUTURE

Solvent-dispersed thermal cure release coatings (condensation and addition cure) will be an important segment of the silicone coating market for years to come, particularly in Far East and Third World countries, where environmental and regulatory pressure to reduce the use of solvents is not a great factor.

New coating capacity in North America and Europe will be predominantly solventless, with radiation cure capability built into new coating lines or retrofitted to existing 100% silicone solids equipment as acceptance of new silicone curing technology becomes widespread. Silicone suppliers will accelerate development and commercialization of new and improved radiation cure and "zero temperature" cure products. An interesting area of research is room temperature cure silicones, consisting of one-part packages with cure that is triggered by moisture rapidly permeated through thin release coatings after they are applied.

Addition cure approaches to lower temperature ure 100% solids silicones have focused on more active catalysts and more volatile inhibitors; combining thermal and UV addition cure via UV-light-activated platinum hydrosilation catalysts or UV-sensitive inhibitors is another promising area of development. Few potential radiation cure silicone chemistry options have evaded development efforts, but vinyl ether-functional silicones (in concert with appropriate iodonium salt catalysts) seem to be capable of combining extremely fast noninerted UV cure with EB curability, too.

Interesting and exciting new silicone chemistry is undergoing rapid acceptance by the traditionally conservative converting industry. The result is that yet-unknown liner materials and applications of silicone release agents will expand pressure-sensitive tag and label markets. Research and development by silicone manufacturers will drive this growth.

SILICONE HARD COATINGS

Introduction

Silicone (polysiloxane) hard coatings are finishes of superior abrasion resistance and inertness to hostile chemical and environmental conditions. They consist of several monomers and other ingredients, and the makeup of the formulations varies from manufacturer to manufacturer. Among the highly varied components are monomeric silanes, dimerized silanes, silane hydrozylates, silaceous materials, leveling agents, flow control agents, cross-linking agents, and catalysts of various types. Silicone coatings are solvent-borne coatings. Some of the possible solvents are alcohols and glycol ethers. This includes such alcohols as isopropanol, propanol, ethanol, n-butanol, isobutanol and methanol.

Polysiloxane coatings are applicable to many substrates, but the majority of applications are on nonmetallic surfaces, especially plastics. Silicone coatings can be dyed or pigmented, but for the most part these coatings are used as clear top coatings. They have excellent light transmission and actually improve the optical properties of the material that is coated.

Some of the plastics that are used with polysiloxane coatings are polycarbonate, acrylic, polyarylate, polysulfone, vinyls, nylons, polyester, cellulose acetate, cellulose acetate-butyrate, and polyolefins, just to name a few. Despite this plethora of materials, acrylics and polycarbonates are generally the plastics of choice.

Silicone hard coatings do not have automatic adhesion to all plastic surfaces. Materials such as polycarbonate and acrylics will, in many cases, have tape adhesion without need of surface treatment, etch, or primer coat. Many of the other substrates cannot be coated successfully without pretreatment.

Many of the materials require either a corona discharge or a plasma treatment. For long-term outdoor stability, polycarbonate requires a primer coat to be applied first.

Plastic, to be coated, comes in many forms. Silicone coatings are used on cast sheet stock, extruded sheet stock, molded parts, lenses, windows, films, etc. The application methods are also quite varied. Some of the methods are spraying, flow coating, spin coating, and dip coating with various withdrawal speeds and, with some formula modifications, pad coating and roller coating. The coatings are not suitable for screen or gravure application.

Hard silicone coatings are useful as antifog, antistatic photochromic, color-dyed, pigmented, UV absorbing, UV stabilized for exterior exposure, chemical resistance, 5-min curing, and tinted coatings.

Many of these properties can be combined in one coating. The coatings are used in such diverse areas as the automotive, electronic, computer hardware, architectural and architectural glazing, recreation, sporting goods, protective eyewear, safety, and optical industries.

Table .1: Cure Schedules and Pretreatment of Some Substrates

Plastic Substrate	Typical Cure Temperature (°F)	Typical Cure Time (min)	Prime Coat, or Surface Treatment (Recommendation)
Polycarbonate	250	5–60	Primer (exterior and some interior use) None
Acrylic	190	20–120	None
Polyarylate	190–325	20–60	Primer or solvent addition
Vinyl	140	60–180	Corona or plasma discharge
Cellulose acetate	150	30–60	Alkali etch
Corona or plasma discharge	145–200	30–60	Alkali etch
Polyester	180–375	4–60	Corona or plasma discharge

The coatings cure thermally at baking schedules that vary with the heat stability of the substrate. The baking schedule can be influenced by such process variables as airflow, nonair media, and uniformity of heating. There are many types of silicone coatingd, and because of this most data given will have large variances. The cure times of these coatings vary from 5 min to 3 h. Since these are solvent-borne coatings, if they are placed in heat immediately after coating, the higher area of the coating strata will cure first, leaving solvent entrapped in the film. Solvent entrapment in the coating entails such possible problems as lower abrasion resistance, weaker adherence to the surface of the substrate, chemical resistance, and loss of some durability.

The parts must be free of dust, dirt, grease, and mold release. When not in use, this type of coating should be stored at a temperature of 40°F. Storage of the coatings at low temperature increases the life of the coating by three or four times that of the same coating under ambient conditions. The storage life for polysiloxane coatings is anywhere from 2 weeks to about 17 months at room temperature. With this wide range of stabilities, the stability of the coating chosen should be confirmed by querying the manufacturer.

In many cases, it is wise to have the coating shipped by refrigerated truck or by air freight.

The price of the coatings may vary from $70 to $200 per gallon. The coverage can vary from 300 square feet per gallon to as much as 2000 square feet per gallon, depending on drip reusability, dilution, and coating viscosity.

Substrates

The Table .1 suggests how silicone hard coatings are used. The coatings are applicable for many other substrates. Solvent addition will change coating properties and give additional adhesion in some cases. With treatment, the use of these coatings on metals is quite possible. The solvent flash-off time will also have an effect on the ultimate cure schedule. The more complete the removal of solvent, the quicker the time of cure. Hardness is increased by longer cure times.

Uses

Table .2 gives a number of current uses for polysiloxane coatings. These very versatile coatings are alcohol based, which makes them quite mild even on sensitive substrates. The biggest problem in determining a new use is the heat stability of the material to be coated. The length of the cure time will also help to determine whether the process is feasible.

Application of the Coating

A typical application procedure is summarized as follows:

1. Remove all protective masking and wrapping material.
2. Check the parts for optical quality.
3. Clean the parts. Some of the methods are solvent cleaning with alcohols and other mild solvents, ultrasonic cleaning with aqueous cleaning solution or trichlorotrifluoroethane, vapor degreasing, and roll cleaning machines that remove particulate with a high tack, nondeposit surface.
4. Before coating, treat the parts to prevent any possible static problems. This can be accomplished in a number of ways.
5. Perform annealing, which may be necessary in some parts, depending on substrate, coating, and the stress in the ultimate part.
6. Consider such other factors in the coating operation as temperature at which the material must be maintained, temperature of the room, maximum and minimum humidity, material of construction for coating vessels, and need of filtration.
7. Be sure that the parts have their solvent flash-off under very clean conditions. Baking should be performed for the required time period and temperature.

Train windows	Plane windows
Glazing for windows	Car headlight
Car taillights	Auto parts
Computer screens (low gloss)	Electronic components
Gauge covers	Instrument cover
Electronic sensors (antifog properties help continuous readability)	
Computer housings	Instrument housings
Protected printed parts	Racquetball glasses
Ski goggles	Swimming goggles
Sunglasses	Safety glasses
Prescription glasses (may require tintable coating)	
Chemical splash glasses	Fireman's face shields
Lenses for gas masks	Full face shields
Scuba masks	Plano lenses
Shields for self-contained breathing apparatus	
Bus windows	Truck windows
Motorcycle shields	Watch crystals
Surgical shields	Outdoor lighting
Operating theater lights	Dental lights

The following coating processes may be used:

Dip Coating

The part is immersed in the bath and withdrawn by hand or by a mechanized method. Slow withdrawal produces even coating thickness from top to bottom and no drip buildup at the bottom. Faster withdrawal methods produce parts faster, but with less coating uniformity and a greater buildup at the bottom of the part.

Flow Coating

This consists of running a filtered stream of coating across the top of a part and letting the excess material drip off at the bottom. This method is good for coating one side of a part.

Spray Coating

This easily automated method is suitable for coating one or both sides of a part. The coating might have to be diluted to be sprayed.

Conclusions

Hard polysiloxane coatings have become important in many industries. In the overall framework of modern coating technology, these coatings are in their infancy. Their tremendous hardness and excellent properties in very hostile environments make them candidates for expansion into other areas of use.

SOLGEL COATINGS

Introduction

Solgel processing is now well accepted as a technology for thin films and coatings. Indeed, the solgel process is an alternative to chemical vapor deposition, sputtering, and plasma spray. Not only have solgel thin films proved to be technically sound alternatives, they have been shown to be commercially viable, as well.

The technology of solgel thin films has been around for over 30 years. The process is quite simple. A solution containing the desired oxide precursor is prepared with a solvent and water. It is applied to a substrate by spinning, dipping, or draining. The process is able to apply a coating to the inside and outside of complex shapes simultaneously. The films are typically 1 μm, uniform over large areas and adherent. The equipment is inexpensive, especially in comparison to any deposition techniques that involve vacuum. Coatings can be applied to metals, plastics, and ceramics. Typically, the coatings are applied at room temperature, though most need to be calcined and densified with heating. Both amorphous and crystalline coatings can be obtained.

The Solgel Process

The solgel process is the name given to any one of a number of processes involving a solution or sol that undergoes a solgel transition. A solution is truly a single-phase liquid, while a sol is a stable suspension of colloidal particles. At the transition, the solution or sol becomes a rigid, porous mass by destabilization, precipitation, or supersaturation. The solgel transition to a rigid two-phase system is not reversible.

The first step is choosing the right reagents. To illustrate this, silica will be used as the model system. Of the available silicon alkoxides, tetraethylorthosilicate (TEOS) is used most often, because it reacts slowly with water, comes to equilibrium as a complex silanol, and in a one-quarter hydrolyzed state has a shelf life of about 6 months. The clear TEOS liquid is the product of the reaction of $SiCl_4$ with ethanol. The colorless liquid, $Si(OC_2H_5)_4$, has a density of about 0.9 g/cm^3, is easy to handle safely, and is extremely pure when distilled.

The other ingredients are alcohol and water. Ethanol serves as the mutual solvent for TEOS and water. As soon as TEOS is introduced into ethanol with water, the chemical reactions of hydrolyzation and polymerization begin. The chemical reactions are approximately as follows:

An acid is usually added to control the rates of these reactions. This mixing takes between 1 and 3 hours, while the solution viscosity increases. Complete hydrolysis of $Si(OC_2H_5)_4$ to $Si(OH)_4$ would give silicic acid. This does not occur. Instead, a condensation occurs between silanols and ethoxy groups to give bridging oxygens or siloxane groups.

$$\text{Hydrolyzation:}\quad -\overset{|}{\underset{|}{Si}}-O-C_2H_5 + H_2O = -\overset{|}{\underset{|}{Si}}-OH + C_2H_5OH$$

$$\text{Polymerization:}\quad -\overset{|}{\underset{|}{Si}}-O-C_2H_5 + Si-OH = -\overset{|}{\underset{|}{Si}}-O-\overset{|}{\underset{|}{Si}}- + C_2H_5OH$$

In the systems described above, the solgel transition is reached when the one-phase liquid becomes a two-phase alcogel, solid plus liquid. Alcogel is used to differentiate gels prepared with alkoxides from those prepared from ion-exchanged solutions or colloidal sols. Those gels are called hydrogels. Once through the solgel transition, the solvent phase is removed by ordinary evaporation to create xerogels or by hypercritical evacuation to create aerogels. At this point, the dried gel is a microporous oxide.

When it comes to coatings, the solution is said to go through its solgel transition while the coating is on the substrate. What is observed is that a liquid film undergoes a sharp increase in viscosity. If the substrate has been handled properly, a tacky gel covers the surface uniformly.

The majority of formulations are for oxides. For example, silica is used as an amorphous coating that may be applied for passivation of a surface or improving surface perfection. Titania is an example of an amorphous or crystalline coating that may be used for its high index of refraction or its electrical properties. Alumina is generally a crystalline coating that finds use in harsh environments. Of course, there are other specialized applications that call for multicomponent oxides. Increasingly, there are formulations for nonoxides, carbides, nitrides, and fluorides, along with organic–inorganic coatings.

THIN FILM APPLICATIONS

When considering where solgel coatings find use, several broad areas come to mind. These include optical coatings, electronic coatings, abrasion coatings, protective coatings, porous coatings, and composites.

Optical Coatings

One of the first applications was for rearview mirrors for cars from Schott Glaswerk. These coatings consist of titania-silica-titania interference filters that give the effect of total reflectance and replace conventional metallizing. Another early application was solar reflecting films for windows. These coatings consist of Pd-containing Titania films that show selective absorption.

In addition, there are antireflective coatings of several types. Both borosilicates and Titania silicates have been developed for this application. Antireflection in the ultraviolet (UV) range has been accomplished with thoria and hafnia films.

Broadband antireflection has been developed for laser optics, using silica broadband antireflective coatings with reduced UV scattering.

Electronic Coatings

Thin films have been applied from solution for numerous electronic and magnetic applications. Indium tin oxide (ITO) has been developed for electrically conducting transparent films for displays. Optoelectronics and PZT have been prepared. Iron oxide layers have been used for magnetic films. Vanadia layers have been used for antistatic films. Barium titanate has been prepared in thin layers for multilayer capacitors. Tantalate has been used for capacitors as well. The advantage of the gel thin films is that the thinner the layer, the higher the capacitance of the device, and the higher the density of devices possible on a chip.

Tungstate films and niobate films have been prepared for electrochromic displays. More recently, there has been work on devices, such as Titania thin films for electrodes, antimony-doped tin oxide films for transparent heat mirrors, and reaction-formed silicon oxynitride films for oxidation masks in microelectronic device fabrication.

Abrasion Coatings

Alumina coatings have been developed because they provide higher hardness on the surface of many substrates, in particular, stainless steel. In porous and dense conditions, a simple application of alumina on the surface can provide scuff resistance. Ceramics in general have higher hardness than metals and plastics. This means that ceramic coatings have fewer tendencies to show scratches. There is utility in applying such coatings to windows, especially polycarbonate windows.

Protective Coatings

Early on, phosphate coatings were applied to silicate glass panels to improve their chemical resistance to attack by water. This improvement was accomplished in 0.2 μ m. These points out one of the primary advantages of solgel processed films — the material use is efficient. Any excess material is recovered and can be used again. Even relatively costly components such as zirconia can be considered.

Generally, because the gel is amorphous, meaning no stoichiometric rules, any amount of metal oxides can be added to a composition. Being thin, solgel films do not delaminate when the substrate is bent. Finally, because the gel is porous, it is undetectable when a cover coat is applied and fired.

Porous Coatings

Of course, fully dense oxides can be obtained from gels with appropriate heat treatment. On the other hand, microporous films find unique applications modifying a physical property such as thermal conductivity or in performing physically as a membrane.

Composites

Microporous gel can be infiltrated with organic monomers and polymers to form organic–inorganic coatings or hybrid coatings. The low temperature of the solgel process enables the combination of organic and inorganic in ways not possible with conventional processes. The new field of organic–inorganic nanocomposites is in its early stages.

ADVANTAGES

The advantages of the solgel process in general are high purity, homogeneity, and low temperature. Because alkoxides can be multiply distilled, the products can be relatively free of impurities. Because the mixing is accomplished in solution, components mix on the atomic scale in relatively short times. If the plan is to use this process to make a dense material, the porous dried gel can be compacted at temperatures lower than conventional films.

In addition to an energy savings for this low temperature process, there is a reduction in volatilization. In fact, this process can make some compositions that cannot be made by conventional means because of phase separation or devitrification.

Another way to look at this process is that initially it creates an amorphous material that is porous. The porosity provides access to the nanostructure, allowing the design of a variety of nanocomposites.

RADIATION-CURED COATINGS

Introduction

Curing coatings by means of radiation represents one of the new techniques that is replacing the use of conventional or low solids, solvent-borne coatings. Radiation-cured coatings offer a manufacturer several important features. These include the following:

- High solids — usually 100% solids
- Low capital investment (with certain specific exceptions)
- Low energy curing costs — low power requirements and elimination of solvent costs
- Rapid cure speeds
- Ability to cure a variety of substrates, including heat-sensitive substrates such as plastics and parts for the electronics industry

- Increased productivity
- Shorter curing lines and decreased floor space requirements for operating line and for liquid coating storage
- A variety of different chemistries from which to select, and thus broad formulating latitude from the wide variety of formulation ingredients available

The main sources of actinic energy for curing coatings by radiation are electron beam and ultraviolet light.* It 1984, Pincus indicated that there were four suppliers of electron beam (EB) equipment and more than 40 suppliers of ultraviolet light (UV) equipment.

These figures include laboratory, pilot, and production units. With the industry growing at about 10 to 15% per year, it is very reasonable to expect that these numbers had increased by the end of the decade. Equipment for EB curing is significantly more expensive than that for UV curing, and it represents the exception listed above in the "low capital requirement" advantage of radiation curing.

Even a laboratory-sized EB unit will run into six figures, whereas simple conveyorized UV systems can be obtained for a few thousand dollars. Furthermore, energy consumption is high with EB units and low in UV systems.

In 1980 about 80 million pounds of formulated radiation-curable product were sold in the United States. This number increased to 127 million pounds in 1985 and was expected to increase to 216 million pounds in 1990.

About 88% of the 1985 total was cured by UV light and the remainder by EB. Thus, it is readily apparent that radiation curing is a specialty area, but it is one that is growing much faster than the overall coatings market. A study indicated that the market will be 336 million pounds by 1995, which represents a 9.2% compounded growth rate from the prediction for 1990. There are two basic technologies involved in the radiation-cured coatings market. The older and more well-known technology involves free radical chemistry, and the cure of compounds containing unsaturation (i.e., acrylates, styrene/unsaturated polyester, and the like). The other technology involves cation chemistry and the cure of cycloaliphatic epoxides and compounds that will copolymerize with them under the influence of Brønsted or Lewis acids.

EQUIPMENT

Electron Beam

With EB curing, energy transfer is caused by energetic or accelerated electrons.

Therefore, photoinitiators usually are not used in EB cure formulations and there are no photoinitiator fragments, which might have an odor or otherwise detract from properties in the cured coating. The electron source is a filament that is heated inside a vacuum tube. The electrons are accelerated to a high energy (~10 5 V) by an impressed electrical field with the degree of acceleration increasing with increasing applied voltage.

High voltages (150,000 to 300,000 V) are involved. The accelerated electrons pass through a metallic foil window and are sent on to the compound, which is capable of absorbing their energy and thus undergoing polymerization. The ability of the electrons to be absorbed by the compound (i.e., the radiation-curable formulation) depends on the material's density, and therefore depth of penetration is inversely proportional to density. The electron bombardment of the formulated coating abstracts hydrogen atoms from some of the molecules, thereby generating short-lived free radicals that can initiate the polymerization of acrylates and similar compounds that will interact and react with such species. Clear coatings of up to 20 mils (500 μ m) and pigmented coatings of about 15 mils (400 μ m) can be cured with EB equipment.

Ultraviolet Light

In general, UV light units operate with electromagnetic radiation that is in the optical region of 200 to 760 nm. They do produce infrared radiation of 760 nm to 1.0 mm, but this energy is thermal and acts either to anneal the cured coating and relieve internal stresses and strains or to enhance cure rate in cationic cure systems. Of course, in certain formulations, this thermal energy can have a deleterious effect by causing volatilization of reactants, and it is minimized in some equipment. In other cases, such as cationic UV curing, the thermal energy can be highly beneficial by kinetically enhancing the cure rate of the compounds.

Basically, in the case of UV curing, compounds susceptible to rapid polymerization are contacted with initiating species obtained by photolysis of a photoinitiator. In the UV cure* of such compounds, a photoinitiator that is capable of photolyzing or degrading to an active species is added to the formulation.

When light of the proper wavelength strikes the photoinitiator, the active species is generated and polymerization rapidly takes place. Certain photoinitiators generate free radicals, and these are used to cure acrylates. Other photoinitiators generate cations, which are used to cure cycloalipathic epoxidebased systems. It should be readily apparent that matching the output of the UV light source with the absorption spectra of the photoinitiator is an important aspect of this technology.

There are four different UV light curing technologies. These are as follows:

- Medium pressure mercury vapor lamps (in certain countries, low pressure and high pressure mercury vapor lamps are also used)

- Electrodeless vapor lamps
- Pulsed xenon lamps
- Lasers

Medium pressure mercury vapor (MPH) lamps have been used commercially for about 20 years. The bulb is an evacuated quartz tube that contains metallic mercury and has electrodes at each end. Electrical energy is supplied through the electodes, and an arc is struck between them. This heats mercury in the tube to plasma that emits UV, visible, and infrared radiation. Output ranges from about 100 to 400 in., and a warm-up time of 10 to 15 min is often recommended for the bulb to be fully operational. The systems can be doped to alter the emission spectrum, but usually only mercury is used. Electrodeless lamps have been in use for about a decade and currently are very popular in the industry.

The vacuum tube UV bulb is manufactured from quartz, which is invisible to the UV radiation. The bulb contains either mercury or other proprietary metals and gases. The system is activated by microwave or radiofrequency energy (i.e., it does not have electrodes). Because of the nature of this activation system, it essentially has instant on and off operation. A variety of doped bulbs are available with various output spectra that can be "matched" to that of different photoinitiators. Bulb lifetime is usually 5 to 10 times longer than that of standard MPH lamps. Xenon lamps are quartz tubes filled with doped xenon. The lamp is powered by pulsed electrical current.

These units offer very low heat output along with short time, extremely high peak intensity output. (Some types have as high as 8000 to 10,000 W output, although most operate at much lower intensities.) The output spectrum is continuous with this source rather than that of the discrete line types as from the MPH lamps. However, in certain instances, mercury is added to the tube contents to enhance curing in the region of mercury's spectral lines. The output has also been modified with other metals such as iron and beryllium.

Argon ion and nitrogen lasers in combination with specific photoinitiators that have a strong absorbance at the emission line of the laser have been used to cure multifunctional acrylates. Although the studies are interesting and may hold promise for the future, at present, this is considered to be a research area of potential interest for the electronics industry.

Chemistry

Most of the following discussion deals with UV light technology. The reason for this is that photoinitiator breakdown is important to cure. In a general sense, the same compounds that will cure with the free radical photoinitiators will cure with EB.

PHOTOINITIATORS

Free Radical Type

Free radical generating photoinitiators are of two general functional types. The first type involves a mechanism known as hemolytic fragmentation, in which a compound such as benzoin alkyl ether undergoes a photochemically induced fragmentation into highly active free radicals as described in Figure .1.

The second type of free radical generation functions through a mechanism that is termed Electron transfer. This mechanism involves a photolytic excitation of the photoinitiator from a found state singlet to an excited triplet state. This is followed by electron transfer to a hydrogen atom donor, such as dimethylethanolamine (DMEA), and the formation of highly excited free radicals as described in Figure .2.

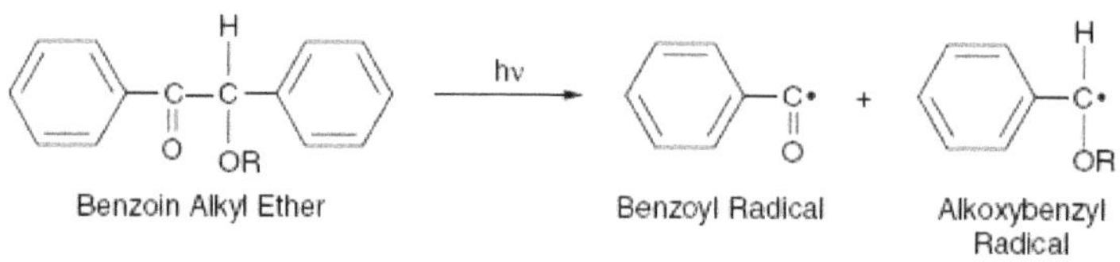

Figure .1 — Homolytic fragmentation.

Typical commercial photoinitiators include compounds such as 2,2-diethyoxy-acetophenone, 2,2-dimethoxy-2-phenyl acetophenone, hydroxycyclohexylphenyl ketone, benzophenone-triethylamine, 2-methyl-1-4-(methylthio)-2-morpholino-propane-1, 1-phenyl-2,2-propane dione-2-(o -ethoxycarbonyl) oxime, and benzoin methyl, isopropyl, isobutyl, and other alkyl ethers.

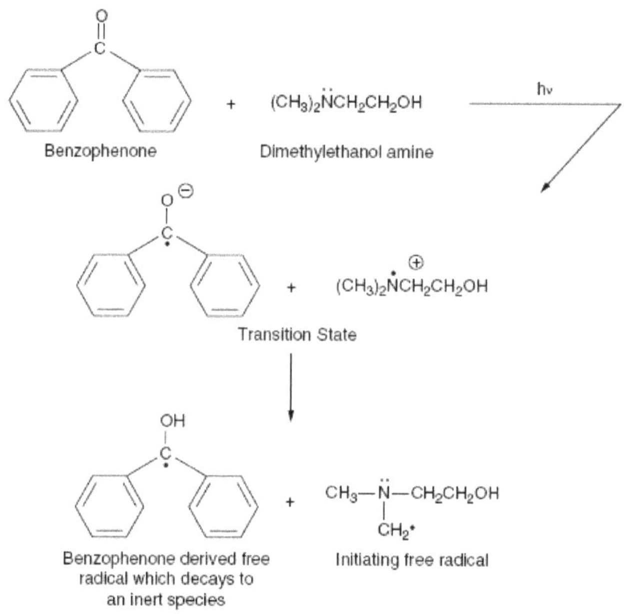

Figure .2 — Electron transfers.

Free radical generating photoinitiators of the foregoing types are inhibited or inactivated by oxygen as a result of a complex that forms between the light-activated photoinitiators and molecular oxygen. This effect can be overcome by inerting the coating with nitrogen during cure, by adding waxes to the system, or by using excess photoinitiator. Air that has been dispersed in the coating system during formulation contains oxygen, and it acts as a stabilizer. However, formulations containing very active photoinitiators of this type have a tendency to polymerize during storage if this oxygen is depleted over a period of time. Compounds that will help prevent such instability include phenothiazine and Mark 275 stabilizer.

Tertiary amines will act as photosynergists, and they greatly enhance curing rate of compounds such as those described above. Ureas and amides also have been described as synergists for benzophenone. Compounds that have been used to accelerate cure rate of pigmented systems include isopropylthioxanthone, ethyl-4-dimethylaminobenzoate, and 2-chlorothio-xanthone.

Cationic Type

Although there are various types of photoinitiators that photolyze to yield a cationic species capable of polymerizing cycloaliphatic epoxides and active hydrogen compounds of the hydroxyl type or vinyl ethers, only the arylsulfonium salts are commercial at present. These types include aryldiazonium salts, aryliodonium salts, iron-arene complexes, aluminum complex-silanols, and the commercial arylsulfonium salts.

Aryldiazonium hexafluorophosphates and tetrafluoroborates decompose under the action of UV light and yield Lewis acids such as BF and PF, nitrogen, and other fragments. These photoinitiators were used in the infancy of cationic UV cure of cycloaliphatic epoxides. Although they were quite active for first-generation products, the disadvantages of thermal instability, which led to short shelf life, and of nitrogen evolution, which led to pinholes and bubbles in films thicker about 0.2 mil, inhibited commercial use and led to their replacement by the onium salts in the marketplace.

The polymerization of epoxides with aluminum complex-silanol photoinitiators has been described. The technology is not being practiced in the United States, but it may be in use in Japan. The iron–arene complexes represent a new type of cationic photoinitiator that was recently described. When photolyzed, these compounds degrade to yield both Lewis acid-type catalysts and free radicals. Because these compounds are relatively new, detailed information about them is not available. Various investigators studied the onium salts of iodine or the Group VI elements.

Currently, the arylsulfonium salts are commercially used as photoinitiators. These compounds do not have the deficiencies of the diazonium salts because there is no nitrogen evolution on photolysis and, if protected from UV light, the systems can have ambient-condition shelf lives in excess of 2 years. When UV light interacts with the onium salts, an excited species is formed. This species undergoes hemolytic bond cleavage to yield a radical cation, which extracts a hydrogen atom from a suitable donor and generates another free radical species. The new compound then gives up the proton for formation of a strong Brønsted acid. The Brønsted or protic acid that is the polymerization catalyst is of the form HMF 6 where M is a metal such as antimony, arsenic, or phosphorus. This catalyst is long-lived, and the cationic polymerization of the epoxide system can continue in the "dark" after initial exposure to UV light until the available epoxide is exhausted or the polymerization is terminated by some other mechanism. Thus, the onium salts generate both cationic species and free radicals and can be used in radiation-activated, dual-mechanism systems.

Note that the onium salt photoinitiator is a blocked or latent photochemical source of the strong Brønsted acid that acts as a catalyst/initiator for the formulated system. Because of the acidity of the UVgenerated catalyst or initiator, it is necessary to keep the formulated system (substrate, coating equipment, etc.) free from basic compounds that would neutralize the acid and either negate or slow cure rate. Even very weak basic compounds will react or interact with the strong acidic species.

Dual-Mechanism Curing

Since the cationic photoinitiators generate both free radicals and Brønsted acids when exposed to UV light, it is possible to combine acrylates that will cure with free radicals and epoxides that cure with the protic acids. Free radical generating photoinitiators such as 2,2-diethoxyacetophenone can be added, if an additional source of free radicals is necessary. Experience has shown that this usually is not necessary.

Of course, the benzophenoneamine systems described earlier should not be used. Little can be found in the literature about this interesting topic, but dual-mechanism curing should prove to be a useful technique in the future and merits further study.

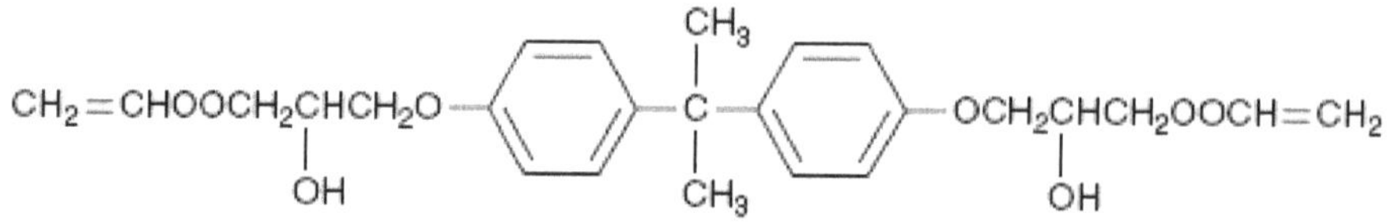

Figure .3 — An "epoxy acrylate."

Dual-mechanism systems that involve free radical chemistry coupled with thermal chemistry are also known. Dual-cure plastisols and dual-cure pigmented coatings have been reported. The combination of UV and infrared radiation for curing coatings, which can be beneficial for either of the dual-mechanism systems, has also been discussed.

Formulation

The subject of formulation involves both free-radical-curable ingredients and how they are used, and cationic-curable ingredients and how they are used. If one takes an extremely simplistic approach, only a single species that will cure under the influence of radiation is needed for a coating. However, most formulations contain a variety of ingredients to achieve the property balance that is needed to meet a given set of performance criteria.

Most formulations contain a base oligometric or multifunctional compound, a flexibilizer (which may be the base compound) a low viscosity, reactive diluent (which may be mono- or multifunctional), for viscosity control, a multifunctional acrylate for high cross-link density, a photoinitiator (unless electron beam cure is used), and usually a surfactant or flow and leveling agent. Combinations of more than one material of each type are usually used to optimize properties. The performance characteristics of specific end uses may also require use of a slip agent, nonreactive organic or inorganic fillers and/or pigments, adhesion promoters, flattening agents, an augmenting flow and leveling aid, or some other additive-type ingredient.

Free Radical Systems

In general, systems based on acrylates are used in free radical cure systems. The base material is usually an epoxy acrylate or a urethane acrylate. Epoxy acrylates are the reaction products of acrylic acid with various diglycidyl ethers of bisphenol A. The epoxy acrylates contain acrylate and hydroxyl functionality, but they do not contain epoxide functionality and should not be confused with the epoxides to be discussed in the cationic cure systems section. The term "epoxy acrylate" is a widely used and accepted misnomer; as can be seen by the structural formula in Figure .3, "acrylated epoxy" would be a better term to use for these compounds.

Urethane acrylates are often prepared by end capping a polyether, polyester, or caprolactone polyol with a diisocyanate and then reacting this isocyanate prepolymer with hydroxyalkyl acrylate. The chemistry is more complex than this simple description, and the order of component addition is important to minimizing viscosity. Both epoxy acrylates and urethane acrylates have a high viscosity (≥ 10 6 cp) at ambient temperatures.

To facilitate manufacture, handling, and later formulation, the compounds are often made in a low viscosity mono- or multifunctional acrylate, which will later serve as a reactive diluent and/or cross-linking agent in a formulated system.

In either the neat or the diluted form, these base compounds are diluted to application viscosity with more mono- or polyfunctional acrylate. Polyfunctional acrylates increase cross-link density and improve solvent resistance, increase hardness, and increase glass transition temperature. Care is always taken to avoid skin and eye contact and inhalation of the acrylates because of the potential for skin irritation, sensitization, and toxicological reaction. It is important to read the Material Safety Data Sheets supplied with these and all other compounds. The manufacturer's precautions should be carefully read and heeded when working with these and other chemicals.

Acrylates are available from a variety of sources. In addition to the urethane acrylates and epoxy acrylates described above, polyfunctional acrylates such as trimethylol propane triacrylate, pentaerythritol triacrylate, 1,6-hexane diol diacrylate, tripropylene glycol diacrylate, and tetraethylene glycol diacrylate, are available. There has been a trend to increase molecular weight by alkoxylation of compounds used to make multifunctional acrylates, to give them better handling and health characteristics. Monofunctional compounds useful as reactive diluents include N -vinyl-2-pyrrolidone, 2-ethylhexyl acrylate, dicyclopentadiene acrylate, hydroxyalkyl acrylates, hydroxylactone acrylates, and ethoxyethoxyethyl acrylate.

Specific formulations are highly varied, and performance requirements guide or dictate ingredient levels. Many formulations can be found in the cited literature or other literature available from material manufacturers.

Cationic or Epoxy Systems

The most important formulating ingredient in a cationic UV cure system is a cycloaliphatic epoxide of the 3,4-epoxy cyclohexylmethyl-3,4-epoxy cyclohexane carboxylate or bis(3,4-epoxy cyclohexylmethyl) adipate type.

Systems usually contain from 100% to about 30 to 49% cycloaliphatic epoxide. When this epoxide is used alone or at very high concentrations, strong, hard, and brittle coatings that are useful on rigid substrates result. These rigid coatings can be flexibilized and toughened in various ways. Although commercial, compounded flexibilizers/tougheners exist for these systems, various polyols such as the propylene oxide or caprolactone polyols can be used. Polyester adipates can be used, but the relatively high acidity of these polyols can lead to shortened shelf life because the cycloaliphatic epoxides are wellknown acid scavengers and will readily react with any carboxylic or other acid groups in this system.

This will either increase viscosity or cause gelation. Other flexibilizing agents include epoxidized soybean and linseed oil epoxides and epoxidized polybutadiene. Care should be exercised when incorporating these compounds in the formulation because they can cause significant softening, along with flexibilization, and little or no increase in toughness.

Relatively small amounts (~1 to 20%) of the diglycidyl ethers of bisphenol A can be added to systems. However, the light absorbing characteristics of these compounds lead to a decrease in cure rate and in depth of cure. In addition, the compounds cause rapid increases in viscosity. Novolac epoxides appear to cure well in cationic systems, but their high viscosity is rapidly reflected in formulation viscosity.

Low molecular weight epoxides available under trade names can be used as reactive diluents. Although somewhat slower in reactivity than many other cycloaliphatic epoxides, limonene mono- and diepoxide can be used as reactive diluents. Vinyl ethers can act as reactive diluents and cure rate enhancers in cationic cure, cycloaliphatic epoxide based systems.

These compounds have not been fully investigated, but the available evidence suggests that they have formulating potential. Since nonbasic, active hydrogen compounds react under cationic conditions with the oxirane oxygen of cycloaliphatic epoxides to form an ether linkage between the compound and the ring and a secondary hydroxyl group on the epoxide ring, low molecular weight alcohols, ethoxylated or propoxylated alcohols such as butoxyethanol, and similar compounds can be used as reactive diluents in cationic systems. However, since these compounds are monofunctional, they can act as chain stoppers — although they do generate the secondary, ring-attached hydroxyl group, which can further propagate polymerization or chain extension — and can be used only in limited amounts, about 1 to 10%, that are dependent on molecular weight. Low molecular weight glycols (diethylen glycol, 1,4-butanediol, etc.) can also be used. Such compounds may enhance cure rate by providing a source of active hydrogen; but, when used at permissible low levels, the glycols do not enhance toughness. In certain instances, inert solvents such as 1,1,1-trichloroethane are used to decrease viscosity and/or increase coverage from a given volume of coating. However, most end users prefer systems that only contain reactive components.

As mentioned above, the reaction mechanism of epoxides and hydroxyl groups is such that a new hydroxyl group is generated for every hydroxyl group that is present. Thus, the initial hydroxyl content of a formulation is conserved after the reaction is complete. Although low levels of hydroxyl groups will often enhance adhesion, too many of these groups can detract from performance characteristics and cause adhesion loss under wet, moist, or high humidity conditions.

Specific formulations are highly varied, and performance requirements guide or dictate ingredient levels. Many formulations can be found in the cited literature or other literature available from material manufacturers.

End Uses

Radiation-cured coatings, which are often taken to include inks, adhesives, and sealants, are used in a large number of ways for rigid and flexible metal, plastic, glass, paper, and wood substrates. Particular end-use arenas include the communication, construction, consumer products, electronics, graphic arts, medical/dental, packaging, and transportation markets. Specific end uses for radiation-cured compounds are numerous and include coatings for appliances, beer and beverage can bodies and ends, book covers, bottles and bottle caps, catalogs, closures, compact discs, cosmetic cartons, credit cards, decorative and functional foils and films, decorative mirrors, electronic components, flocked fabric, furniture, labels, magazines, magnetic tape, natural and simulated wood paneling, optical fibers, orthopedic casts and splints, photoresists, plastic cups and containers, printed circuit board assemblies (conformal coatings), record album jackets, solder masks, steel can ends for composite paper–metal cans, toys, transfer letters, and vinyl flooring.

FIRE-RETARDANT/FIRE- RESISTIVE COATINGS

Conventional Paints

Non-flame-retardant coatings usually give a low flame spread rating over asbestos-cement board, steel, or cement block. When the coatings are tested over wood and other flammable materials, flame spread ratings similar to those of the substrate are obtained.

The fire-retardant effectiveness of paints is highly dependent on the spreading rate or thickness of the coating as well as the composition. When conventional paints are applied at the heavy rate common for fire-retardant coatings, they give flame spread indices comparable to those of fire-retardant paints. For example, coating of latex and flat alkyd paints applied to tempered hardboard at an effective spreading rate of 250 ft /gal reduced the flame spread index of the uncoated substrate by factors of 3 and 5, respectively.

Fire-Retardant Paints

Fire-retardant coatings are particularly useful in marine applications. Ships are painted repeatedly to maintain maximum corrosion protection. As the layers of paint build, they pose a fire hazard even though the substrate is steel. In the event of fire, the paint may catch fire, melt, drip, and cause severe injury and damage to the vessel. Coatings are therefore formulated that do not sustain combustion; they should not spread the flame by rapid combustion nor contribute a significant amount of fuel to the fire.

Polyvinyl chloride containing 57% by weight chlorine is self-extinguishing. However, it is not a good vehicle for a flame-retardant coating because of its high melting point. This can be lowered substantially by copolymerization with other vinyl monomers such as vinyl acetate. To make these copolymers useful, addition of plasticizers and coalescing solvents is often necessary to give suitable application and per-formance properties. These additions dilute the overall concentration of chlorine thereby reducing the flame retardancy.

Table .1 — Formulation for a Fire-Retardant Latex Paint

Ingredient	Parts by Weight
Palyvinyl acelate copolymer emultion (55% solids)	20
Chlorinated paraffin	5
Antimony oxide	11
Titanium dioxide	22
Maca	9
Thickeners, etc.	3
Water	30

Table .2 — Formulation for an Alkyd-Based Paint

Ingredient	Parts by Weight
Long oil alkyd	22
Chlorinated paraffin	7
Antimony oxide	6
Titanium dioxide	29
Micronized talk	7
Whiting	11
White spirit drier, etc.	18

Fire-retardant coatings are based primarily on chlorinated alkyds, alumina trihydrate, or a combination of chlorinated paraffins and antimony trioxide. Table .1 and Table .2 give typical formulations. Flame spread test results depend both on the substrate and the thickness of the film.

FIRE-RETARDATION MECHANISMS

The combustion of gaseous fuel is believed to proceed by a free radical mechanism:

$$CH_4 + O_2 \rightarrow CH_3 + H + O_2$$

$$H + O_2 \circ OH + O$$

$$CO + OH \rightarrow CO_2 + H$$

The H, OH, and O radicals are chain carriers and take part in a number of reactions in the flame zone. The function of halogen containing compounds as flame retardants has been explained by the radical trap theory and takes place in the gas phase. In the foregoing reactions, liberated HCI or HBr competes for the radical species that are critical for flame propagation:

$$CH_4 + X \rightarrow HX + CH_3$$

$$H + HX \rightarrow H_2 + X$$

$$OH + HX \rightarrow H_2O + X$$

$$O + HX \rightarrow OH + X$$

The active chain carriers are replaced with the much less active halogen radical, slowing the rate of energy production and helping flame extinguishments.

Antimony oxide is known as a flame-retardant synergist when used in combination with halogen compounds. Volatile antimony oxyhalide (SbOX) and/or antimony trihalide (SbX 3) form in the condensed phase and transport the halogen into the gas phase.

Phosphorus compounds are also used as primary flame retardants. The flame-retardant mechanism for phosphorus compounds varies with the type of compound, the polymer, and the combustion conditions.

For example, some phosphorus compounds decompose to phosphoric acids and polyphosphates.

A viscous surface glass forms and shields the substrate from the flame. If the phosphoric acid reacts with the polymer (e.g., to form a phosphate ester), subsequent decomposition results in a dense surface char. The coatings that form serve as a physical barrier to heat transfer from the flame to the substrate and to diffusion of gases; in other words, the substrate is isolated from heat, flame, and oxygen.

This is the mechanism for fire-resistive intumescent coatings discussed below.

Triaryl phosphate esters are thermally stable, high boiling (>350°C) materials. They can volatilize without significant decomposition into the flame zone, where they decompose. Flame inhibition reactions, similar to the halogen radical trap theory, have been proposed.

$$H_3PO_4 \rightarrow HPO_2 + PO + \text{etc.}$$

$$H + PO \rightarrow HPO$$

$$H + HPO \rightarrow H_2 + PO$$

$$OH + PO \rightarrow HPO + O$$

Alumina trihydrate (ATH) or magnesium hydroxide inhibits ignition by absorption of heat due to decomposition, releasing large volumes of water of hydration (>30%).

FIRE-RESISTIVE INTUMESCENT COATINGS

Intumescent paints and mastics swell and char when exposed to heat and flame, giving carbonaceous foam that insulates the substrate from heat, air, and fire. This may delay the onset of combustion of a wood or plastic substrate or delay the heat buildup and tensile loss of structural steel. In the latter case, thick coatings or mastics are used. These coatings are considered to be fire resistive and significantly more effective than fire-retardant coatings. The former offer protection to the substrate. Intumescent coatings require three basic ingredients: a carbonific or carbon producer such as pentaerythritol or other polyol, an acid releasing agent such as a phosphate, and a spumific or gas producer such as melamine.

Polyols such as starch and dipentaerythritol (less water sensitive) are carbonifics. Other acid producers and their temperatures of decomposition are ammonium polyphosphate (215°C), monoammonium phosphate (417°C), and melamine phosphate (300°C); spumifics include urea (130°C), dicyandiamide (210°C), and melamine (300°C).

When the coating or mastic is heated, the decomposing phosphate forms phosphoric acid. This esterifies the hydroxyl groups of the polyol, which subsequently decomposes, forming water and a carbonaceous char, regenerating the phosphoric acid. As the char forms, the paint binder softens and the spumific decomposes, liberating nonflammable gases. These gases expand the softened binder into foam. Rigid carbonaceous foam is formed as charring of both carbonific and binder resin is completed.

Table .3 and Table .4 give examples of intumescent systems. When used as a thick mastic coating on structural steel, such materials can prevent the steel from reaching the failure point for 3 hours, as indicated in Table .5.

Table .3 — Formulation for a Typical Intumescent Emulsion Paint

Ingredient	Part by Weight
Polyvinyl acetate latex	18
Ammonium polyphosphate	22
Pentaerythritol	12
Starch	3
Dicyandiamide	16
Titanium dioxide	4
Water	25

A unique intumescent coating containing no water-sensitive polyols is composed of p,p′ -oxybis(benzenesulfonamide) in a vinyl chloride–vinylidene chloride copolymer binder. The addition of melamine pyrophosphate improves the homogeneity of the system. Dipentaerythriol can also be added to improve the homogeneity.

Table .4 — Formulation for a Typical Intumescent Solvent–Thinner Paint

Ingredient	Part by Weight
Vinyl toluene–butadiene copolymer	7
Ammonium polyphosphate	28
Dipentaerythritol	8
Melamine	9
Chlorinated paraffin	9
Titanium dioxide	6
Mineral spirits	33

Table .5 — Rating on Steel Column

Coating Thickness (in)	Rating (h)
.031	1
0.39	1.5
0.58	2
0.78	2.5
0.97	3

Pitt-Char (PPG Industries) coating is a two-component epoxy-polyamide produce designed for structural steel. This intumescent epoxy coating requires no reinforcement and provides up to 3 hours of fire protection.

Miscellaneous Coatings

Magnesium oxychloride mastics will protect steel by releasing large quantities of water upon heating and by relatively high temperature calcining (500°C). This material can be painted or sprayed and can be used in exterior applications.

Polystyrene foam block can be protected against fire for specified time periods. Building plastics containing gypsum and perlite and an expanded vermiculite, Portland cement, and limestone coating are particularly effective in protecting polystyrene foam block against the heat of a simulated mine fire.

For an insight into potential methods of flame retarding polyurethane and epoxy coatings, an analogy with plastic flame retardants can be made. The most common way to achieve flame-retardant polyurethanes is by the addition of halogen, phosphorus, chlorophosphate, or chlorophosphonate compounds.

Reactive halogen and phosphorus diols have also been used. Epoxy resins are generally flame retarded using the reactive tetrabromobisphenol A. Use of phosphorus–halogen compounds has been reported.

LEATHER COATINGS

Introduction

Tanned leather is usually coated with a thin pigmented or lacquer coating. One of the purposes of such a coating is decorative. The coating may also change some physical properties of leather: it may decrease water and air permeability, increase its rigidity, etc. Such changes depend on the coating type, especially on the polymer used as a film former. The properties also depend on coating formation technology: the coating may penetrate deeply into leather, or it may remain only on the surface. The coating technology chosen depends on the leather structure and the degree of its surface damage.

Tanned leather is the midlayer of an animal's hides — the derma, which is processed chemically and mechanically. During processing, leather becomes resistant to bacterial and fungal attack; its thermal resistance and its resistance to water increase. The derma consists basically of collagen protein having a fibrous structure. Collagen in the derma is in the form of a fibrous mat, and the fibers extend at varying angles with respect to the leather surface. The fiber diameter is 100 to 300 μ m. Other proteins (albumin, globulin) and mucosaccharides are located between fibers and bond the proteinaceous materials into multifiber ropy structures. Such a multicomponent leather structure determines its capability to deform — its elasticity and plasticity.

Leather is used for many applications: footwear, gloves, clothing, purses, furniture upholstery, saddles, and a variety of other uses. Leather is processed differently for each application: different chemicals are used; their quantity and processing conditions may also be different. Thus, leathers of different physical-mechanical properties are obtained: very soft, thin, and extensible for gloves and clothing, more rigid for footwear, and hard and stiff for soles. Often leather is dyed during processing. Dyeing may take place by the immersion of leather into a dye solution bath (usually in a rotating drum), or by covering the dry leather surface with a colored liquid coating. The latter technique confers a protective leather coating.

There is also another, but rarely used, method to form a surface coating: lamination of a polymeric film to the leather surface. In such cases, the surface is covered by a film, which is caused to adhere to the surface by pressing with a hot plate.

In general, there are several combinations of finished leather: undyed leather, dyed in a bath without a coating (aniline leather), surface dyed by applying a coating, and both bath dyed and surface coated. If the leather surface has many defects, these may be removed by grinding. In such cases, the coating is thicker and forms an artificial grain.

Coating of leather and coloring it by coating are accomplished by applying several coating layers. Initial grounding layers penetrate the surface of the derma and help to bond the leather. The midlayer is the most important; it has a protective and a decorative function. The top layer determines the gloss level, brings out the aesthetic qualities of the main layer, and often protects it from the effects of moisture.

The number of separate coating layers required to produce the desired leather surface quality depends on their composition and on the condition of the leather surface, which may be natural or polished. The formation of a leather surface coating generally consists of the following steps.

1. Application of a single nonpigmented seal coating. In the case of leather with a natural top grain, this coating is applied by spraying; in the case of polished surface leather, by flow coating. This coating has a good penetration and it helps to bond collagen fibers and to form the ground for the formation of a new leather surface.
2. After some drying and light pressing (pressure 0.6 to 0.9 mPa and temperatures 55 to 60°C), the leather is coated with a pigmented ground coating. This coating may be applied by brushing.
3. After drying and sometimes light pressing, the main pigmented coating is applied. Depending on the color, it may consist of two or three separate applications; in the case of light colors, even four applications may be required.
4. The top coating is applied by spraying in a single or double application. Sometimes drying and pressure may be applied between the coats.

All these steps are not necessary in all cases. Depending on the leather's surface condition, sometimes the first ground coat may be omitted; quite often the second pigmented ground coating is omitted, and its function is performed by the main coat.

CHARACTERISTICS OF LEATHER COATINGS

Main Coating

Leather coatings may be water or solvent-borne. The main advantage of aqueous dispersions is the absence of flammable, volatile, and polluting organic solvents. Therefore, aqueous dispersions are used for leather coatings more often than solvent-borne coatings. That is especially true for footwear leather.

Many different coating compositions are used. The main coating ingredient is the film former or binder. Pigments are usually predispersed and added as a paste. Plasticizers are added to increase the flexibility, softness, and elasticity. Surface active agents improve wettability and therefore improve the adhesion. Waterborne coatings may employ thickeners to increase the viscosity. Cross-linking agents are used sometimes to improve temperature and chemical resistance. In the formulations for ground coatings, penetrating agents may be used; they usually are mixtures of surface active agents, organic solvents, and water. For solvent-borne coatings, organic solvents and diluents are used.

Various synthetic and natural materials are used as film formers. They are mainly aqueous emulsions of acrylic, and also of vinyl polymers and copolymers; polyurethanes and nitrocellulose, available as solvent solutions and aqueous dispersions; casein and its derivatives; and waxes and other materials capable of film forming and exhibiting a good adhesion to leather. The desired appearance of such coatings is also important in determining the choice of a binder.

Coatings are classified according to the binder used: thus we have emulsion, nitrocellulose, polyurethane, and casein coatings. Casein coatings, having the milk protein casein as the binder, are used less and less frequently. Casein as well as nitrocellulose is being replaced by various synthetic polymers.

Unpigmented Ground Coatings

The dermal structure consists of two main layers: the upper or gland layer and the lower or net layer. The gland layer is less dense and has large pores, and the fibers are bound less strongly together. The gland layer is further weakened and damaged when the leather surface is ground. Therefore, before applying a pigmented coating over a ground leather surface (and sometimes one that is not ground), an unpigmented ground coat is applied. This coating penetrates the gland layer of derma, bonds its structural elements, increases the bond between the gland and net layers, and in this manner, eliminates leather breakage in the finished product. In addition, the unpigmented ground coating provides improved adhesion of the main coating.

Unpigmented ground coatings are prepared from various latexes, which are diluted with water and contain penetrating agents, which increase the penetration of the ground coating into leather. Penetrating agents consists of organic water-miscible solvents, such as alcohol, nonionic or anionic surfactants, and water. These components are mixed in various ratios: solvent 25 to 40%; surfactants 10 to 15%; the remainder, water. Polymer concentration in the ground coating is 7 to 10%. Depth and rate of penetration depend on the coating's capability of wetting leather, the density of the leather upper layer, its viscosity,

and the chemical nature of the polymer. Penetrating agents improve the wetting capability. The penetrating capability of ground coatings can be characterized by the time required for a ground drop (0.1 ml) to be absorbed by leather. Absorption time of 3 to 8 sec is acceptable. A ground coating penetrates to one-third of the leather thickness. The composition of ground coatings is selected empirically, because the relationship between the coatings's reinforcing properties, its adhesion, rate of penetration, wetting capability, and emulsion particle size is not easily defined.

A ground coating recommended by Reichhold Chemie is of the following composition, in parts by weight:

- Durlin-binder 5318, 200
- Durlin-binder 5534, 70
- Durling-penetrator RCH-30, 130
- Water, 600
- Stall recommends the following ground coating:
- Resin 6509, 70
- Resin 6582, 30
- Penetrator 6507, 60
- Water, 280
- The consumption of ground coatings is 150 to 200 g/m 2.

After application of the ground coat, leather is allowed to relax and to dry. Drying conditions are very important. Leather is dried horizontally at a slow rate. The temperature should not exceed 30°C, and only a mild air circulation should be used. At such conditions, water evaporation from the surface and its migration into internal capillaries of the leather proceed slowly and do not cause the migration of dispersed polymer to the surface. The polymer remains uniformly distributed in the gland layer of the leather.

In addition to aqueous polymer emulsions, polyurethane solutions, capable of forming elastic films, are used. If polyurethane has free isocyanate groups, formation of chemical bonds between isocyanate and functional groups in collagen may develop. This increases the interaction between the ground coating and fibrous structure of the gland layer of the leather.

Aqueous Pigmented Coatings

Aqueous pigmented coatings are the most widely used coatings in leather manufacturing. In these coatings, the film former is dispersed in water. They are prepared by blending pigment concentrates (paste consistency pigment dispersions, including a binder, which acts as a protective colloid with latex), various additives, and water to dilute to the required pigment concentration. Aqueous pigmented coatings are complicated colloidal systems; therefore, their stability is especially important. They must not coagulate when diluted with water, when subjected to a mechanical force (mixing, pumping, applying by brush), and under the addition of some electrolytes. It is important that the pigments be protected from flocculation and that they do not precipitate. The protective colloid and emulsifier must be compatible with the latex film former and must form a uniform structure throughout the coating volume. The protective colloid function in various pigment pastes is performed by ammonium or sodium caseinates, methyl cellulose, carboxymethyl cellulose, or acrylic carboxylated copolymers. Film formers are acrylic and diene copolymer emulsions.

When formulating coatings, it is important to select components to ensure that the coating is elastic and resistant to aging, and that the pigments are uniformly distributed.

Acrylic and Diene Latexes

Latex may be blended by employing polymers that form soft and tacky coatings with polymers that form stronger and harder coatings. The elasticity temperature range may be expanded into lower temperatures by blending acrylic copolymers with diene latexes. However, diene copolymer latex films are less resistant to light. Therefore, acrylic latexes are more suitable for white coatings.

Diene copolymer latexes are prepared by copolymerizing various diene monomers with acrylic or methacrylic acid esters. Such useful copolymers are methyl methacrylate-chloroprene (30:70), methyl methacrylate-butadiene-acrylic acid (35:65:1.5), piperylene-acrylonitrile-methacrylic acid (68:30:2), and many other copolymers. Films from these copolymers retain their elasticity at least down to –20°C and are useful for blending with acrylic latexes to extend their low temperature flexibility.

Casein

Casein is a protein prepared from milk. It is soluble in dilute alkalies. It is used as a binder in the preparation of pigment concentrates and also in casein and combined casein–emulsion coatings. Modified casein is a methylacrylate and ammonium caseinate emulsion polymerization product used as an additive in coating compositions with other, usually acrylic, latexes. Films of modified casein are elastic (elongation of 600 to 900%), strong (tensile modulus at failure 6 to 8 mPa), and soluble in water. However, they may be easily rendered hydrophobic by treatment with formaldehyde or solutions of polyvalent metal salts. Butadiene–ammonium caseinate copolymer latex has similar properties.

Wax Emulsions

Wax emulsions are water-dilutable dispersions at pH 7.5 to 8.5 and are stabilized with nonionic surfactants. The basis is usually montan or carnauba wax. Emulsions are used as additives to pigment and top coatings.

Pigment Concentrates

To obtain well-colored leather, it is important that pigments be well dispersed in the binder and that a strong bond is formed between pigment particles and the binder. Direct pigment dispersion in the coating is difficult. Aqueous polymer emulsions used for leather coatings are not sufficiently viscous to maintain a uniform distribution of pigments. Furthermore, emulsions might not be stable enough to allow a direct addition of pigment. Therefore, the pigment is dispersed separately in the binder solution, yielding a stable dispersion that can be safely blended with film forming emulsion and other additives to ensure the stability of the heterogeneous system. The pH should be similar in the two dispersions.

Pigment concentrates, depending on the binder used, can be of varying composition: casein, where the binder is an aqueous alkaline casein solution, or a synthetic polymer base, mainly acrylic.

A pigment concentrate in casein may have the following composition (in parts by weight):

- Pigments, 14 to 60
- Casein, 3.8 to 8.6
- Oil of alizarine, 2.4 to 4.0
- Emulsifiers, 0.5 to 1.0
- Antibacterial agent, 0.5 to 0.9
- Water, up to 100

To prepare such a composition, casein glue is made up first (18 to 20%); then antibacterial agent is added, followed by other additives. The pigment is dispersed employing suitable equipment until a stable dispersion is obtained. Casein binder is suitable for the dispersion of all pigment types.

In addition to casein-based pigment concentrates, pigment dispersions based on acrylic polymer are used. Pigments are dispersed in a thickened acrylic emulsion. If the film former in the coating is acrylic latex, it mixes well with such pigment dispersions. Acrylic dispersions also are more effective in improving the coating elasticity, as compared to casein dispersions. The composition of pigment dispersions in acrylic latex may be as follows:

- Pigments, 11.2
- Blend of two or three acrylic emulsions, 85.3
- Ammonium hydroxide (25%), 2.3
- Oil of alizarine, 0.7
- Surfactants, 0.5

Pigment concentrates based on acrylic emulsions are of a viscous paste consistency; they are easily dilutable with water, and they blend well with all aqueous emulsion film formers. In addition to pigments, dyes may be used in coatings. In the preparation of pigment or pigment–dye blend coatings, attention must be paid to the pigment properties — their resistance to light, their opacity, and the elimination of such side effects as bronzing.

Nitrocellulose-Based Compositions

These products are used for nitrocellulose coatings, but most frequently, for top coatings over coatings of other types. Nitrocellulose solutions (lacquers) in organic solvents, or solution dispersions in water, are used. Nitrocellulose lacquer is a solution of nitrocellulose in organic solvents and diluents compounded with plasticizers. Nitrocellulose is available in alcohol-soluble and -insoluble forms. The latter is used for leather coatings. Each type is available in several viscosity grades, depending on the molecular weight of the nitrocellulose. A compromise is usually made between the coating's physical properties, which improve with increasing molecular weight, and coating solids, which decrease with increasing molecular weight for a solution of required viscosity.

The solvents used are ethyl and butyl acetates, acetone, and methyl ethyl ketone. Alcohols (ethyl and isopropyl), while not solvents by themselves; enhance the solubility of nitrocellulose in other organic solvents. Diluents are miscible organic liquids that do not dissolve nitrocellulose but decrease the solution viscosity. They are also less expensive than true solvents. Such diluents are toluene, xylene, and some aliphatic–aromatic hydrocarbon blends. The choice of solvents/diluents for nitrocellulose lacquer is determined by economics and by such properties as sufficiently low volatility, lack of water absorption, or capability to form azeotropic blends with water. For film formation it is important to have an optimum amount of alcohol, which has a relatively low volatility.

Nitrocellulose is brittle, and therefore plasticizers are used in compounding nitrocellulose coatings. Plasticizers used are alkyl phthalates, castor oil, camphor oil, and others. Nitrocellulose lacquer is a clear, water-white, easily dilutable, viscous liquid containing 15 to 18% solids. The tensile strength of nitrocellulose film is 1.5 to 1.8 mPa; elongation at break is 50 to 60%.

Aqueous nitrocellulose dispersions also contain some organic solvents, which facilitate the coalescence of nitrocellulose lacquer particles. Film formation from nitrocellulose dispersions that do not contain any solvent is difficult. Both types of dispersion are used: oil in water and water in oil. Nitrocellulose coatings are used as top coatings over aqueous emulsion coatings. The mechanical properties of nitrocellulose films obtained from aqueous dispersions are poorer than those obtained from solutions.

Leather that does not require vapor and air permeability (e.g., leathers used for applications other than footwear or clothing) may be coated entirely with nitrocellulose, starting with the ground coat and ending with the top coat. For the ground coat, aqueous nitrocellulose coatings are mainly used; the main coat consists of pigmented nitrocellulose enamel, and the top coat is a clear nitrocellulose coating.

Polyurethane Coating Compositions

Coatings described here are used for all polyurethane coatings and also as top coats for coating of other types. Polyurethane solutions in organic solvents and aqueous dispersions are used. Coatings of this type have a good adhesion to leather (because of the urethane groups). Polyurethane lacquers may be clear and colored (pigmented or dyed). Many different polyurethane structures are available: tridimensional, branched, or linear. Thus, coating properties may be obtained that exhibit variations in elasticity, resistance to solvents and abrasion, and thermal and light resistance. The main disadvantage of these coatings is that they mask the appearance of the leather surface and contribute to the artificial appearance of the product. This is mainly the case for coatings consisting entirely of polyurethane, but not for top coatings.

Polyurethane coatings are also used to prepare varnished leather. Polyurethane leather coatings are made from ester oligomers (molecular weight of about 2000) and isocyanates. Polyurethanes may vary widely in their properties. If oligomers have more than two hydroxyl groups, the resultant polyurethanes are three-dimensional structures and have a high chemical and temperature resistance. If ether oligomers or polyethers (polypropylene glycol, polytetramethylene glycol) are used, the resultant polyurethane film has a higher resistance to hydrolysis and better low temperature properties, retaining the elasticity at lower temperatures.

The synthesis of polyurethane is carried out in a solution of organic solvents, such as methyl ethyl ketone, dimethylformamide, butyl- and ethyl acetate, and their mixtures. Two main polyurethane polymerization processes are used: single stage and two stages. The two-stage process begins with the reaction between ester oligomers and isocyanate, which yields a low molecular weight prepolymer. In the second stage, under the effect of a chain extender (1,4-butanediol, diamines), the chains grow. In the manufacture of varnished leather, the second stage is usually carried out after coating application to the leather surface.

Polyurethane coatings can be applied by spraying or by flow coating. The coating is dried for about 16 hours at 30 to 40°C and low relative humidity (40 to 50%), with the leather sheets in the horizontal position.

Coatings with a metallic sheen may be prepared by the addition of metallic particles, prior to the application of a clear top coating.

A single-component polyurethane lacquer is a solution of thermoplastic, high molecular weight, linear polyurethane in organic solvents. Such lacquers are produced from linear polyesters and diisocyanates, and in dilute solution they are useful as top coats. Thermoplastic polyurethane films are elastic, their elongation at break reaches 1000%, and the elastic properties are retained over a wide temperature range (–30 to 120°C).

Aqueous polyurethane dispersions are obtained by the dispersion of polyurethane in water, employing emulsifiers. Aqueous dispersions are useful for pigmented coatings: ground coats, main color coat, and also for top coats.

Mixtures of polyurethane and nitrocellulose dispersions are also used as top coats. Such top coats dry faster, but their mechanical and aesthetic properties are poorer.

TECHNOLOGY OF DECORATION OF SKINS OF LARGE HOOFED

Animals with Artificial Grain

The processing technology becomes more complex when it is required to form an artificial grain. The natural grain is ground down and a coating is applied in several applications. An artificial grain, including one imitating natural leather is formed.

This process starts by grinding the grain surface two or three times. Grinding is performed by employing special grinding–polishing equipment and using #5 to #6 abrasive paper. After grinding, the leather surface is coated with an unpigmented ground coat of the following composition (in parts by weight):

- Acrylic dispersion (40%) solids, 100
- Penetrator, 40
- Water, 300

An unpigmented coating is applied by a curtain coater. Its consumption is about 0.2 to 0.3 kg/m 2 depending on the density of the leather's glandular layer. The leather is then stored for 8 to 24 hours, dried, stretched, pressed, and ground again. Pressing before grinding improves the grinding–polishing quality. The leather is dyed by running it twice through special equipment, which also involves drying, initially at 110°C, and then at 80°C at the end of the drying tunnel. The drying cycle lasts 1 to 2 min.

In the next step the leather is pressed at a plate temperature of 50 to 70°C and a pressure of 14.5 mPa. One or two additional color coatings are applied until the surface is sufficiently well covered. Light and intense colors may require three or four coating applications. The leather is dried at 65 to 75°C after each coating and pressed at a plate temperature of 65 to 75°C and a pressure of 14.7 mPa. After this smooth press, the grain is pressed in by a hydraulic press at a temperature of 70 to

80°C and 14.7 mPa pressure. The process is ended by spraying one or two layers of top coat. Nitrocellulose lacquer (or emulsified lacquer) or polyurethane lacquer is used for the top coat.

SOME NONSTANDARD COATING APPLICATIONS

Rohm & Haas Company suggest employing foamed color coatings. Such coatings consist of few components: latex film former, pigment dispersion paste, and foaming agent. These coatings contain 50% solids and are applied to a coating weight of 0.4 to 0.5 kg/m 2. The application technology is not much different from the standard application techniques: drying, pressing, and coating spraying are employed. The coating is elastic and resistant to flexing (endures 20,000 to 40,000 flex cycles). The method is more economical than traditional techniques.

Another method is casting on a silicone elastomer matrix with a negative leather grain. A polyurethane coating is applied (a prepolymer containing free hydroxyl groups and an isocyanate cross-linker). Coating thickness varies from 0.15 to 1 mm. Leather (which also could be split leather) is pressed against the uncured coating and after a contact time of 10 to 15 min, it is dried in a drier, where a well-adhered coating is formed. After such curing, the leather is removed from the matrix.

METAL COATINGS

Metallizing

The objective of metallizing techniques is to place metal on the substrate for appearance or protection of some sort. The classes of metallization are many and complicated, but may be separated by their process details. Processes that apply metal to surfaces may use metal as individual atoms or ions, as the fluid molten metal, or as the solid metal. We deal with each separately.

Liquid Metal Processes

Galvanizing

The metal item (iron or steel) is dipped into a molten bath of zinc, and then withdrawn, and the excess zinc allowed dripping off. The item is cooled and the zinc crystallizes on the surface (giving an appearance called "spangle"), with the cooling rate determining the size of the zinc crystals showing on the surfaces. This process can be made continuous for rod, wire, coiled sheet or pipe, and semicontinuous for reinforcing rod, pipe, and cut sheet.

The process is not quite as simple as it sounds. There are iron/zinc compounds that form at the molten interfaces. The time of heating a molten zinc–iron interface governs the thickness of the interface containing these compounds, and the ratios of iron and zinc in the compounds at the interface. In addition, sal ammoniac (ammonium chloride) is used as a flux in the molten zinc, but it can appear as a blue coating or streak on the galvanized item, to the item's detriment. That "sali"-contaminated item should be recoated.

The zinc layer on the item acts as a physical barrier to corrosion. However, as soon as there is physical damage to the zinc layer, exposing the iron, the zinc acts as a sacrificial anode, giving the iron electrolytic corrosion protection, as well. Since zinc is soft and easily corroded, it will "wear" away, showing the white zinc oxide surface as the zinc layer becomes thinner. The white zinc oxide has substantial water solubility and washes away in rain. The zinc surface is often painted to give protection against these losses.

Flame/Plasma Spray

Although all the flame and plasma spray processes project liquid metal droplets through the air to the substrate, each starts with a solid metal wire or solid molten powder. The solid is taken into a device which heats it to the molten state, breaks it into microscopic droplets, and propels the droplets at the substrate. Longo edited a reference book on this topic.

The "flame" or "thermal" spray uses a modified oxyacetylene torch as the heat source. After the flame has been adjusted to its hottest, additional compressed air is blown into the flame. The wire or powder is then fed in via a funnel, and the blast of liquid metal particles is pointed at the substrate. The "plasma" spray uses an electric arc as the heat source to melt the wire or feed powder into the compressed airstream.

The distance between the molten metal's origin and the substrate will determine the type of coating obtained. The close approach of the nozzle means that most particles will hit the substrate surface as a liquid. Many will stick, and transfer their heat to the substrate, but some particles will bounce off. If the nozzle is further away, the smaller liquid particles will solidify and bounce off the substrate, while the larger particles will still stick. If the nozzle is too far from the substrate, none of the particles will stick because they will have cooled too much. Losses from bounce-off or from premature cooling will be on the order of 25% of the weight of the metal sprayed.

The metals most commonly sprayed include copper, zinc, iron, and aluminum. Alloys and even metal oxides can also be sprayed, provided there is enough heat in the flame to make the particle soften. Since the sprayed metal is of lower density than the solid metal, there is occluded air, and even porosity.

However, adhesion to most substrates is good, and is often assured by cleaning and roughening the surface (e.g., sandblasting). The inorganic coating may be built up enough to be machined, and may be strong enough to be a bearing surface.

Other Liquid Metal Coatings

Any solid can be dipped into a molten metal.

Solid Metal Processes

Sherrodizing

An item of steel or iron is put in a drum with powdered zinc and steel balls, and the drum is rotated. The zinc powder is essentially hammered onto the surface of the steel/iron item as individual spots. The length of time in the rotating drum, the amount of zinc powder, the number of steel balls, and the number of items to be treated all govern the actual amount of zinc that ends up on the surface of the item to be treated. The zinc coating on the iron/steel acts as a corrosion (rust) inhibitive protection for the item.

"Detaclad" Process

The construction of metal laminates, such as the coinage products now used by the U.S. Treasury, is a simple process involving a layer of metal applied to another metal surface by the force of an explosion on one surface. If silver is needed as an outer surface over copper, the explosive is on the side away from the copper in the setup, as shown in Figure .1.

VAPOR PROCESSES

Vacuum Evaporation

The simplest of the vapor processes is vacuum evaporation. The items to be coated are put on racks that circle a central set of trays formed from electrical resistance heaters. A bell jar is lowered over the whole array, and sealed to be pumped down to 1 mm Hg of vacuum. The resistance heaters are fired off to evaporate the metal powder or slug in the tray, and the individual atoms of metal fly off in a straight

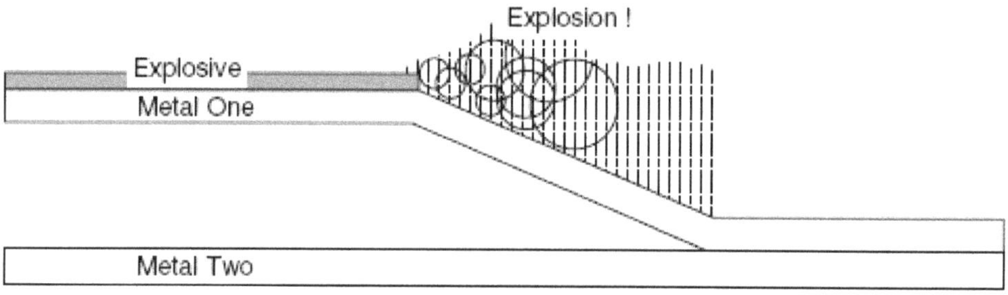

Figure .1 — Explosive bonding of metal

line toward the bell jar in all directions. The items to be coated are hit by these atoms in a line of sight from the resistance heater tray. The atoms condense on all surfaces facing the heater trays, forming discrete globules at first, and covering the whole surface as the globules coalesce.

The items to be coated must be treated to make sure the metal wets and adheres. Such treatments might be simple washing or precoating with a polar lacquer, or as complicated as a plasma etch or a corona discharge surface treatment. Even nonpolar materials like poly(propylene) or acrylonite-butadiene- styrene resins can be successfully coated with good adhesion by vacuum evaporation. Common items so treated include toy cars and lipstick cases.

Good continuous film coatings can be formed on just about any surface, at thicknesses ranging from micrometers to millimeters. Indeed, the transparent gold coating put on aircraft windshields for deicing/ defogging is about 10 nm thick. This coating MUST be continuous, as any microhole in the coating would cause a blue “burn” spot as soon as current was applied.

Almost any metal may be vacuum evaporated, though some metals sublime, rather than undergo the preferred heat/liquefy/boil process. The subliming metals do not do quite the line of sight depositions desired, and the vacuum fittings can become plugged. The most common are aluminum, copper, gold, and silver.

Since the metal deposited by this process is in the bare, unprotected, metal atom form, it may corrode quickly if it is so disposed in the electrochemical series. One must apply topcoat for corrosion protection immediately. Sorg described vacuum metallized parts for automobiles, while others compared a number of different approaches for automotive bright work.

Sputtering

The key elements in the sputtering process make it different from simple vacuum evaporation. The metal is vaporized by an electron beam, which charges the metal atoms or droplets with a negative charge. The substrate is made the cathodic target (or is directly in front of a cathodic target). Thus, electromotive force is the driving force coating the substrate. It is used quite successfully to coat Mylar films for sunshade applications with copper, aluminum, and even stainless steel. Sputter plating onto actual polymers has been described from plumbing fixtures.

Chemical Vapor Deposition

A reference book on plasma deposition has been edited by Hochman and others. The substrate is placed in a container that is then evacuated, and a gas or gases are drawn in. The container has a cylindrical electrode just within the inner surface, and this is to react with each other or the surface of the substrate.

Monomers used in plasma deposition of polymer include methyl methacrylate, propylene, acrylic acid, acrylonitrile, norbornadiene, hexamethyldisiloxane, tetrafluoroethylene, ethylene, and styrene. Nonmonomers (such as pentafluorobenzene, methane, or tetrafluoromethane) also may deposit polymeric coatings. Colloidal metal may be included within the polymer film to generate a colored coating.

The advantage of plasma coating is the very thin film formed, with no pinholes and no contaminants (i.e., the additives needed in solvent or waterborne coatings).

Plating

Electroplating

Plating on plastics has been reviewed by Saubestre and by Muller and associates. Springer and associates described details of polymer treatments and morphology as important in electroplating polymer surfaces. Safrenek detailed the properties of electroplated metals and alloys. A zinc–nickel alloy electroplated onto steel is reputed to have better corrosion properties than galvanized steel.

Electroless Plating

The handbook cited earlier is a good reference for electroless plating as well. The idea is simply to put a metal solution on a surface, and let it deposit the zero valent metal on the surface because of an added chemical. It turns out not to be quite so simple, because addition of the reducing agent to a solution does not guarantee that the metal will deposit on the surface, and it does not mean that the deposit will adhere if deposited. Hence, there are sequences of "washes" and "activators" that prepare the substrate to accept the final plating formulation. In the case of plating on plastics, several oxidative techniques (chromic acid solutions, plasma etching, nitric acid washes, etc.) are used to prepare a surface. One process uses three deposition steps, with clean water rinses between, to end up with copper on plastic.

The three solutions are tin chloride, palladium chloride, and a formulated copper solution that has reducing agent along with rate modifier and surface modifier chemicals as well. Rigorous rinsing between the metal solutions is needed to assure that the last solution does not become slurry of colloidal copper, because of "dragout" or contamination by preceding metal ions. The electroless plating deposits are continuous, conductive, and bright, but are thin — micrometers in thickness.

Coating on Metals

Coatings for metals may be divided into two classes by main function: those that are decorative, and those that are protective. That is not to say that a protective coating cannot be decorative, but that the coating's main function is protective, while it is chosen to be decorative as an additional option.

Decoration

The decorative aspect of coatings lies in several features that are dealt with elsewhere in this book. Among those aspects are color, gloss/flatness, and texture. Each entails specific approaches the formulator uses to attain the desired appearance, mainly involving choices of vehicle (the adhesive that sticks the pigment particles together and then to a substrate), pigments, the ratio of pigments to vehicle, and (occasionally) the additive that confers a certain property when pigment or vehicle cannot.

Yet another aspect of decoration is pattern. If and when there are multiple colors on a surface, the shape described at the color interfaces can be important. The camouflage paints strive to have what appear to be random splotches, because a regular geometric pattern (especially a straight line) attracts the eye to a potential target. The multicolor paints have the dots, splotches, smears, and lines simulating natural mineral or rock formation surfaces. And the pattern is a main point in signage or artwork.

Protection

The metal substrate is generally thermodynamically unstable and is easily converted to the more thermodynamically stable oxide. However, the protection provided to the metal substrate is often designed to guard against sorts of attack other than chemical oxidation. Mechanical damage may aid the oxidation, by providing sites for the oxidant to work. Electrical exposure or damage can ease the chemical degradation.

So, work on protecting a metal substrate must include consideration of the insults or attacks to be expected. In the instance of corrosion, the protection may aim to prohibit contact of the coating with oxygen or water — each a necessary element of the corrosion reaction. The aim may also be to prohibit contact with corrosion catalysts — salts, acids or alkalis.

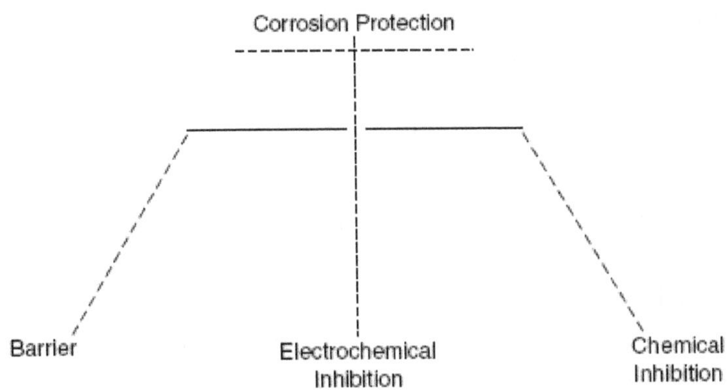

Figure .2

At the National Coatings Center, we divide a "pitchfork" diagram (Figure .2) to describe the three major corrosion protection schemes. The main emphasis was on improving ways to give corrosion protection through combinations of some of the corrosion protection schemes.

Examples of the barrier coatings are fairly straightforward. As noted earlier, the zinc in galvanizing is initially a barrier, and any impervious coating is a barrier, be it wax, asphalt, coal tar, polyethylene, or whatever. There are selective barriers that aim to protect against corrosion by blocking a specific corrosive element. The wax, polyethylene, coal tar, or asphalt is specifically aimed at prevention of water permeation to the metal surface, as water is a specific corrosive substance. Other coatings (PVDC, Barex-type nitriles, etc.) aim to prohibit oxygen permeation. In both cases, the coatings are diffusion barriers, and the key is to have a coating that dissolves as little as possible of the permeant within the coating, and also inhibits permeation. A highly polar coating material is always more water permeable than the nonpolar material, because water dissolves so well in the polar material that the polar material acts as a pipeline rather than a barrier. It is a chuckle to hear of silicones or acrylates described as waterproofings, but you have to understand the implied statement that they protect against liquid water, while water vapor goes through them 100 to 1000 times faster than it would go through a hydrocarbon barrier.

The electrochemical protection schemes are bound up in converting the iron or steel into a cathode, since the corrosion reaction is an anodic oxidation of the zero-valent iron metal to ionic forms (usually ferrous). The easiest way is to simply contact the metal with something that is oxidized more easily (zinc, magnesium, aluminum, etc.) and let the iron be the cathode while the other metal is corroded as the sacrificial anode. Indeed, there is a substantial market in bars of magnesium or zinc that are attached to iron (pipelines, underground tankage) to act as sacrificial anodes. We already noted that galvanized iron with the zinc surface damaged to penetrate to the iron is still electrochemically protecting the iron. The "zinc-rich" paints (having about 85% by weight of zinc metal powder and 15% binder) are also cathodic protectors of steel. But the sacrificial anode does send its corrosion products into the surroundings. In a tidal area, the sacrificial anode specified by some regulatory agencies feeds metal ions into the underground water while it is protecting the underground storage tank.

An alternative in cathodic protection is to impress a current onto the potentially corroding metal to make it cathodic. Here a battery or a step-down transformer is used with another power source. This technique is used on shipboard for iron hulled vessels, and the battery can be recharged when needed. Some pipelines are cathodically protected with impressed current. The advantage is that there are no ions lost to the surroundings, hence subsequent contamination of water.

The chemical inhibitors are many and varied items of commerce and may work in many ways. Most are sold because they have been shown to work, though no mechanism is proposed. Some of the surfactant-type materials (oleyl sarcosine, for instance) may only add a physical barrier by adsorbing to the surface and blocking approach of oxygen, water, or catalytic ion. Other materials may adsorb on the metal and act as a pH modifier or buffer, as an amine would inhibit acid-catalyzed attack. Some inhibitors modify the electromotive potential at which corrosion occurs and are said to have "passivated" the surface.

Most of the chemical inhibitors are low molecular weight compounds and can be washed away or otherwise rendered ineffective by chemical attacks or reactions. They are effective over short periods of time, and can be stabilized against erosions by formulation to some degree. For instance, there are oils into which corrosion inhibitors are formulated.

There are corrosion-inhibitive pigments. Seldom is there discussion of the mechanism by which these chemical inhibitive pigments work, and such characterization could extend their utility. It has been shown that nominally corrosion-inhibitive pigments do not work in some formulations. Indeed, Nadim Ghanem (American University of Cairo) told of a basic lead carbonate formulation study in which the solvent-borne vinyl binder gave no corrosion protection, while a waterborne vinyl gave excellent protection. His hypothesis was that the coating needs to be permeable to water to get the lead ions to migrate to where they need to be to act as corrosion protectors. That may have also been the message in the Los Angeles Society for Coatings Technology work with aminosilane-treated talcs in latex formulations. Though many demonstrations of the effectiveness of the corrosion-inhibitive pigments exist, the mechanisms should be more thoroughly described to aid the formulator.

OPTICAL FIBER COATINGS

Introduction

The concept of high speed data transmission on a beam of light was demonstrated by Alexander Graham Bell in 1880. His "photophone" used mirrors and sunlight to transmit low quality voice messages. Today's technology has developed to a point of passing extremely high data rates (>2 Gbit/s) unrepeated through a high purity quartz fiber for distances exceeding 100 km. Optical fibers offer many advantages over copper conductors and are gaining acceptance for the following reasons:

1. They carry more signals for greater distances without repeaters. A fiber with a 1 GHz bandwidth can transmit several thousand simultaneous telephone calls through use of state-of-the-art multiplexing techniques. Undersea cables have repeaters spaced as widely as 30 to 55 km, whereas copper cables require signal regeneration every 2 to 5 km.

2. Electromagnetic interference does not occur.
3. Cross-talk is eliminated because the optical signal is maintained in fiber and "short-circuits" do not occur.
4. Weight and volume of fiber cables are greatly reduced (for the same signal carrying capacity).

These advantages have led to the development of numerous optical fiber telecommunication networks, with an increasing awareness that the initial growth of this technology for long-distance trunk lines will continue to accelerate as fiber replaces copper wire in the subscriber loop (residential and business installations).

Optical Waveguide Principles

In any transparent material, light travels more slowly than in a vacuum. The ratio of these speeds is the index of refraction. When light, traveling in a material of one refractive index, strikes a material having a lower refractive index, the light is bent back toward the higher refractive index material. This phenomenon can be used to guide a light down a high purity glass rod, if the rod is clad with a material having a lower refractive index. Current telecommunication fibers employ a doped (germanium, phosphorus, etc.) silica core in combination with a lower refractive index (n = 1.46) silica cladding. The core will generally vary in thickness from 8 to 100 μ m, whereas the outside diameter of the cladding is usually 125 μ m. When made from pure silica, the resultant fibers have excellent strength (to 14,000 N/mm 2) 1, but they rapidly degrade as a result of the development of microscopic scratches, which are subject to growth, resulting in breakage when stressed (stress corrosion).

Coating

The application of a protective coating is required to preserve the optical fiber's strength and to protect it from lateral deformation, which can result in a reduced light signal (attenuation).

Attenuation is one of the major concerns of a fiber user. This is a measure of the light lost in passing through the fiber and is described in electronic terms as noise (i.e., decibels per kilometer: dB/km). This loss results from light scattering due to fluctuations of glass density, absorption due to impurities, and radiant signal loss due to microbending, the result of unequal forces or distortions of the glass fiber core.

Long-term durability of the fiber and its ability to transmit a signal of high fidelity is of paramount importance. Lifetimes of 25 to 40 years are generally expected, even though the anticipated environments vary from tropical climates to arctic winters to undersea water pressures. In addition, the fiber may be exposed to groundwater (having a pH of 2 to 12), solvents, steam, stress, and rapidly fluctuating temperatures.

Under conditions of high stress and humidity, silica glass is known to undergo rapid stress corrosion through a process of hydrolysis, resulting in premature breaks of the communication link. To avoid this difficulty, manufacturers minimize flaws and conditions during fiber manufacture. To protect the glass, a protective buffer coating is applied within 1 or 2 sec of drawing, or as soon as the glass has cooled to below 100°C. The buffer coating is designed to protect the glass from being scratched. However, it is mandatory that the coating be free of any particulate matter, which may cause microscratches on the glass surface, forming a site for crack propagation.

Handleability

To keep up with increasing demand, the coating or coating system must lend itself to high production speeds. During the commercial development period of the late 1970s, typical production speeds varied from 0.5 to about 1 m/sec. At these slow speeds several coatings were found to be acceptable. Twocomponent polysiloxanes were found to offer the best overall properties, namely good strength, and minimal temperature dependence. (Their flexibility and modulus remain largely unchanged over a temperature range of - 55 to 85°C.) Their durability is good, and their low modulus properties provide a soft coating, which protects the glass core from physical distortions and microbending. Major deficiencies of the silicones include relatively limited application and cure rates, high hydrogen generation, limited pot life (when catalyzed), and a high coefficient of surface friction, which requires overcoating. Other early coating systems included cellulose-based lacquers, Teflon extrusions, and polybutadiene rubbers.

COATING REQUIREMENTS

In 1977 the demand for increased volume and productivity resulted in the development of ultraviolet (UV) light curable acrylate coatings, which met the needs for strength, longevity, and performance while also providing production speeds of 5 m/sec and faster. 2 - 4

By using combinations of acrylate-functional oligomers and monomers, it is possible to formulate coatings having a wide range of useful properties, including the following:

- Modulus, 1 to 1000 MPa (125 to 150,000 psi)
- Elongation, 10 to 250%
- Hardness, 35 Shores A to 70 Shores D
- Tensile strength, 0.5 to 40 MPa (70 to 6000 psi)
- Low water absorption (<3%)

It has long been theorized that an optimal fiber coating system should consist of a very soft, low modulus, particulate-free coating against the glass to provide for surface protection and microbending resistance. Microbending is the attenuation of the light signal caused by small bends of the core, resulting from nonuniform stresses on the fiber. The primary coating is followed by a harder, tough, secondary buffer to provide mechanical protection and water resistance, while offering a hard, slick surface that is compatible with subsequent cabling operations.

It is desirable that these coatings be applicable in-line, at maximum possible draw rates, and at several different outside diameters. Coating thicknesses may be from 3 μ m. to nearly 400 μ m. The properties usually desired and generally provided by multifunctional acrylate coatings include the following:

1. Primary buffer coatings

Good adhesion to the glass. Buffer removal or stripping can be accomplished with the use of solvents such as methylene chloride, or short lengths of coating can be removed mechanically.

Another construction might have minimal bonding to the glass to facilitate rapid, safe mechanical stripping of the buffer during installation.

Low tensile modulus over a range of operating temperatures (55 to 85°C). Long stability with excellent resistance to oxidation, hydrolysis, saponification, and UV light.

A satisfactorily high index of refraction, so that the coating can be applied using laser forwardscattering for concentricity control, and an index of refraction sufficiently higher than the cladding that light launched into the glass cladding will be stripped from the fiber (mode stripping) ($n > 1.48$).

Low glass transition temperature (g) to allow the primary coating to function without going through a glass transition phase that will cause differential expansion and contraction, resulting in microbending. Low generation of hydrogen during the fiber's installed lifetime.

2. Secondary buffer coatings

Low surface free energy (i.e., low coefficient of friction and tack) as to allow for easy handling in the cable plant.

Low surface tack, to prevent binding or coupling of the coated fiber to the walls of a "loose tube" cable.

Good intercoat adhesion characteristics when overcoated with other components of a cable system (i.e., tape or extrusion coatings such as nylon, Hytrel Tefzel, etc.).

Excellent resistance to environmental factors such as water, acids, bases, solvents, and fungus growth.

High modulus to resist external lateral forces that will deform the light path. High T g to avoid differential expansion and contraction during thermal cycling. Low volatility to resist outgassing during hot extrusion applications to tightly buffered cable processes and longterm changes in physical properties.

The ability to be colored or to accept coloring, which permits individual fiber identification. Resistance to the chemical gels used to fill "loose tube" cables.

Composition of Fiber Coatings

More than 75% of the world's quartz optical fiber is coated with acrylate-based materials and cured with UV light. The classical reaction chemistry uses photoinitiators to absorb photons generated by medium pressure mercury, xenon, or doped mercury lamps. For full cure in less than 1 sec, 1/cm 2 of a wavelength of 300 to 390 nm is generally sufficient. Free radicals, which are produced by this reaction, cause rapid addition polymerization of the unsaturated acrylate groups. The proprietary chemistry is in the design and choice of the acrylate oligomers and reactive monomers. The oligomers are generally based on a range of polyesters, polyethers, and epoxy polymers, which have been esterified with acrylic acid (coreacted with a diisocyanate and a hydroxyfunctional acrylate). The reactive diluents are both mono- and multifunctional. Solvents are used only rarely for viscosity control. Stabilizers, inhibitors, flow aids, and a wide assortment of additives are used to gain required properties. Secondary buffers may include pigments or soluble dyes for color identification.

Two-component polymethylsiloxane coatings are used in decreasing quantities for quartz telecommunication fiber. A phenyl variation provides the high refractive index required for a primary buffer, but at a significant cost disadvantage. The usual low refractive index makes silicones the coatings of choice for most all-plastic fiber and high loss, "plastic clad silica" (PCS) fiber. Physical properties of the silicones are excellent, but line speeds are generally limited to speeds well below 2 m/sec. Silicones are also difficult to mechanically remove from glass fiber for splicing, because they leave a residue.

Fluorine- and silicone-modified acrylates are used for specialty applications, as are polyimides for high temperature resistance. A new class of inorganic, hermetic coatings is being commercialized to provide stronger, more durable fiber. These include various vapor-deposited coatings such as reactive silica and titanium nitrides and carbides, as well as molten aluminum. Gold and indium have also been used.

PHARMACEUTICAL TABLET COATING

History

The coating of solid pharmaceutical dosage forms began in the ninth century B.C., with the Egyptians. At that time the primary solid dosage form was the pill, a hand-shaped spherical mass containing drug, sugar, and other diluents. A variety of materials were used to coat pills, such as talc, gelatin, and sugar. Gold and silver were also used. Many of these coatings proved to be impervious to chemical attack in the digestive tract; as a result, the pill never released its active ingredient and was thus ineffective.

The candy-making industry was the first to develop and enhance the art of coating. It is most likely that the pharmaceutical industry adopted sugar coating technology for its own use. The first sugar-coated pills produced in the United States came out of Philadelphia in 1856. Coatings resistant to enteric or gastric fluids were developed in the 1880s. In 1953 the first compression-coated tablet was introduced, and in 1954 the first film-coated tablet was marketed.

Reasons for Coating Tablets

There are many reasons for coating tablets; some aesthetic, some functional. One important reason is to enhance drug stability; that is, to protect the drug from oxygen, moisture, and light, the three key causes of drug degradation. Coating can also be used to separate reactive components in a tablet formula.

Another important reason for tablet coating is identification. Tablet coatings may take on a variety of colors. A coated tablet may also be imprinted with a symbol or word. In the case of the film-coated tablet, the tablet core may be embossed with a symbol or word that remains visible after the coating process. The definitive identification of a coated tablet has saved patients and health care professionals alike. Additionally, coating is used to uniquely identify a branded product. Tablet coating is done for aesthetic reasons as well. Often the appearance of the tablet core is mottled or otherwise unattractive. Coating masks this. Many times, too, the drug itself has a bitter taste. Coating masks this as well.

Tablet coating can also be used to control the duration and site of drug release. Overall, tablet coating, through an additional step in the manufacturing process, is often vital.

Types of Coating

There are two main types of tablet coating done today: sugar coating and film coating; film coating is the more popular. Coated tablets fall into three main subcategories depending on how the drug is released: immediate release, enteric release, and sustained release: Immediate-release coating systems, as the name implies, allow immediate release of the drug compound to the body.

Enteric coatings are soluble only at a pH greater than 5 or 6. Thus, the drug is not released in the stomach but in the small intestine. Enteric coatings are by far the most unreliable because of the wide and unpredictable variance in gastric pH profiles. Gastric pH varies considerably based on stomach content, age of the patient, and disease state.

Sustained-release coatings permit drug to dissolve slowly over a period of time. This helps to reduce dosing intervals and improves therapeutic reliability.

Film coating can be carried out using either an organic solvent system, such as ethanol or methylene chloride, or by using water as a solvent. The solvent film coating systems are fast disappearing because of cost, environmental, and safety concerns. Most film coating carried out today is done with aqueous systems.

The Sugar-Coated Tablet

The sugar-coated tablet is the most elegant solid dosage form produced today. Its glossy appearance, slippery feel, and sweet taste are unmatched by any other coated tablet. The sugar-coated tablet is also the most difficult and time-consuming to produce. The tablet consists of a core upon which layer after layer of coating material is slowly and carefully built up. In some cases this is done by hand and in other cases automatically. In any event, there is still an art to sugar coating.

To successfully accept a sugar coating, the tablet cores must be robust. They are subjected to wetting and rolling in a coating pan with 50 kg or more of other cores. Generally the coating pan is spherical and has a solid exterior surface. Temperature-controlled air is introduced and removed from the pan via external ducts. The following procedure is used for the manual sugar coating of tablets.

The first step is to slightly waterproof the tablets by applying a coat of pharmaceutical-grade shellac. This prevents the cores from dissolving prematurely in the presence of the other coating liquids that are to be applied.

The second step is subcoating: a solution composed of acacia, gelatin, and sugar is applied to the tablets. The wetted cores are then dusted with dicalcium phosphate or calcium sulfate and allowed to dry. This step is repeated many times until a smooth rounded tablet form has been achieved.

The third step is the grossing coat. The cores are wetted with a sugar solution and dusted with titanium dioxide powder. This creates a very white base coat on which color may be applied. The fourth step is the color coat. In this instance an insoluble opaque color solid is suspended in sugar syrup and applied to the tablet. No dusting of the cores takes place. The tablets are simply air dried.

The fifth step is the shutdown coat. In this step diluted sugar syrup is applied to the tablet and allowed to dry. This produces a very smooth finish in preparation for the last step.

The last step is polishing of the tablets. The tablets are placed in a canvas-lined drum. Beeswax or carnauba wax is dissolved in methylene chloride, and the solution is applied to the tablets, which are tumbled until the solvent evaporates and tablets achieve a very high shine.

In all, 40 or more separate layers are applied during the manual sugar coating process. The process takes between five and eight 8-hour shifts to complete.

Automated sugar coating is generally faster. For example, the various syrups used in the coating process have the dusting powders suspended in them. The syrups are applied by spray. This process can be automated to reduce the number of operators required. Perforated coating pans, which greatly enhance air throughput, are used almost exclusively. With greater air throughput, water evaporates more quickly, thus speeding the process. Using automated techniques, tablets can be sugar coated in about 16 hours.

The Film-Coated Tablet

The film-coated tablet consists of a core around which a thin, colored polymer film is deposited. Thus, a film-coated tablet gains about 3% of total tablet weight upon coating. The sugar-coated tablet undergoes a 100% weight gain. Overall, film coating is a much faster procedure, and much less prone to error.

The basic film coating formula consists of a film former, pigment dispersion, a plasticizer, and a solvent. A variety of polymeric film formers can be used to coat tablets. By selecting the solubility properties of the polymer, one can produce an immediate-release, an enteric-release, or a sustainedrelease tablet.

The most popular immediate-release film formers are the water-soluble cellulose ether polymers. The two most common are hydroxypropylcellulose (HPC) and hydroxypropylmethycellulose (HPMC). The low viscosity grades of these polymers are employed in the coating formula to maximize polymer solids concentration. Both these polymers are water soluble.

Water-insoluble film formers can also be used to prepare immediate-release coatings. These products fall into two categories: cellulose ethers and acrylate derivatives. The most common cellulose ether is ethylcellulose. This material is commercially available in two forms: as pure polymer and as an aqueous dispersion. The pure polymer is generally dissolved in an organic solvent; the dispersion is delivered out of an aqueous media. In both cases, a certain amount of water-soluble component (up to 50% of the total polymer solids) is included in the coating formula, to provide immediate drug release.

The ethylcellulose and acrylate compounds are also used to formulate sustained-release products.

Again, a water-soluble component is included in the coating formula. However, the level is very low: usually about 3% of total polymer solids. When the coated dosage form is exposed to water, the watersoluble component dissolves. This leaves a porous film surface through which drug diffuses.

The third class of coatings, the enterics, resists the attack of gastric fluids. As a result, drug is released only in the small intestine. Enteric coatings are prepared by using a polymer with pH-dependent solubility properties. Cellulose esters, substituted with phthalate groups, are the primary polymers used in this application, especially cellulose acetate phthalate. Polyvinyl acetate phthalate is also used. Acrylate derivatives are also capable of providing enteric release.

Compression Coating

Compression coating is a technique wherein a large tablet either completely or partially surrounds a smaller tablet. Essentially, a small tablet is compressed first and is then surrounded by powder, which undergoes compression. This type of coating technique requires the use of special tableting machinery and it is used to produce sustained-release tablets.

CORROSION AND ITS CONTROL BY COATINGS

Introduction

Energy Transfer

The metallic state is in most metals an unstable condition resulting from the smelting operation, in which energy is imported by the ore as the metal is derived. After extraction, most metals undergo a slow deterioration process during which they shed this energy and return to a more stable condition in which they are combined with some element of their environment, such as an oxide, a sulfide, or some other corrosion product. This energy conversion process is known as corrosion.

The Electrochemical Nature of Corrosion

Corrosion is most usually driven by some electrochemical inhomogeneity in the metal or its environment. In this process, different areas of the metal, having different levels of free energy and therefore different corrosion potentials, become the electrodes of an electrochemical cell in contact with a common electrolyte.

The electrochemical couples are set up with areas of more active electrochemical potential acting as anodes of the cell, while more passive areas act as cathodes (Figure.1). Corrosion takes place at the anodes as metal dissolves into the electrolyte as ions, so releasing electrons, which pass through the metal to the adjacent cathode areas where they react with the environment. This flow of electricity, the electron passage from anode to cathode, and the accompanying charge transfer back through the electrolyte from cathode to anode, make up the corrosion current. The rate of the current flow, i.e., the magnitude of the corrosion current (I) that develops, is a measure of the amount of degradation and is related to the potential difference (V) between the anodic and cathodic sites by Ohm's law:

$$I = \frac{V}{R}$$

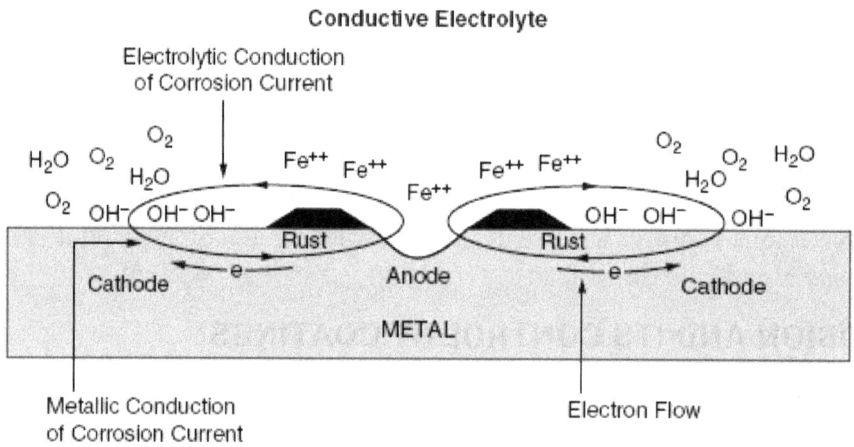

Figure .1 — Electrochemical corrosion cell

In the normal corrosion cell, R is in fact made up of several individual resistances: the resistances of the two electrodes (R a and R c), the resistance of the internal electrical contiguity between them, R w, and the electrolytic resistance of the common electrolyte across their external faces, R e.

Ohm's law may thus be rewritten

$$I = \frac{V}{R_a + R_c + R_w + R_e}$$

As the anode, the cathode, and their contiguity are all metallic, these resistances are negligible compared to the resistance of the electrolyte, which is ionic, and which may be exceedingly high in electrolytes of low ionic content (e.g., deionized water). The resistances of salt solutions, however, are much lower. It is for this reason that corrosion proceeds much faster in seawater, and only very slowly in freshwater.

Because of these things, the relationship shown in Equation (.2) becomes practically equivalent to minimization of the conductivity of the electrolyte is, therefore, an important maxim in corrosion control.

$$I = \frac{V}{R_e}$$

All metals have different electrode potentials (see Table .1), and large potential differences are set up where the electrochemical cell is composed of different metals. The electrical contiguity of a more noble metal and an active one in contact with the same electrolyte can produce severe galvanic corrosion in the active metal. On the other hand, the deliberate electrical fixation of a more active metal (such as zinc) to one of more passive potential (steel) is a powerful stratagem that is widely employed to eliminate corrosion in the passive material in the presence of a common electrolyte. Zinc, magnesium, and rarely aluminum anodes (generally ingots or in the case of zinc often films) are widely used to protect steel in this manner. Table .1 lists many common metals in terms of their electrochemical potential. Those having extremely positive electropotentials (gold, platinum, etc.) are very stable and do not corrode. They are found in nature in the metallic state.

Electrode Reactions

The chemical reaction at the anode (the dissolution of the metal) may be written

Table .1 — Electrode Potentials

Electrode Reaction			Potential (volts)
Active			
Na	⇔	Na^{+}+e	−2.71
Mg	⇔	Mg+$^{2+}$2e	−2.38
Al	⇔	Al^{+3}+3E	−1.66
Zn	⇔	Zn^{+2}+2E	−0.76
Fe	⇔	Fe^{+2}+2E	−0.44
Pb	⇔	Pb^{+2}+2E	−0.13
H	⇔	H^{+}+E	0.00
Cu	⇔	Cu^{+2}+2E	0.34
Ag	⇔	Ag^{+}+E	0.80
Pt	⇔	Pt^{+2}+E	1.2
Au	⇔	Au^{+3}+3E	1.4
Noble (Passive)			

$$M = M^{n+} + ne$$

Where n is the valency of the metal.

In the case of iron, this equation becomes

$$Fe \rightarrow Fe^{2+} + 2e$$

The exact nature of the reaction at the surface of the cathode (in which electrons released in anodic dissolution are, in turn, consumed) depends upon the nature of the environment. Under neutral and alkaline conditions, the reaction involves oxygen and proceeds

$$2H_2O + O_2 + 4e = 4OH^-$$

Under acidic conditions, if oxygen is present, the reaction may proceed

$$O_2 + 4H^+ + 4e = 2H_2O$$

Where oxygen is not available, hydrogen gas may form under acidic conditions:

$$2H^+ + 2e = H_2$$

Migration of the oxidative product (M n+) from the anode and the reduction product (OH -) from the cathode occurs until they combine to form the oxide, which precipitates. In the case of steel this may be Fe(OH) 2 , ferrous hydroxide, or, depending upon the nature of the environment, one of several precursor products, such as ferrous hydroxy chloride in salt water. Ferrous products are readily soluble, and this favors migration, so that oxide formations are not intimately associated with the anode but are loosely adherent and porous. Given sufficient oxygen, a second oxidation reaction will occur in steel corrosion, which converts the divalent ion to the trivalent ferric state,

$$Fe^{2+} \rightarrow Fe^{3+} + e$$

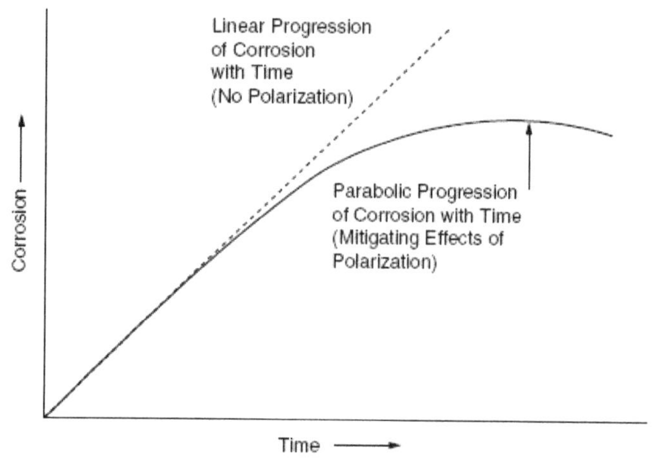

Figure .2 — Polarization and its effect on corrosion rate.

The solubility of the trivalent corrosion product is much less than that of the ferrous product. Under normal circumstances, however, where the secondary oxidative process occurs gradually after the ferrous ions have migrated away from the anode, the corrosion product is no more tightly adherent than is the ferrous product from which it is formed, and films of rust, hydrated ferric oxide (Fe 2 O 3 × Ç 2 Ï) , are usually loose and crumbly.

Polarization

The accumulation of ions at the anode tends to insulate the metal from the electrolyte and reduce the amount of ions going into solution. In effect, the anode deposit or film increases the resistance of the electrochemical cell, and the cell is said to be anodically "polarized." Similarly at the cathode, cathodic films are possible. In oxygen-deprived systems, for example, hydrogen gas may accumulate at the cathode, preventing further access of hydrogen ions and the consumption of electrons. The cathode reaction is therefore stifled, and the metal is said to be cathodically polarized. (Where oxygen levels are high enough, the cathode reaction occurs irrespective of the amount of H + ions, and in this case, the oxygen is said to have depolarized the cathode, allowing increased corrosion.)

The presence of these electrode films has a great effect on the rate of corrosion current transfer (i.e., the rate at which corrosion occurs). This can be illustrated in the resultant modification of the Ohm's law Equation (.3). Thus,

$$I = \frac{V}{R_{af} + R_{cf} + R_e}$$

Where R af and R cf represent the resistances of the anode and cathode films, respectively. These resistances may be very high, and as such polarizing films build up, the rate of corrosion is greatly diminished (see Figure .2).

ELECTRODE FILM BREAKDOWN AND DEPOLARIZATION

The permanence of such films has great significance to the control of corrosion of a specific metal. On aluminum, for example, under near neutral conditions, the naturally formed anode oxide film is very dense and insoluble (if thin) and quite resistant to mechanical removal. R af becomes very high, and I becomes negligible. Aluminum, therefore, is quite resistant to corrosion under neutral conditions.

Where the environment is sufficiently acidic or alkaline, however, the oxide film on aluminum is chemically dissolved. Thus, the metal is stripped of polarizing film (depolarized) and corrodes rapidly, i.e., the cell reaction resumes its linear relationship between corrosion and time. On other metals such as steel, the oxide film is normally less adherent and easily dislodged mechanically.

Such physical removal of the oxide film has the same result as chemical removal, the corrosion rate becoming more linear until additional corrosion product can repair the site of removal, and repolarization occurs. Mechanical agitation can also depolarize the cathode. If, under acidic conditions, the cathode, polarized with hydrogen gas, is rapidly agitated in the electrolyte, the physical shearing stresses will dislodge the bubbles of hydrogen, and the cathode will again be accessible to hydrogen ions from the electrolyte. Electrons will again be consumed, and corrosion will, as a consequence, increase.

If the secondary oxidative process on iron can be engineered to occur rapidly and close enough to the anode, a more adherent, denser, and more permanent barrier film of hydrated ferric oxide will be formed across the anode. This effectively prevents further dissolution of the iron into the electrolyte and greatly increases the resistance of the anode film. Under these conditions, the metal behaves as if it were more noble than it actually is. This phenomenon is known as passivation, and the thin anode films involved are known as passive films.

Passivation and Depassivation

Certain materials such as chromates, phosphates, and molybdates are known to induce passivity artificially. 6 – 10 These materials, known as corrosion inhibitors, if present as soluble ions within the electrolyte, will adsorb onto the metal surface, reinforce the natural oxide layer, and facilitate the passive state. Other materials, most commonly soluble chlorides and sulfates, effectively compete with inhibitive ions for adsorption onto the metal surface and tend to prevent passivity thereby.

These salts are known as depassivating agents. The ratio of passivating ions to depassivating ions in any interfacial electrolytic environment will dictate the effectiveness of induced corrosion control.

Other factors will also affect the rate of corrosion. The passive state is more readily achieved under alkaline conditions, and, at high enough pH, oxygen alone will induce passivity without benefit of other inhibitors. Corrosion will increase on the other hand as the temperature of the system increases.

Area Effects

If the rate of corrosion is largely dependent upon the rate at which electrons from the anode are consumed by the cathode reaction, then the intensity of attack is dependent upon the ratio of the areas of cathode and anode sites (Figure .3).

Where an anode is large compared to a given cathode, the effects of the corrosion current generated will be spread over the large anode area, so that corrosion is more general and attenuated. This type of corrosion, while undesirable, is not particularly dangerous. When the anode is small compared to the cathode, however, the current fall will be concentrated on a small area, and here corrosion will be intense. If the anode is small enough, this may lead to pitting attack, which is very dangerous and can perforate metal in a short time frame, especially where there is appreciable difference between anode and cathode potentials.

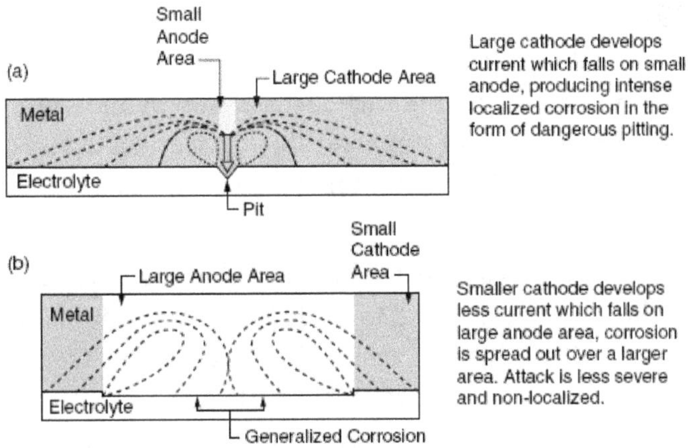

Figure .5 — Area effects

Control of the area ratio is a most important consideration in corrosion engineering, and coating films may be effectively employed to modify this ration. Diminishment of the cathode area must always receive first priority in any design. Holiday detection, to eliminate anodic pinholes at which pitting may occur, is a critical quality assurance process in the application of barrier coatings.

Table .2 : Basic Techniques Employed in Corrosion Control by Coatings

Barrier coatings	Resistance inhibition: oxygen deprivation
Ihibitive pigments	Modification of interfacial environment (establishment of passive firms)
Zinc-rich coating	Prevention of current discharge from metal to environment by affixation of anode film

COATINGS

Corrosion Control by Coatings

Coatings play an important role in corrosion control, used either alone or as part of an overall strategy of corrosion control.

While theoretically, coatings may be employed in a variety of ways to modify, hinder, or thwart the corrosion reaction, three or four devices find practical application (see Table .2). These devices are utilized in the three types of coatings employed in corrosion control.

There are barrier coatings, functioning via resistance inhibition and/or oxygen deprivation, Inhibitive coatings, functioning by modifying the interfacial chemical environment against the metal surface, and zinc-rich coatings, which prevent current discharge from the steel to the environment by employing a more anodically active metal (almost exclusively zinc) as a pigment at loadings high enough to ensure electrical contiguity with the substrate and low electrical resistance across the film.

Barrier Coatings

It is a practical impossibility to prevent enough moisture transmission through even the densest films to control the cathode reaction by moisture deprivation. It is, however, possible by the use of impermeable barrier coatings to suppress both oxygen transmission and the transmission of ionic solutions to a sufficient extent that corrosion may be precluded.

The prevention of oxygen access to the metal deprives the cathode reaction of critical fuel, and the film's resistance to ionic transfer renders any moisture that does access the substrate so high in resistivity that current transfer between the anodic and cathodic areas is frustrated (Figure .4).

Designing devices to utilize these precepts involves the use of molecularly dense films (highly uniform cross-linked thermosets and relatively crystalline thermoplastics). These coatings are formulated with nonhydrophilic components and lamella pigments to give coating systems of high electrical resistance that maintain adhesion even under wet conditions. They are typified by fluoropolymers, highly crosslinked epoxies, and coal tar modified systems, pigmented with pigments such as aluminum flake.

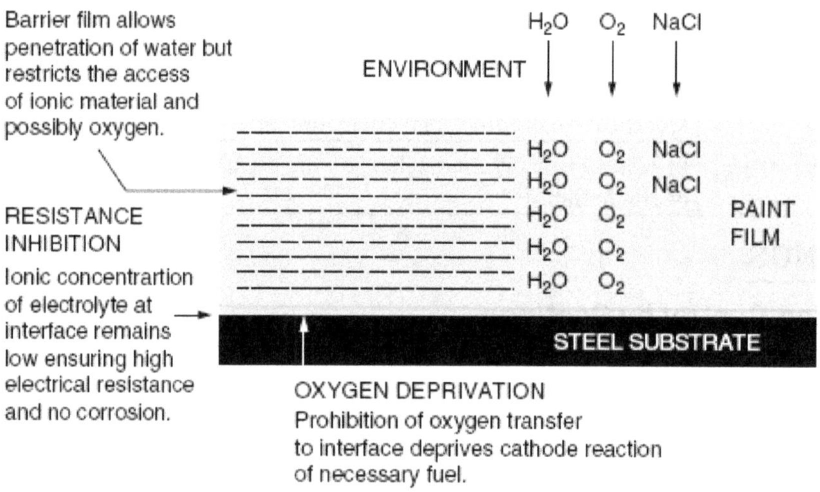

Figure .4 — Corrosion control by barrier coatings

Barrier coatings can be used alone or in combination with impressed current driven cathodic protection systems, where they act to minimize exposed cathode areas and thereby reduce the cost of required electricity that maintains the direction of current flow. Barrier coatings can also be used to minimize cathode-anode area ratios in mixed metal systems where insulation is impractical, thereby eliminating pitting at exposed anodes.

Barrier coatings are widely used for extreme service such as the lining of water, fuel, food, and chemical storage tanks, for marine coatings, etc.

Inhibitive Coatings

Inhibitive metal primers reduce corrosion by chemically modifying the interfacial condition against the metal. This is done by including pigments in the primer film that release oxidizing ions and other passivating moieties into the aqueous phase against the interface. Reviews of pigments used in this manner may be noted. In turn, this reduces the levels of electrolytic oxygen necessary to establish passive films on the metal5 beneath the coating (Figure .5). In order to minimize blistering and provide sufficient inhibitive ions to achieve long-term passivity without rapidly exhausting the available ionic reservoir (provided by the pigment), selection of both binder and inhibitive pigment is critical. Control of the PVC/CPVC is also important in order to maintain permeability so that there is sufficient moisture to access and dissolve enough inhibitive ions from the pigment in the film, but not so much that the film becomes too porous. Highly porous films will allow the transmission of depassivating ion species (chlorides and sulfates) into the film and through to the metal, there to compete with inhibitive ions for adsorption and so prevent passivation.

Inhibitive coatings are generally oleoresinous systems, oils, alkyds, epoxy esters, phenolic varnishes, etc., although epoxies and many latex primers also function by this mechanism. Inhibitive coatings are generally used for light to moderate duty atmospheric service, although they may be part of relatively high performance systems such as automotive primers, aircraft coatings, and coil coating primers. They are generally not favored for long-term freshwater immersion service, however, or for continuous use under high-temperature, highly humid conditions.

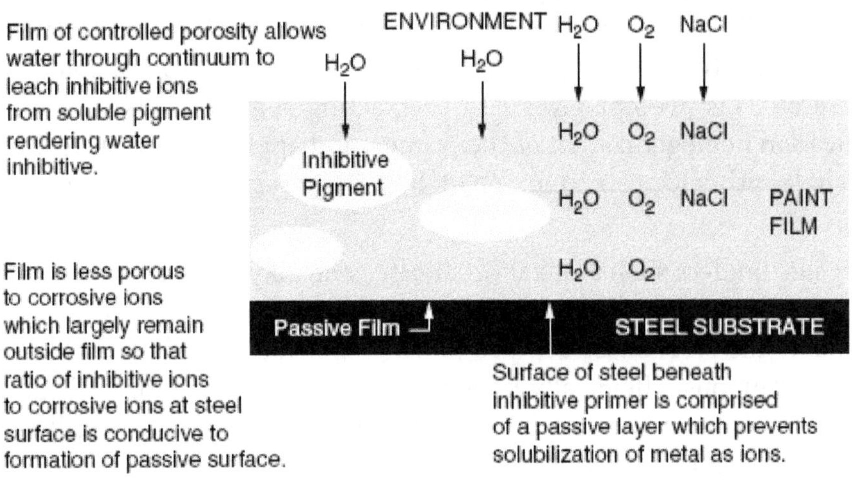

Figure .5 — Corrosion protection by inhibitive primer

The formulation of inhibitive metal primer is considered in several texts and papers.

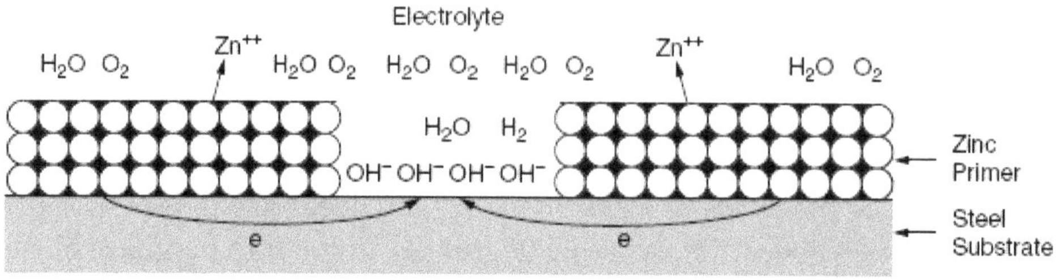

Figure .6 — Fundamentals of zinc-rich protection. The presence of strongly electronegative zinc-pigmented coating short-circuits all local cell activity on steel. The steel becomes totally cathodic to the zinc anode. The zinc corrodes, but the steel will not corrode even at bare spots. It is mandatory that the zinc coating be in electrical contact with the steel surface; therefore, the steel must be stripped of all contamination.

Zinc-Rich Coatings

Zinc-rich coatings contain high loadings of zinc dust as a pigment and serve to prevent corrosion galvanically by acting as the anode of an electrochemical cell of which the steel substrate is the cathode (Figure .6). All local cell activity on the steel is thus overridden by the electrically proximate anodic primer, which itself corrodes sacrificially to protect the steel. Eventually the zinc surface becomes partially polarized with zinc corrosion product, and protection shifts to a quasi-barrier mechanism, as the corrosion product seals the porosities of the highly pigmented zinc film surface. Where organic binders (epoxies, high-styrene systems, moisture curing urethanes, etc.) are used, PVC/CPVC ratios must necessarily be adjusted to unity (or a little higher) in order to maintain the required conductivity (from pigment particle to pigment particle within the film and from pigment particle to substrate). This precept gives rise to exacting requirements in formulation, preapplication homogenization of the primer, and application itself. Too low PVC will preclude cathodic protection; too high PVC will result in weak films with poor mechanicals.

Inorganic binders such as alkaline silicates and alkyl silicates rely on chemical reaction of the binder with the zinc atoms on the surface of the zinc dust pigment, and probably the steel substrate also (during application and curing) to produce a wholly continuous, chemically bound, inorganic matrix of zinc silicate. While PVC/CPVC levels are somewhat less critical, high levels of zinc are still necessary to provide good cathodic protection over the long term. Although these films are more porous than are the organic zinc-rich systems, again the presence of water within the primer, even against the metal surface, is not threatening to steel, for zinc corrosion is its only consequence. As long as there remains sufficient zinc within the film, all oxidation must occur only within the film itself. Eventually zinc corrosion product packs the porosities and seals the film off as was the case with the surface of the organic zinc-rich film.

Zinc-rich primers are widely employed as nontopcoated systems in environments where the zinc corrosion product is not easily dissolved (i.e., in neutral environments). In such applications, inorganic zinc primers have protected steel for half a century and more. In other systems, especially where environments are either highly acidic or highly alkaline, and therefore reactive with the metal, both organic and inorganic zinc-rich primers are recoated with barrier topcoats (epoxies, vinyls, chlorinated rubbers, and urethanes) for additional protection.

WIRE-WOUND ROD COATING

Introduction

Wire-wound metering rods have been used for more than 75 years to apply liquids evenly to flexible materials. They were the first tools used to control coating thickness across the full width of a moving web. The 1980s saw a new popularity in rod use because of improved quality and the industry trend toward shorter converting runs. Wire-wound rods are used in a wide range of applications but find their greatest appeal in the manufacture of tapes, labels, office products, and flexible packaging. The first rods were made of ordinary carbon steel, wrapped with music wire. Today's metering rods use precisionground core rods made of stainless steel, tightly wound with polished stainless steel wire at high speeds, on custom-designed winding machines. The resulting product is a laboratory-quality precision tool that can control coating thicknesses accurately within 0.0001 in. (0.1 mil). A typical wire-wound rod station is shown in Figure .1.

Also called applicator rods, Mayer bars, equalizer bars, coating rods, and doctor rods, this equipment has found uses in a wide variety of production applications, from the manufacture of optical films to wallboard panels. Wire-wound metering rods look deceptively simple. A stainless steel rod is wound with a tight spiral of wire, also made of stainless steel. The wire can be so small that it is almost invisible to the naked eye, or so large that the windings look like the coils of a hefty spring. Today, the industry has standardized on stainless steel rods because they can be used with almost every coating liquid. Earlier problems with rust and corrosion have been virtually eliminated. Where abrasive wear is a problem, some converters use chrome plating to prolong the life of the rod because of the hard surface presented by chromium. Chrome has its drawbacks, however, as it builds up unevenly at the extreme tops of the wires, changing the shape of the wires and the resulting coating thickness. Also, if not applied properly, chrome can acquire pitting marks or can flake off, contaminating the coating bath or causing uneven coating. Several new products introduced since 1985 have further expanded the market for rods. Where streaking or rod cleaning is a problem, rods with a Teflon surface are available. Particles that might wedge between stainless steel wires tend to slide through, preventing buildups and subsequent streaks in the coating.

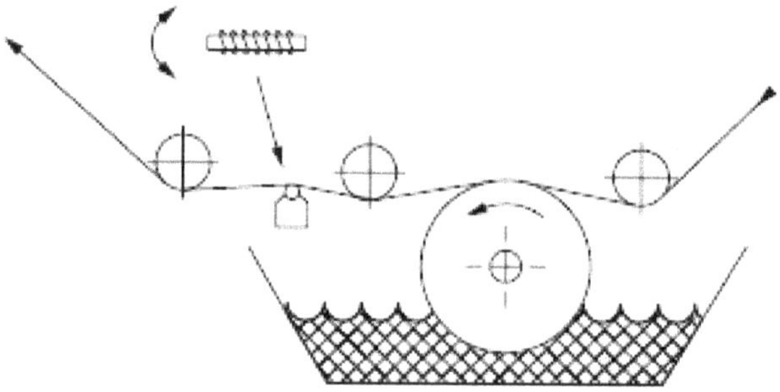

Figure .1 — Typical coating station

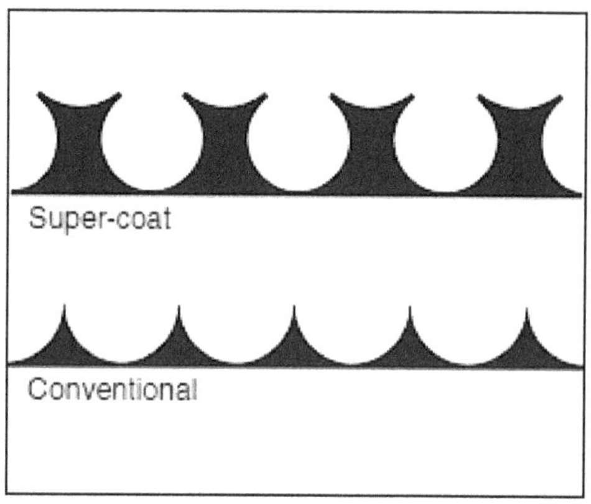

Figure .2 — Super-Coat flow pattern

Other users find the Teflon surface easier to clean, and a few claim increased rod life up to eight times that of stainless steel rods. Super-Coat* rods are conventional rods with a second, smaller wire wound around the primary wire of a normal rod. The shape of the grooves between the wires is altered drastically, which allows certain coaters to produce controlled coatings more than twice as thick as with conventional rods (see Figure .2). Others users find that for any given coating thickness, the deeper grooves between the wires are spaced closer together and provide a smoother finish on the product.

Today's coating machines have wider webs and run at higher speeds, outproducing earlier equipment many times over. Meanwhile, the market for coated material requires more and more specialized products, and hence, more frequent changeovers and setups. Compared with other coating methods, changing coating rods requires little downtime for changeover and makes precise control of coating weight without changing formulations a practical consideration for many coaters.

History

In 1905, Charles W. Mayer founded the Mayer Coating Machines Company in Rochester, New York. The firm made equipment for manufacturers of carbon and wax papers, two new and growing industries. These machines used "equalizer bars" or "doctor rods," the forerunners of today's precision metering rods. The rods were made of carbon steel wound with different sizes of music wire. Mayer was issued a series of patents on his coating machines, including one in 1912 that covered his equalizer bars. Coaters of carbon and wax papers found that they could easily change the thickness of their coating by switching rods, so they began ordering rods with different wire sizes. Also, the early rods tended to rust and wear out, so Mayer found a ready and growing market for replacement rods, which became an important part of his business as more machines were shipped. Mayer's success in the coating machinery business became the target of federal antimonopoly laws, and he was forced to release his designs and patents to others in the 1930s. New companies sprang up, and their machines used doctor blades and roll coating methods, in addition to wire-wound rods. Under the federal guidelines, Mayer was not required to expose his proprietary techniques for making equalizer bars. The Mayer factory continued to produce these rods on special winding machines that were developed for this purpose. As his machinery business declined, rod manufacturing continued to expand.

Theory and Principle

Coating thickness is governed by the cross-sectional area of the grooves between the wire coils of the rod, as shown in Figure .3. The geometry of this system creates a wet film thickness directly proportional to the diameter of the wire used. For example, you can double the coating thickness by doubling the wire size. The groove between the wires determines the precise amount of coating material that will pass through. The initial shape of the coating is a series of stripes, spaced apart according to the spacing of the wire windings. Almost immediately, normal surface tension pulls these stripes together, forming a flat surface, ready for drying in air or under heat. Wet film coating thicknesses, accurate to 0.1 mil, can easily be achieved using metering rods. Mathematically, the average thickness of the area between the wires, and therefore the coating thickness, is 0.173 times the wire diameter. From a practical standpoint, most users use a 1:10 ratio to select the best theoretical rod size for a given wet coating thickness. A rod wound with a 0.015 in. diameter wire, for example, will ideally produce a wet film that is 0.0015 in. (1.5 mil) thick.

In production coating, other physical factors can influence the actual coating thickness. The most important of these is the phenomenon of shearing action of the liquid. In effect, not all the coating material passes through the gaps; some liquid adheres to the surface of the wire. The impact of this is usually small, but it can be significant when small (0.025 to 0.005 in.) wires are used or when viscosities are high. In high volume coating operations, the speed of the web, web tension, penetration into the base material, and other factors will also affect the coating thickness. The sum of all these variations seldom exceeds a 20% difference from the theoretical coating weight; however, the differences can be large enough to suggest a trial-and-error procedure for selecting the ideal wire size for each coating application. Because of the relatively low cost of rods, and the ease of changing from one wire size to another, the selection process is both economical and practical. Most coatings laboratories have a set of "lab rods," which are shorter versions of production coating rods in a variety of wire sizes. To test a coating for coverage, adhesion, opacity, or other characteristics, drawdown samples are made and examined.

A piece of the substrate material is attached to the flat surface, and some coating liquid is puddle near the top of the substrate sample. A lab rod with a preselected wire size is chosen and pulled manually through the liquid, meeting a precise thickness of coating on the substrate.

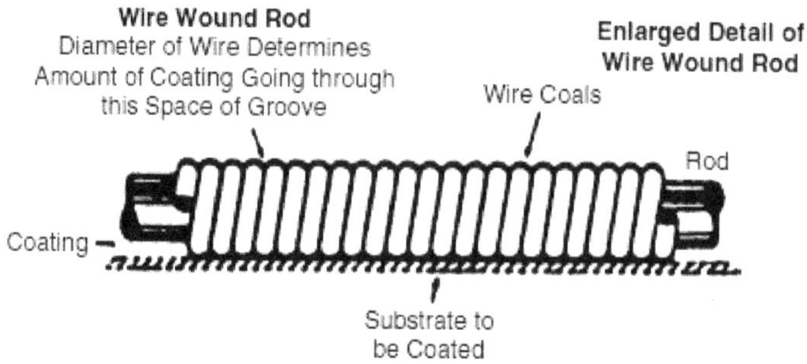

Figure .3 — Rod design

Table .1 — Wire Size Selection Chart

Wire Size Wire	Wire Diameter (inches)	Wet Film Thickness (mill)	Wet Film Thickness (microns)	Wire size	Wire Diameter (inches)	Wet Film Thickness (mill)	Wet Film Thickness (microns)
#0	None	0.00	0.0	#38	0.038	96.5	96.5
#2	0.0025	0.25	6.4	#40	0.040	101.6	101.6
#3	0.003	0.3	7.6	#42	0.042	106.7	106.7
#4	0.004	0.4	10.2	#44	0.044	111.8	111.8
#5	0.005	0.5	12.7	#46	0.046	116.8	116.8
#6	0.006	0.6	15.2	#48	0.048	121.9	121.9
#7	0.007	0.7	17.8	#50	0.050	127.0	127.0
#8	0.008	0.8	20.3	#52	0.052	132.1	132.1
#9	0.009	0.9	22.9	#54	0.054	137.2	137.2

Wire Size Wire	Wire Diameter (inches)	Wet Film Thickness (mill)	Wet Film Thickness (microns)	Wire size	Wire Diameter (inches)	Wet Film Thickness (mill)	Wet Film Thickness (microns)
#10	0.010	1.0	25.4	#56	0.058	142.2	142.2
#11	0.011	1.1	27.9	#58	0.058	147.3	147.3
#12	0.012	1.2	30.5	#60	0.060	152.4	152.4
#13	0.013	1.3	33.0	#62	0.062	157.5	157.5
#14	0.014	1.4	35.6	#64	0.064	162.6	162.6
#15	0.015	1.5	38.1	#66	0.066	167.6	167.6
#16	0.016	1.6	40.6	#68	0.068	172.7	172.7
#17	0.017	1.7	43.2	#70	0.070	177.8	177.8
#18	0.018	1.8	45.7	#72	0.072	182.9	182.9
#19	0.019	1.9	48.3	#74	0.074	188.0	188.0
#20	0.020	2.0	50.8	#76	0.076	193.0	193.0
#22	0.022	2.2	55.9	#78	0.078	198.1	198.1
#24	0.024	2.4	61.0	#80	0.080	203.2	203.2
#26	0.026	2.6	66.0	#82	0.082	206.3	206.3
#28	0.028	2.8	71.1	#84	0.084	213.4	213.4
#30	0.030	2.0	76.2	#86	0.086	218.4	218.4
#32	0.032	3.2	81.3	#88	0.088	223.5	223.5
#34	0.034	3.4	86.4	#90	0.090	228.6	228.6
#36	0.036	3.6	91.4	a			

The quality of test samples made by hand is largely dependent on the skill of the technician. Variations in rod pressure, angle of stroke, and speed can cause differences in the thickness and consistency of drawdown samples.

Film Thickness

A wire-wound rod applies a finite thickness of liquid to a web of base material. In production coating operations, it is difficult to measure the actual thickness of a wet coating before it starts to dry through evaporation, although there are gauges made for this purpose. In addition, many substrate materials are absorbent, so there may be little or no actual increase in thickness after coating. If the coating is applied to a hard, nonabsorbent material and is allowed to dry, its dry thickness can be measured accurately with micrometers or other dry thickness gages. Table 1 compares the calculated wet film thickness to the wire diameter of metering rods. The 10:1 ratio between wire diameter and wet film thickness allows an easy selection of the required rod.

The Rod Coating Station

In production coating, the web must pass through a wetting station and then to the metering rod, where a measured thickness of coating is allowed to pass between the wires, and the excess liquid is returned to the reservoir. The coating liquid may be applied at the wetting station by several different methods.

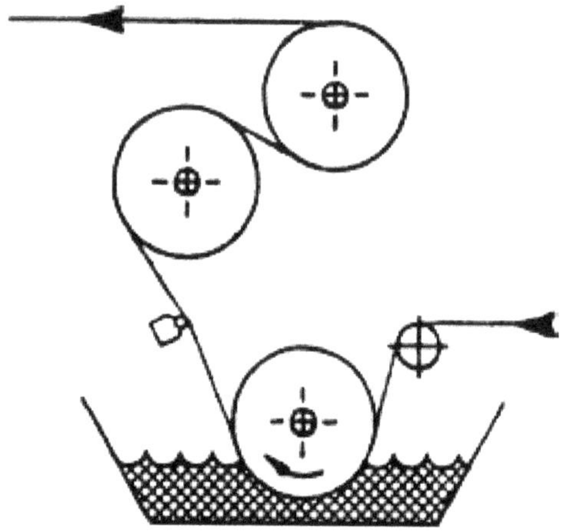

Figure .4 — Web immersion method

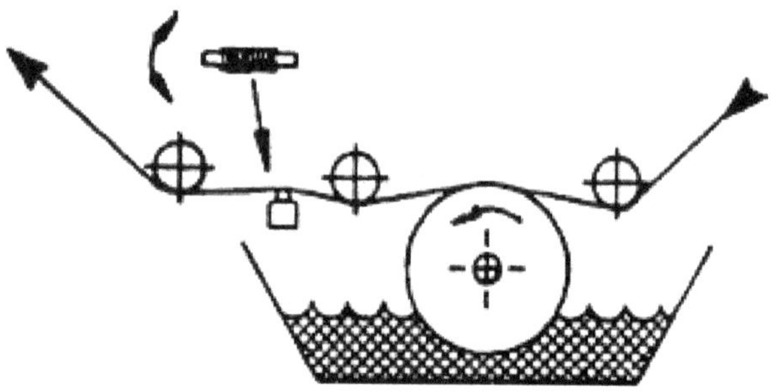

Figure .5 — Applicator roller method

The web can be immersed directly into a tank (Figure .4); or an applicator can be rotated in the reservoir to transfer the liquid to the web at the top of its rotation (Figure .5). It is important to apply an excess of coating liquid at this station, to let the metering rod do its job. When an applicator roller is used in a rod coating system, the speed of the applicator is not a critical factor. In addition, the machine operator can adjust the applicator roller speed within a side range, even while the machine is running. The web passes over the metering rod, which may be stationary or may be rotated slowly. The rotation may be either in the same direction as the web or in the opposite direction.

The choice of stationary rod depends on movement with different coaters and with different products. Establishing the ideal speed of rotation will also be different from job to job, and converters experiment to find the best procedure for each run. The most common procedure, however, is to rotate the rod slowly in the opposite direction to the movement of the web. The rotation flushes the coating material between the wires, keeping the wire surfaces wet, and preventing setting up and hardening of some liquids. The rotation also distributes any abrasive wear evenly on the wires and prevents flat spots from forming. The purpose of the metering rod is to remove excess coating liquid, allowing a measured amount to pass between the wire windings. The web should pass above the rod, to allow the excess liquid to fall back into the tank. The web, however, need not be perfectly horizontal, as long as the surplus coating can return to the tank through gravity. Metering rods for production coating can be made in a wide variety of sizes. The most common core rod diameters are quite small (3/16 and 1/4 in.), although sizes up to 1 in. diameter are also used. The main advantages of small-diameter rods are their low cost and ease of storing and handling.

These thin rods must be supported in the coating machine, because they are not rigid and will deflect with pressure from the web. There are several types of rod holder in common use; the simplest is a square of a rectangular steel bar, with a "V" groove machined into it. This rod holder is mounted between the side frames of the coating machine, and the metering rods are placed in the grooves.

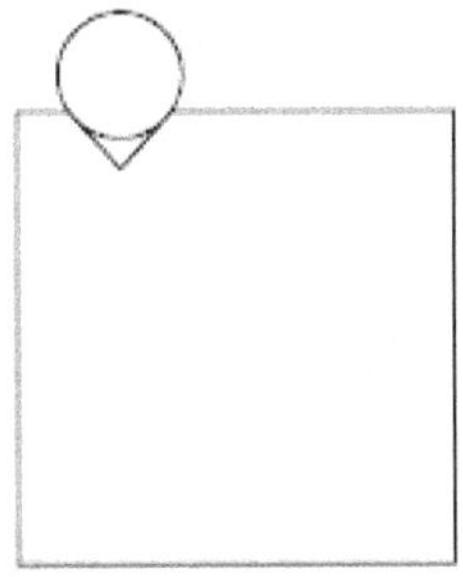

Figure .6 — Rod holder.

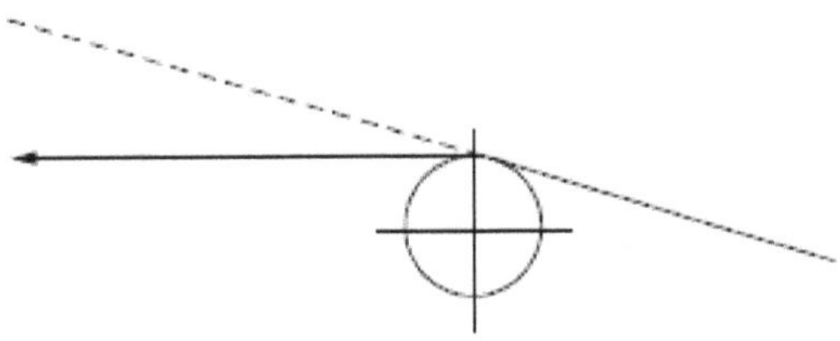

Figure .7 — Wrap angle.

The "V" groove should be ground and polished to minimize wear on the rod and should be mounted accurately, at a right angle to the direction of web travel and parallel to the idler rollers of the machine (Figure .6). The design of a rod coating station should ensure that the web makes intimate contact with the wires of the metering rod. The wrap angle, the angle between the web direction as it approaches the rod, and its direction as it leaves the rod, should be 15° for a heavy web tension or up to 25° for a light web tension (Figure .7).

Web tension is a critical factor in the design of a rod coating station. With a wrap angle of 15 to 25°, the web must be tight enough to ensure intimate contact with the metering rod, yet not so tight that the web is deformed by the wires. Adhesives and some other liquids can solidify between the wire windings of the rod whenever the coater is stopped. Many coating machines have a "throw-off" feature, a mechanical means of separating the web from the rod automatically, whenever the machine is turned off. This allows for quick removal of the rod for flushing and cleaning before the coating material has a chance to congeal between the wires (Figure .8).

This automatic releasing feature also simplifies rod changing between production runs. One method used by coating machine manufacturers is a rocker arm throw-off system. A series of idler rollers presses the web against the metering rod while the coating machine is running. Whenever the coater is stopped, the idlers automatically rise, lifting the web up, away from the metering rod. At the same time, a water flushing system can be triggered to remove coating material from the rods before it can set up. Other techniques involve lowering the metering rod and its holder when the coater is turned off.

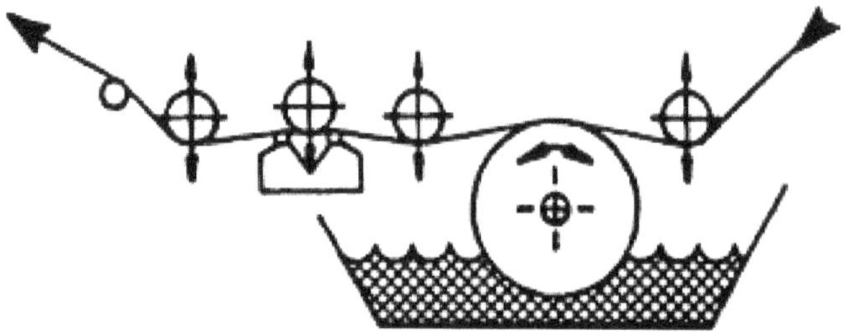

Figure .8 — Automatic throw-off

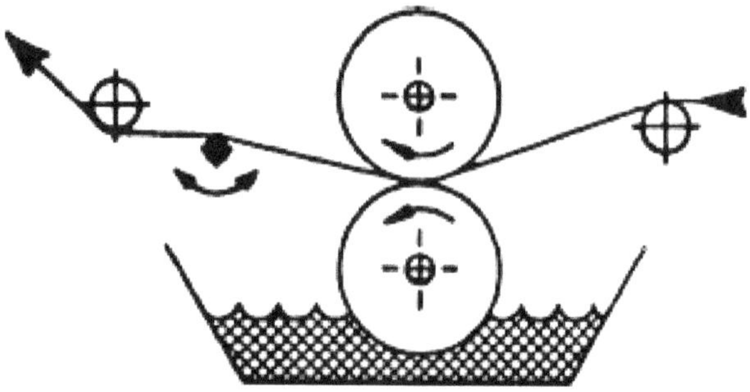

Figure .9 — Smoothing rod

Rod Station Variations

Every coating application is different. Converters have created many unique systems for rod coating that they developed through experimenting in the lab or in actual production. A smoothing rod is a finely polished metering rod without wire, which is used several different ways. It can be used in the web path after printing with a gravure cylinder, to eliminate the etched pattern and enhance the coated surface. It also can be used after a wire-wound metering rod, when the viscosity of the coating liquid prevents it from leveling through the surface tension effect (Figure .9).

When a large amount of coating liquid must be metered off the web, two wire-wound rods are used in tandem, spaced 1 or 2 in. apart. The first rod has a larger wire size for doctoring off most of the excess liquid, and the second, smaller wire rod controls the finished coating thickness.

A typical coating station may use one or two metering rods and also have a position for a smoothing rod, which can be engaged for certain applications. The coating roll applies liquid to the moving web that is metered off by one or two rods, and then it can be flattened by the smoothing rod. The complete coating station design must also consider draining, filtering, and pumping the coating liquid, controlling the wet edge of the web, and disengaging the metering rods when the machine is stopped.

ADVANTAGES AND DISADVANTAGES

Metering rod coating is the third most popular method in use today, behind gravure and reverse roll coating.

Low Cost

Rod coating stations are relatively inexpensive and easy to add to existing coaters. Replacing worn or damaged rods, and changing from one coating weight to another, are both inexpensive and fast. The machine downtime for changing rods or cleaning them can be measured in minutes instead of hours, using much less labor than is required by other coating systems.

Precise Coat Weights

Metering rods can be selected to control the wet coating thickness in 0.1 to 0.2 mil increments, without changing the coating formulation.

Lower Setup Cost

Another important factor in the new popularity of rod coating is the worldwide trend toward shorter and shorter production runs. The faster setups inherent to rod coating allow the converter more productive running time, in addition to lower labor costs for changeovers.

Less Edge Wear

Rod coating offers the converter other advantages because of the method used to control wet edges. In a rod coating system, the dry edge is controlled by wipers or deckle straps on each edge of the applicator roll. Because the wipers are constantly wet with coating liquid, the tendency to scratch the roller is reduced. Even when scoring eventually occurs, it is on the applicator roll, not on the metering rod (which controls the final coating thickness). Also, because the wipers are easily moved, their positions can be adjusted while the coater is operating, with no downtime.

Limitations

Standard metering rods work best with low viscosity liquids, which will flow easily between the wire windings. Two-wire Super-Coat rods can be used where viscosities are higher, however. Rod coating also can be used for a flat web, without tight or baggy edges. In production rod coating, the actual thickness of the coating can be affected by web speed, viscosity, and other factors. Depending on the application, rod coating speeds is usually limited to 1000 ft/min, although some coaters claim web speeds up to 2000 ft/min. The critical factor in the web speed of a rod coating system is the time required for the striations formed by the rod to level. The rod meters liquid by allowing a measured amount to flow through the spaces between the wires. Normal surface tension forces the raised portions to flow out and form a flat, even coating, but there is a time element involved, which is different for each coating material. The web speed must be controlled to allow time for leveling before the web is dried.

EXTRUSION COATING WITH ACID COPOLYMERS AND LONOMERS

Product Considerations

Acid copolymers and ionomers are high-performance resins that offer adhesion, heat seal, and barrier properties markedly superior to those of conventional polyethylenes. Figure .1 shows the structure of acrylic and methacrylic acid copolymers, the two types of acid copolymer commercially available in the United States at this time. The presence of the methyl side group in the methacrylic acid copolymers results in some subtle differences between the two resin types, but they can be regarded as equivalent after allowing for the difference in molecular weight of the comonomers. For example, a 10 wt% acrylic acid copolymer is equivalent to a 12 wt% methacrylic acid copolymer in carboxyl group content.

Ionomers (Figure .2) are derived from acid copolymers by partial neutralization of the carboxyl group with either sodium or zinc ions. Because the neutralization reaction results in a substantial increase in melt viscosity, it is commonly referred to as ionic cross-linking.

In both acid copolymers and ionomers, the melt and solid state properties are strongly influenced by intramolecular hydrogen bonding, as illustrated in Figure .3. The forces involved in hydrogen bonding are almost 10 times stronger than the intramolecular forces in the nonpolar polyethylenes.

The ionomers are distinguished by the combination of hydrogen bonding and interchain ionic forces perhaps an order of magnitude stronger than the hydrogen bonds. As a consequence, the ionomers have a wide spectrum of melt and solid state properties, including better hot tack and grease resistance than acid copolymers of equivalent acid content.

With the acid copolymers, melt index and acid content are the major variables available to the resin producer. Because melt index is a measure of melt viscosity, it is related primarily to the processing characteristics of the resin. The resins now used for extrusion coating applications fall in the 5 to 15 melt index range to accommodate a broad field of processing needs.

Resins with acid contents of 3 to 15% are currently available on the market. The effects of increasing acid content are as follows:

- Better foil adhesion
- Better hot tack

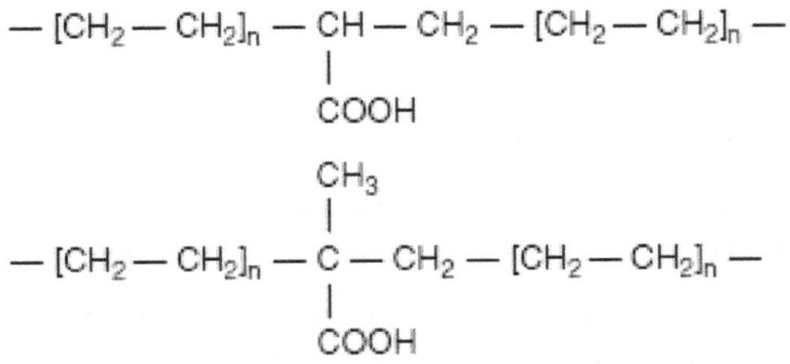

Figure .1 — Structure of ethylene acrylic (top) and methacrylic acid (bottom) copolymers

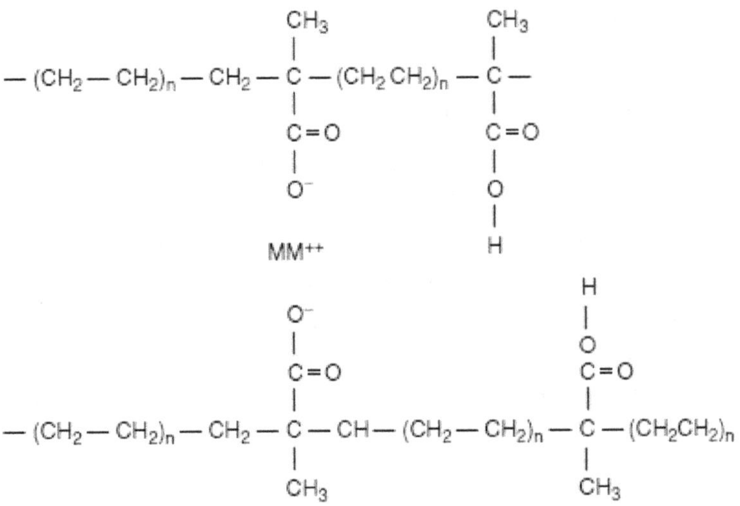

Figure .2 — Ionomer structure

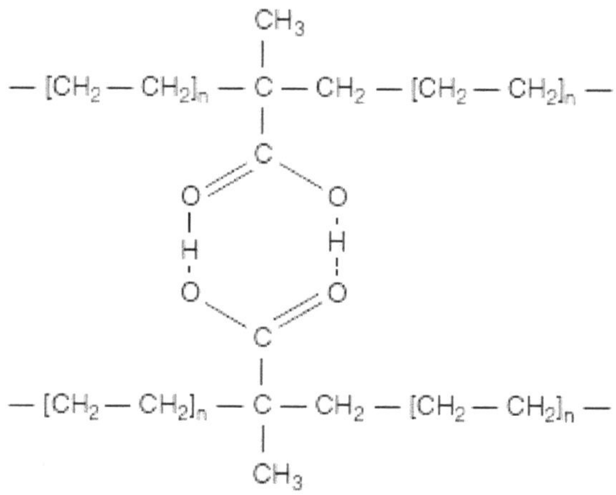

Figure .3 — Intramolecular hydrogen bonding.

Lower seal initiation temperature

- Better oil and grease resistance
- Better abrasion resistance
- Higher gas and moisture permeability

The most important effect is the improved adhesion to unprimed aluminum foil, as most commercial applications of acid copolymers involved foil adhesion.

The ionomers are characterized by additional variables that can be manipulated to obtain a broader spectrum of properties in the final product. These variables include the following:

- Melt index (molecular weight) of base resin
- Percentage of acid in base resin
- Degree of ionic cross-linking (percent acid neutralized)
- Type of ion used for neutralization (Na or Zn)

In particular, the use of the sodium and zinc ions produces families of resins with distinctly different properties. The zinc ionomers are used for the majority of extrusion coating applications because of their excellent adhesion to unprimed foil. The effects of composition on ionomer properties are shown in Table .1. In general, the ionomers based on high acid copolymers and with a high degree of neutralization have the best heat seal and product resistance characteristics. However, these advantages are balanced by a need for relatively high free carboxyl content for optimum foil adhesion.

In practice, most extrusion coating resins used for coextrusion with nylon have a relatively high degree of neutralization. There are limited uses of sodium ionomers in extrusion coating applications that call for maximum grease resistance, but these applications are on either paper or primed foil.

END-USE CONSIDERATIONS

Foil Adhesion

One of the primary reasons for the selection of an acid copolymer or ionomer for use in extrusion coating is adhesion to unprimed aluminum foil. Extensive studies in Du Pont laboratories have found that acrylic and methacrylic acid copolymers of equivalent acid content have equal adhesion to foil. At a particular acid content, a zinc ionomer will have 15 to 20% lower foil adhesion. Resins with high acid content (10 to 13%) have 20 to 25% better adhesion than resins with low acid content (3 to 5%).

The acid copolymers and ionomers have additional utility because of the enduring nature of their adhesion to foil. High levels of adhesion of polyethylene to foil can be achieved with primers or high melt temperatures in extrusion, but the bonding is often deteriorated by extended exposure to aggressive environments. With acid copolymers and ionomers, most hard-to-hold products can be packaged without sacrificing long-term durability of the coating adhesion. Acid copolymers and ionomers of equal acid content are essentially equivalent in product resistance.

Heat Seal Characteristics

The heat seal properties of acid copolymers and ionomers constitute the second major area of superiority over low-density polyethylene (LDPE). Seal initiation temperatures are lower and strengths higher than ionomers, and LDPE. The acid copolymers and ionomers have similar seal initiation characteristics, but the acid copolymers have higher ultimate seal strength. Both have much higher seal strengths than LDPE.

Hot tack strength is the ability of a heat seal to remain together when a force is applied while it is still in the molten state. This is a critical property in vertical form-fill-seal applications, in which the product is loaded immediately after the seal is made. It is also critical in any high speed packaging operation in which the package is exposed to some form of abuse before the seal has cooled.

One method of measuring hot tack strength is the Du Pont spring test. A series of springs of different thicknesses and widths at the narrowest point provides varying levels of spring tension (Figure .5).

Figure .6 shows how the spring is located inside the sample. The sample is heat sealed for 3 sec at a pressure of 40 psi. The 3 sec dwell time allows an accurate measurement of the interfacial seal temperature by using a very fine thermocouple hooked up to a rapid response recorder.

Table .1 : Ionomer Properties versus Composition

Higher Percentage of Acid	Higher Ion Linking	Higher Free Carboxyl: Greater Adhesion to Polar Substrates (Foil, Mylon)	
		Na	Zn
Greater stifffness Better oil resistance Higher gloss Higher tensile and impact strenght Lower haze Lower melt and softening Points	Greater stifffness Better oil resistance Higher gloss Higher tensile and impact strenght Higher toughness Higher hot tack	Better oil resistance Higher gloss Lower haze Higher draw Higher toughness	Less water sensitive Better adhesion (foil, nylon) Les "nech-in"

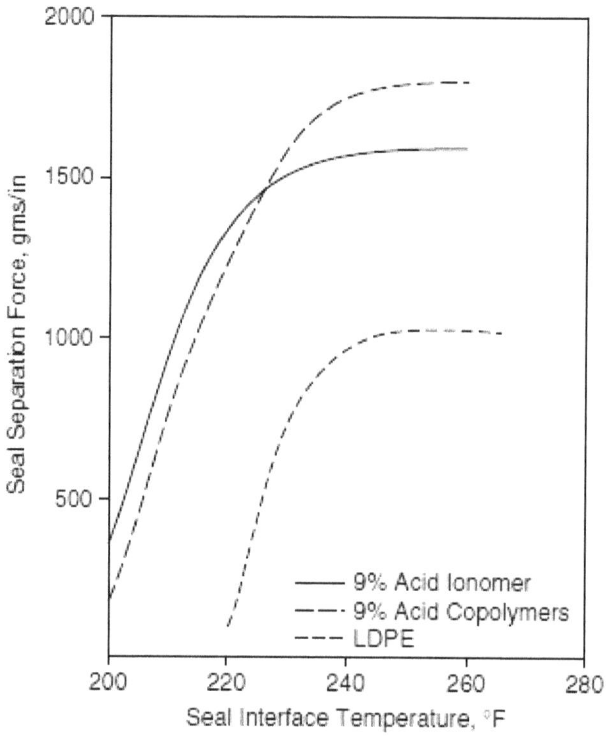

Figure .4 — Heat seal strength versus bar temperature (1 mil coatings on 30 lb kraft paper)

When the heat seal bars are released, the springs apply an instantaneous force to the seal. The separation of the 1 in. wide seal is measured to the nearest tenth of an inch. The results are plotted as the force required obtaining 20% seal separation as a function of temperature.

Figure .7 shows the hot tack characteristics of an ionomer and acid copolymers of equivalent acid content in comparison with LDPE. LDPE, a completely nonpolar material, has very low hot tack strength.

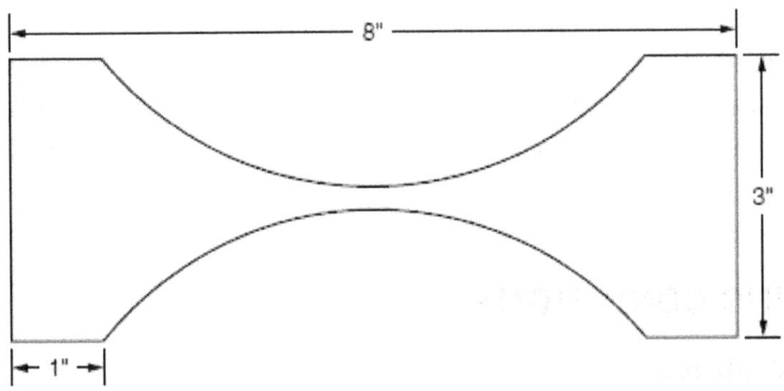

Figure .5 — Spring for Du Pont hot tack tester

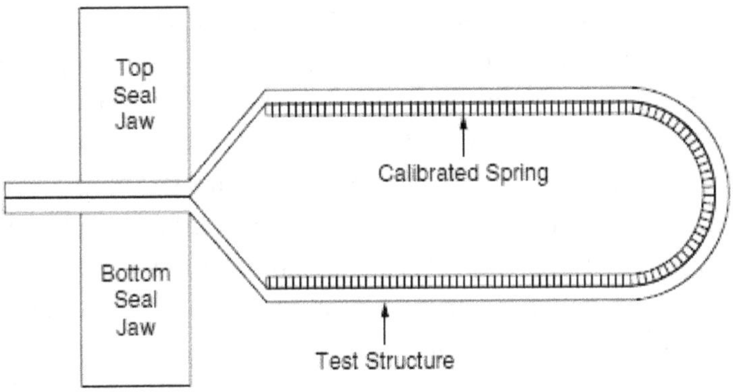

Figure .6 — Hot tack spring test

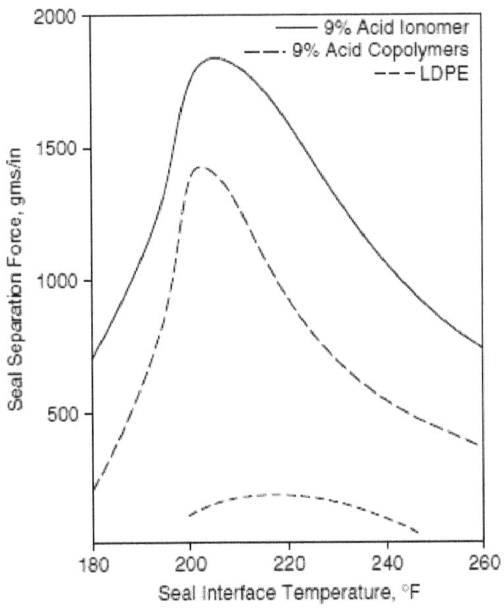

Figure .7 — Hot strength versus interface temperature (1 mil coatings on 30 lb kraft paper)

The acid functionality of the copolymers results in much higher hot tack strength because of interchain hydrogen bonding. The combination of hydrogen bonding and ionic bonding in the ionomers producesan even greater maximum hot tack strength and a broader hot tack range. In practice, this means thatthe ionomers will tolerate the broadest range of sealing conditions and offer the highest speeds onpackaging lines.

PROCESSING CONDITIONS

Melt Temperatures

LDPE is typically processed at melt temperatures in the range of 600 to 630°F. These high temperaturesare necessary to promote surface oxidation of the resin and to ensure adequate adhesion to the substrate.However, it is not necessary to oxidize the acid copolymers or ionomers because the acid functionalityis responsible for adhesion to either foil or paper. These resins should be processed at the lowest melt temperature consistent with adequate product performance. One obvious reason to prefer a lower melt temperature is to reduce the chance of odor or taste problems in packaging sensitive products. Another reason is shown in Figure .8. Both the acid copolymers and ionomers can undergo a process called anhydride cross-linking in which carboxyl groups in adjoining molecules react to form an anhydride cross-link and produce water as by-product.

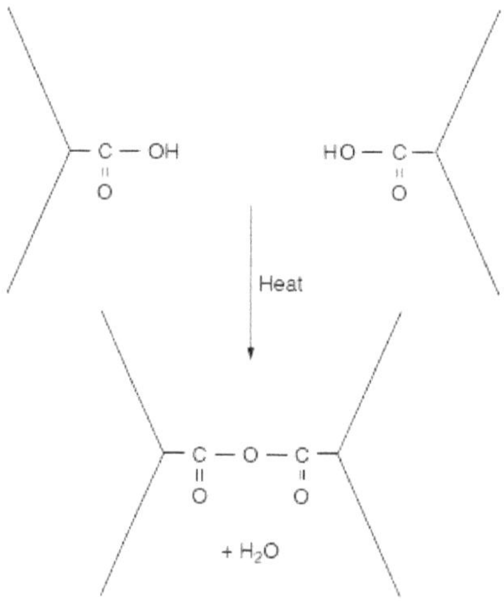

Figure .8 — Anhydride cross-linking

This reaction occurs at a significant rate in the temperature range at which the copolymers and ionomers are normally processed. The cross-linking reaction causes gel formation, while the water causes die buildup problems. As approximate guidelines, the melt temperatures shown in Table .2 are recommended for acid copolymers and ionomers, assuming that desired product properties can be achieved under these conditions.

Table 2 — Melt Temperatures

Product	Melt temperature (°F)
High acid, high melt index	450–500
Intermediate acid and melt index	500–550
Low acid and melt index	550–590

Other Considerations

Both the acid copolymers and ionomers are slightly corrosive, so that metal parts exposed to these products at high temperatures should be plated with nickel or chrome. Another factor to be considered is the moisture sensitivity of the ionomers. The ionomers are hygroscopic and, as a consequence, are packaged in foil-lined bags and boxes. When a box is opened, only a corner of the liner should be cut off to permit insertion of the transfer pipe. If this is done properly, it is easy to reseal the liner of a partially used box by folding it over and taping it securely. With normal precautions, the moisture sensitivity of ionomers does not present an operational problem.

ROTARY SCREEN COATING

Introduction

Not long after the introduction of multicolor printing machines, the one-color printing unit was also introduced, which through the years has found its way to a wide range of application areas such as wall cover printing, production of hard floor covering, technical coatings, artificial leather production, and last but not least, as a one-color printing machine.

Equipment

Screen coating equipment (Figure .1) consists of the following:

- The screen, which is a seamless, perforated, nickel sleeve. The degree of perforation is expressed in the so-called mesh number, indicating the number of holes per linear inch.
- The squeegee, which is mounted in the screen and serves as the supply and distribution pipe of the paste. The squeegee blade, which is mounted to this pipe, pushes the paste out through the wall of the screen.
- The whisper blade smooths the applied coating layer.

The amount of coating to be applied is determined by four factors:

- The choice of mesh number
- The squeegee pressure: that is, the angle formed between squeegee blade and screen (The smaller this angle, the higher the add-on.)

- The viscosity of the paste
- The squeegee setting with regard to the counter-pressure roller

Products

The coatings that can be applied by rotary screen are discussed

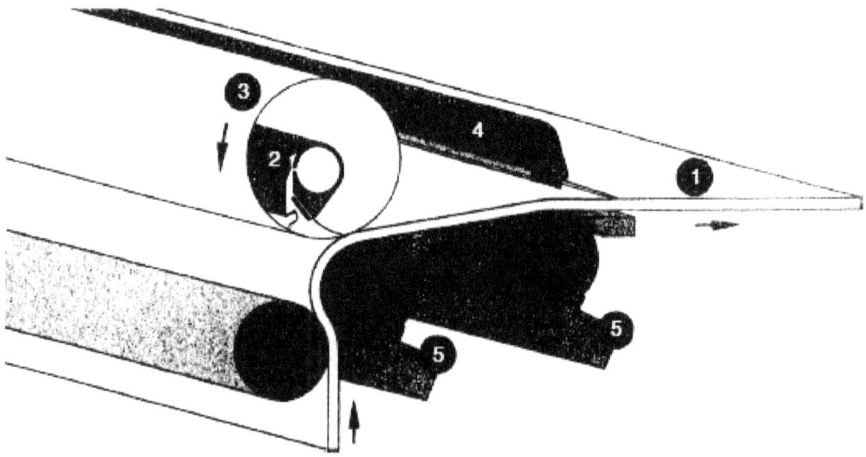

Figure .1 — Screen coater. 1: web; 2: squeegee; 3: screen; 4: whisper blade; 5: doctor blade

Pattern-Type Coatings

In view of the quantities required for pattern-type coatings and flock glue application, frequent use is made of 125 mesh and 40 mesh screens with a repeat size of 25 in. (64 cm). End uses include women's fashions, wall-covering curtains, and partial chintz rainwear.

Dot Coating

Two types of screen are available to accommodate dot coatings: for regular dot coating, the dots are arranged in diagonal rows; for computer dot coating (CP), the dots are arranged at random. For regular dot coating, the 17, 25, and 30 mesh screens find the widest application, and the CP 30, CP 40, and CP 46 (where 30, 40, and 46 designate the number of perforations per square centimeter) are the most frequently used CP screens.

Because of the need to apply an oscillating squeegee on the outside surface of the screen (only used for dot-coating applications), which has been designated to suit 64 cm repeat screens, dot coating can be effected with 64 cm repeat screens only.

End uses for dot coatings include fusible interlinings (woven and nonwoven) and print-bonding of nonwoven materials.

Advantages of screen coating compared to traditional coating by means of an engraved roller are as follows:

- Screen costs are considerably lower than the cost of an engraved roller.
- The time for exchanging screens is extremely short (~10 min).
- The application amount is easy to control.
- Light-sensitive nonwovens (20 g/m 2) or knitted fabrics can be processed without problems thanks to the frictionless application system.
- There is no penetration, resulting in effective use of paste and a good controllable hand.
- The constant paste supply ensures precise level control by means of a past processor.

A new development is the application of foam-dot coating, in which the foam is applied with the so-called closed system (see Figure .2). The foam, which has been made in the foam processor, is applied directly on the substrate via a "closed" squeegee, as shown (Figure .2). The foam processor will be adjusted for the application weight wet.

For example, suppose we have a polyamide paste with 33% solids. To apply a dot coating with a 10 g/m 2 dry add-on, the foam processor is adjusted for 30 g/m 2 wet add-on.

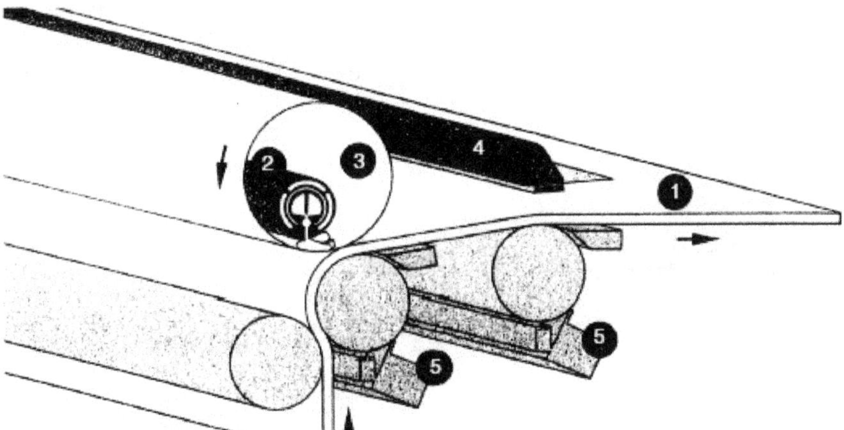

Figure .2 — Foam coater. 1: web; 2: squeegee; 3: screen; 4: whisper blade; 5: doctor blade.

Overall Paste Coatings

Fine or thin coatings are in a range of 5 to 25 g/m 2 of dry application weight. Sorters used are 80 mesh 12% open area, 100 mesh SP (special) 20% open area, and 125 mesh SP 12% open area, 130 μ m thickness (indication only).

End uses include anoraks, skiwear, sportswear, rainwear, umbrellas, pressure-sensitive coatings (medical), and shift-fast coatings.

Medium coatings fall in a range of 15 to 20 g/m 2 of dry application weight. Screen used are 60 mesh LR (long run) 14% open area, 125 μ m thick, and 40 mesh LR 20% open area, 130

μm thickness.

End uses include roller blinds, aluminum-coated curtains, bag material, and ironing bar covers. Heavy past coatings, having a dry application weight of 15 to 130 g/m 2 are frequently applied with the 40 mesh HX (hexagonal) screen with 30% open area and 300 μm thickness.

End uses include vertical blinds, flame-retardant coatings on upholstery and technical coatings, and tablecloths. Screens for paste coating always have a 64 cm (25 in.) repeat size. The squeegee blade is 28 mm×0.2 mm.

Foam Coatings

Only three types of screen are applicable when coatings are applied by means of thermally unstable foam (foam collapses in drier/stenter) or stable foam:

5. 40 mesh HX (50% open area, 300 μ m thick, 91.4 cm [35.8 in.] repeat) for foam coatings having a dry weight of 20 to 60 g/m 2
1. 14 mesh HX (40% open area, 450 μ m thick, 91.4 cm [35.9 in.] repeat) for foam coatings having a dry weight of 50 to 100 g/m 2 3. 25 mesh HX (30% open area, 350 ì m thick, 91.4 cm [35.9 in.] repeat) for average foam coatings. Stable foam is applied by means of a squeegee blade 40 mm wide and 0.2 mm thick.

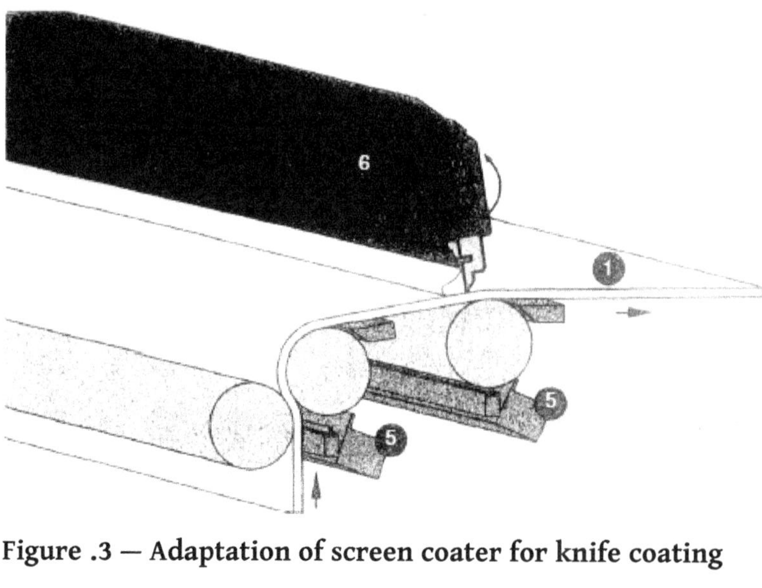

Figure .3 — Adaptation of screen coater for knife coating

End uses include black-out curtains or roller blinds, breathable coatings for anoraks and rainwear, and flock adhesive on curtains and shower curtains.

Thermally unstable foam coatings are applied directly by means of the "closed" system (see Figure .2) via the foam processor and the closed squeegee through the screen on the fabric. The wet application weight can be adjusted on the foam processor.

For example, to apply 30 g/m 2 dry of an acrylic or polyurethane binder (50% solids), simply adjust the wet application weight on the foam processor at 60 g/m 2 .

End uses include velour backcoatings, automotive backcoatings, mattress ticking, and fixation of pile on pile fabrics, flame retardant and cigarette-proof coatings, imitation suede, and antislip coatings.

Advantages

1. Because substrate, screen, and counterpressure rollers have the same speed, coating is done without tension and friction. Thus, virtually all substrates can be processed on this system, includingknitted fabrics, velours, nonwovens, and shift-sensitive materials, such as skiwear, mattress ticking, and Lycra fabrics.
2. The user has penetration control: penetration into the substrate can be completely avoided or, if desired, controlled.
3. Thanks to the low system content, the coating method is clean, and fast changes are possible.

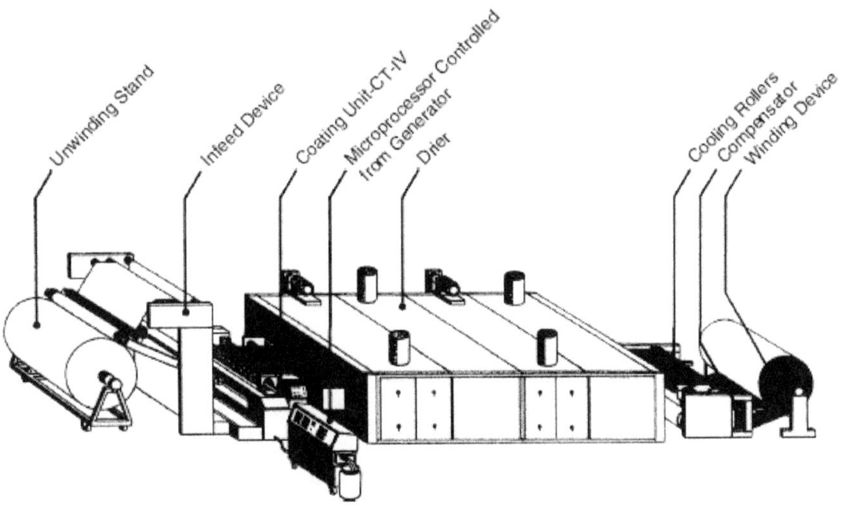

Figure .4 — Screen coating line

4. Coatings are exactly reproducible. As parameters, squeegee pressure, squeegee setting, mesh number, and viscosity can be measured and read off, and any given coating can easily be repeated.

5. Chemical savings (up to 20% of the coating weight) are realized in two ways: (a) through accurately controllable application and because the screen follows the web structure exactly (thus, the textile character is maintained, and an excess of paste, such as occurs with knife coating, is avoided); and (b) through great accuracy, in left/right and longitudinal directions, of the application amount.
6. Application is both tensionless and frictionless.
7. By means of the closed system, the user has total process control.
8. The knife coating option can be attached above the whisper blade roller, mentioned earlier.

This knife coating (see Figure .3) can be used as a knife-on-air system for paste or unstable foam coatings and in the knife-over-roll coater made for foam applications. In both cases, the apparatus is fitted with a paste or foam distribution system over the full working width. In this way, it is possible to apply colored coatings with a totally even appearance. The schematic diagram of a rotary screen coating line is shown in Figure .4.3

10 Paint Manufacturing Process

When paint was first discovered in ancient days, it was produced by mixing a natural resin from a plant or tree with natural colorants from the earth or plants. The ingredients were mashed together in some type of container. It was not very durable, but it allowed people to create art and illustration. Modern paint manufacturing is much more sophisticated, but it still involves a blending process to disperse pigment into a vehicle system.

The vehicle system, typically a resin or a varnish type material, is first charged into a mill with thinner and grinding aids. Pigments and extenders are added. The mill is then operated until the pigment particles are broken down and uniformly dispersed to create a grind paste. The grind paste is extended with additional resins and thinner and dropped into a thin-down tank. There, other materials are added to the tank to make the finished product. At this point, a sample of the material is brought into the quality control laboratory and a number of tests are conducted to compare the batch versus the standard parameters it is designed to meet. Trained technicians evaluate the results and then make adjustments for color matching and other characteristics such as viscosity and gloss. In addition to the normal adjustments made to batches, quality checks are performed throughout the manufacturing cycle.

When the paint in the tank meets all the prescribed specifications, quality control approves it for filling. The paint is then filtered to strain out any large or unwanted particles and filled into containers.

The most critical operation in the manufacturing process is the paste grinding process. This is where the solid pigments are broken down into their finest particle size in the resins, solvents, and additives (the vehicle system).

EQUIPMENT USED FOR PIGMENT PASTE MANUFACTURING

The equipment used to grind the pigment is an important part of the process. Different mills work well with different pigment materials.

High-Speed Mixers (HSD, HLM)

Among the different types of mills used for grinding, the most widely used is the high-speed disperser (HSD). This mill consists of a tank that is jacketed for cooling water, a motor, and a shaft with an impeller that resembles a saw blade with teeth that are bent 90°.

The dispersion is achieved at or near the surface of the impeller by the violent changing of direction of the mill paste. This shearing action is used to handle easily dispersed pigments such as those found in whites and light pastel colors. An HSD is used in conjunction with the continuous mill (COM) to premix the mill paste. Both the sand mill and COM are used for the more difficult dispersing pigments such as those found in blue, green, and red colors.

These mills use a variable speed saw-tooth blade rotating at high speed. Peripheral speeds range from 3000 to 6300 ft/min. HLM mixers also have a slow speed sweep agitator at the tank wall to aid in movement from the tank walls.

- Critical parameters: paste viscosity, RPM of shaft, and level of paste in the tank (optimum range 2.0–2.5 times blade diameter)
- Advantages: requires least amount of labor, easy cleanup, and low maintenance
- Disadvantages: high power consumption, easy dispersing pigments needed, and generates heat
- Typical products: latex, some primers, plastisols, high solids polyesters, polyurethanes, and acrylics

VERTICAL OR HORIZONTAL CONTINUOUS MILL

Another device is a COM. It is also charged with glass or ceramic media and disperses by shear and attrition. In this case, the degree of dispersion is determined by the flow rate of the premixed paste through the COM. These mills have a high-speed shaft with multiple discs of various configurations rotating in a long chamber filled with beads.

- Critical parameters: high-throughput rates, easy to clean, low yield losses, and large batch size
- Disadvantages: requires premix tank, expensive to maintain, and cannot grind tougher pigments in one pass
- Typical products: primers, polyurethane, water reducible (W/R) primers, clean colors, alkyds, and lacquers

Sand Mill

Another type of grinding device is a sand mill. It uses a different type of impeller, and it is charged with glass or ceramic as a grinding media. Pigments are ground by shear and attrition and the finesse of the dispersion is determined by the length of time the paste is in the mill, or residency time. This mill consists of a high-speed shaft with one or two solid disc impellers rotating off center in a cylindrical tank with a cone-shaped bottom, filled with small grinding beads.

- Critical parameters: ratio of paste volume to media volume and paste viscosity; impeller speed = 3500 ft/min
- Advantages: can grind most pigments, low maintenance
- Disadvantages: batch operation—requires many mixes, slow discharge, yield loss high, and cleanup is difficult
- Typical products: high-solids polyesters, plastisols, water-based baking enamels, alkyds (light colors), and lacquers (light colors)

Roll Mill

A roll mill, another milling device, can use two or more rollers to blend the paint ingredients. The material to be milled is placed between the feed and center rolls. Each adjacent roll rotates at progressively higher speeds. For example, the feed roll may rotate at 30 rpm, the center roll at 90 rpm, and the apron roll at 270 rpm. Material is transferred from the center roll to the apron roll by adhesion. The dispersion is achieved by the shear forces generated between the adjacent rolls. The milled material is removed from the apron roll by a knife that runs against the roll.

- Critical parameters: roll pressure, roll clearance, paste viscosity, and tack
- Advantages: can grind concentrated pastes with high polyvinyl chloride (PVC) loadings, no wash needed, very fine dispersions obtained, and no metal contamination
- Disadvantages: requires premix operation, slow process, and expensive to maintain
- Typical products: electrodeposition paste, high-solids polyesters, alkyds, and acrylics

All Mill

One of the older types of milling equipment that is still used is the ball mill. This mill consists of a cylindrical tank mounted horizontally. It is loaded to one-third of its apparent volume with 5/8" steel balls as a grinding media. There are lifter bars located on the inside wall of the mill and as the mill rotates the balls are lifted and then cascaded downward, dispersing the pigment by attrition and shear. These mills are used for the most difficult pigments to disperse such as blacks and organics.

- Critical parameters: ball charge = 33% of mill volume; normal paste charge = 25–40% of mill volume

- Advantages: no premixing required, no solvent loss, low maintenance, and can handle hard grinding pigments
- Disadvantages: long grind times, limited batch size, discoloration from metal, yield losses, and difficult to clean
- Typical products: primers, high solids polyesters, alkyds (dark colors), acrylics (colors), acrylics (colors), and lacquers (dark colors)

After the paint is milled, it is filtered and put into containers. Filtration methods vary, depending on the paint material and the needs of the customer.

Color Matching and Color Control

The appearance of a finished product is often one of the most critical aspects of quality. The way it looks has a significant psychological impact on the perception of quality by the potential customers and on their willingness to buy the product. Color is the most important appearance attribute of an object. There are other attributes such as size, shape, gloss, and texture, but color is by far the most important.

Color, gloss, and texture are imparted to a product by a coating. The coating manufacturer has the responsibility for formulating an accurate color match based on the customer's sample. The applicator also plays a role in accurate color match, since mixing, storage, application, and curing can impact color. It is necessary for the manufacturer to have a quality control program for the desired color and appearance of their product. But it is equally important to be able to detect all sources of error that affect color and appearance.

Color match is usually judged by visual evaluation in standard lighting. Some will use color difference measurement devices as well. An inspector will compare the color of the product to a color standard sample.

If there is some color difference, it will be necessary to judge if it is within acceptable tolerance levels. When there are guideline panels that represent the maximum allowable color difference limit, the task is a little less difficult. Any difference in gloss or texture will further complicate the visual process, because the color match will vary as the angles of view and illumination vary.

Usually, the coating is checked for color by the quality control lab before the customer uses it. A number of sources of error can cause a poor match.

Examples of sources of error:

1. Different color standard
2. Metameric color standard
3. Metameric batch
4. Gloss difference

5. Texture difference
6. Observed metamerism
7. Panel preparation differences
8. Substrate differences
9. Film thickness variance
10. Disagreement between color measuring instruments

In addition to these items, it is possible that the paint needs to be adjusted for accurate shading.

A color difference measurement can provide accurate numerical color match or define a difference from the specification. When measurement equipment is not used, it is hard to describe the visual difference and make an adjustment.

COLOR VARIABLES

Color has three visual variables: hue, saturation, and lightness. Hue is a color variable denoted by red, orange, yellow, green, blue, violet, and purple. Saturation is the variable of color that denotes its purity or its degree of departure from grayness. Lightness is the variable of color that distinguishes black from white, without regard to hue, as represented by a neutral gray scale ranging from black to white. Hue, saturation, and lightness are often called the three dimensions of color, because it takes a three-dimensional space to represent them. Lightness is distance along the vertical axis, saturation is radial distance away from the vertical axis, and variation in hue is angular distance around the vertical axis.

Color matching is the process of selecting colorants to match a given color and adjusting their concentrations until the hue, saturation, and lightness of the trial (batch) match the hue, saturation, and lightness of the given color sample.

Knowing that color has three variables, leads to the four-colorant principles of color matching. To match a color, hue, saturation, and lightness must be adjustable or controllable and can only be done with a minimum of four colorants. For example, if only one colorant—white—is used, there is no way to change it. It can only be white. By adding black as a second colorant and varying the concentration of the white and black, any number of colors, varying essentially only in lightness can be made. By adding a third colorant, for example red, any number of colors can be made, varying in lightness and saturation but all having essentially the same red hue. By adding a fourth colorant, yellow, then all three color variables—hue, saturation, and lightness—can be varied, and a color can be matched exactly.

If a color is formulated with less than three, then the match can only be obtained when all of the colorants are exactly as they were originally. There is no latitude for production variation. If more than four colorants are used, then it becomes more difficult to determine which colorant will correct a hue difference, a saturation difference, or a lightness difference, because two colorants in a five-colorant formula will control the same variable. For example, a green made with white, black, blue, and two different yellow pigments, needs to be more yellow. Which yellow is added? Some yellow pigments will make it more yellow but also less saturated. Other unsaturated pigments may be more effective in correcting the hue without affecting the saturation. To correct the hue, the effect on saturation must be considered.

Metamerism

Another reason why a five- or more-colorant formula should not be used is metamerism.

Metamerism is when the colors match less than one set of illuminants, observer, and viewing conditions, but do match under another. To prevent this, the match must not only contain the same colorants but also the concentration must be the same. In a five or more colorant formula, it is almost impossible to produce batches repeatedly, with the colorants at the same concentrations. Color may not always be controlled. A four-colorant formula is unique, because only one set of concentration of the colorants will produce a match and therefore no batch-to-batch metamerism is possible.

When matching colors by the visual approach, the color matcher selects the colorants by trial and error, experience, and knowledge of subtractive color mixture. It does not take long to learn that mixing blue and yellow together can make a green color. Colorant adjustments are made until the appearance is correct. Blue, black, or some other color is added until a match is made. A quantity of a colorant is added and the change in color is evaluated. If more is needed, the amount is estimated by that change. This process is repeated until a satisfactory match is obtained. With good technique and accurate weighing of colorants, an experienced color matcher can make a match fairly quickly.

The problem with the visual approach is that the matches may be metameric, which may or may not be acceptable. The metamerism occurs because the colorants used are different from those that were used to produce the color sample. There are three types of metamerism: illuminant, geometric, and observer. Illuminant metamerism is what is normally thought of when the word metamerism is mentioned, a change of light that causes a change of color.

Geometric metamerism is when the colors will match only at a certain angle of viewing and illumination. This is caused by difference in gloss and/or texture. A flat color will only match a gloss color at one viewing angle.

Observer metamerism is when colors are a match for one observer and not for another. It is caused by differences in the color vision of the observers and is quite evident when illuminant metamerism also exists. When there is difficulty in getting a match accepted by a customer, it may very well be observer metamerism.

Dr. Henry Hemmendinger is a leader in understanding the breakdown of a color match by change of either illuminant or observer. Hemmendinger and Davidson developed the D&H Color Rule, a device to quantify the extent of observer metamerism, which is still viewed as an indispensable aide in teaching the principles of observer metamerism. Dr. Hemmendinger, also developed methods to use metameric pairs as tools to assess instrument performance. An observer can readily get a match with highly metameric samples. The pairs chosen are noted and compared to the pairs chosen by another observer. If the pairs are the same under the same light source, the observers have the same color vision. If the pairs appear different under the same light source, it indicates a difference in color vision between the observers. Visual color measurement is done in a light booth such as a Macbeth Spectralite.

This light booth provides three light sources that are commonly used to visually evaluate color and color differences. They are North Daylight, Incandescent, and Cool White Fluorescent (CWF). The International Commission on Illumination (ICI) has standardized on North Daylight and Incandescent and given them the designations illuminant D65 and illuminant A, respectively. CWF has no standard as yet. The important feature about the booths is that they supply a light source that is constant. Natural daylight varies with the time of day. In addition, metamerism can be evaluated very readily, because two different light sources are provided.

Two ASTM (American Society for Testing Materials) standard practices—D1729, Visual Evaluation of Color Difference of Opaque Materials and D4080, Visual Evaluation of Metamerism—are methods that recommend the use of light booths. When matching a color where metamerism is obvious, the match under daylight may look like it requires red to correct it, but under incandescent it needs green. When there is no concern about metamerism, color match is done under one light source, preferably the one under which it will be viewed. When metamerism cannot be tolerated, other colorants are selected and tried until the combination does not produce a metameric match. This trial and error method can be very time consuming and is outmoded by methods using color measuring instruments and computer color matching (CCM) systems.

COLOR MEASUREMENT EQUIPMENT

Intelligent use of color-measurement instruments will increase the efficiency of matching colors. They measure color difference in terms that allow description of hue, saturation, and lightness, and also quantify these differences. This makes it possible to estimate the amount of colorant to add to correct the color.

Colorimeters measure color in terms of three numbers called tristimulus values— X, Y, and Z. These values are the amounts of three primary colors—red (X), green (Y), and blue (Z)—needed to match or specify a color under illuminating and viewing conditions as standardized by the ICI. Each color has its own tristimulus values, and identical values are identical colors. Differences in tristimulus values between colors can be analyzed to determine the direction and magnitude of the color difference.

To facilitate the translation of the color difference from numbers to the more meaningful visual descriptive terms and quantify color difference, the XYZ values are converted into the Hunter L, a, b Scale. This transformation is simple, and all modern instruments offer it as an option. L is lightness–darkness, a is redness–greenness, and b is yellowish–bluish. If the difference between the trial values and given color in L, a, b values was -0.5, -0.5, and +0.5 respectively, the trial values indicate that the color is darker, greener, and yellower than the given color. If the color difference is unacceptable, then the trial must be corrected by adding white to make it lighter, adding the bluest colorant to make it bluer or less yellow, and adding the reddest colorant to make it redder or less green.

By keeping records of the readings and the addition of the colorants, it is possible to estimate the amounts of the colorants for a correction. For example, one pound of blue lowered b from +0.5 to +0.25. The goal is to minimize b. Therefore, if one pound of blue changed b by 0.25 units, then another pound of blue would lower b to 0. The other colorants are estimated similarly.

Some instruments indicate color difference in the tristimulus values. The color difference is in terms of XYZ ratios, expressed as percent between the values of the batch and the standard. XYZ ratios of 100, 100, 100 indicate that the values of the standard and the batch are the same and the colors match. Any values other than 100 indicate that a color difference exists. It is much more difficult to translate these values into visual descriptive terms as opposed to L, a, b values.

As an example, green reads 101.5, 100.5, 100.5. The Y ratio, which is a measure of lightness, is greater than 100 indicating the batch is lighter. In order to determine the hue and saturation difference, it is necessary to compare the relative values of XYZ ratios. The hue of a green can only be yellower or bluer so the Z ratio is the hue indicator. However, it must be compared to the X ratio that is the saturation indicator for a green. In the example, Z is lower than X meaning that the batch is yellower.

For saturation, the X ratio must be compared to the Y ratio and since the X ratio is greater than the Y ratio, the batch is less saturated or grayer. Another way of looking at this is that there is more red than needed. Red is the complementary color of green. Complimentary colors when mixed produce gray. The logic is similar for analyzing the remaining category of colors blue, yellow, and red.

XYZ values are the basic measurement of color, and for this reason estimating a correction to a batch is easier and more reliable than other color scales but still requires experience. A green reads 110, 110, 110 and is made with white, blue, yellow, and black. An addition based on experience of one pound of blue, one pound of yellow, and 0.1 pound of black is made. The result of the addition was 105, 104, 105 and another addition is necessary to correct the match. To estimate the amount of blue, the ratio affected the most is used to compute the addition.

Although this can be determined experimentally by adding some blue by itself and seeing which ratio has changed the most, it is more expedient to use the following rule. The bluest colorant will lower the X the fastest, the reddest colorant the Y, and the yellowest the Z. Therefore, X will be used to estimate the amount of blue, Y the black, and Z the yellow. One pound of blue lowered the X ratio by 5 and to get 100, another pound is added. A pound of yellow will lower the Z by 5 so another pound of yellow is needed. The black lowered Y by 6, so 0.06 pounds more is needed.

This method works very well, but experience is needed for the first addition and all colorants must be added simultaneously or too much of the colorants will be estimated. Once a batch is shaded using this method, and the readings and colorant additions are recorded, the information can be used for succeeding batches. This has been called the "historical" method and has proven to be very successful.

This method would not have been possible without color measurement. The most difficult decision to make when matching a color is deciding when to stop. Viewing colors that differ and judging whether or not it is a close enough match is very subjective. No two people see color alike. It is a rare occasion when total agreement will occur, and this is especially true when colors are metameric. Every individual has his or her own idea as to what is an acceptable match, and this may change from day to day because of deadlines imposed to get the job done. This is where instruments perform well. Measurements are objective and consistent and not influenced by external pressures.

Color Difference Equations

Quantification or color difference makes it possible to establish a meaningful measurement of the closeness of a color match. The basic unit is a just noticeable difference (JND). One JND is a very close match whereas three would be easily perceptible.

Color difference equations calculate the color difference in units of E, the error of the color match.

Tristimulus Filter Colorimeters

The Hunterlab D25 and the Gardner XL 805 are useful for color correction (shading) and quality control, where the colorants are known and there is no metamerism of any sort. The tristimulus data they provide is for only one illuminant (usually C), and therefore, metamerism cannot be identified or checked.

Computerized Color Matching

CCM is the best tool for matching colors. In the hands of a skilled color matcher, the quality of matches and gain in efficiency can be quite dramatic. It is a big improvement over visual trial and error and the instrumental methods mentioned previously. CCM systems are available from Diano, Macbeth, ACS, Universal Color Systems, X-Rite, and Instrumental Color Systems. All are good and all have advantages and disadvantages as far as user friendliness is concerned. None of them will do exactly what you would like them to do, but this is true of any software that is not custom designed. Several companies, DuPont, Sherwin-Williams, PPG, and Benjamin Moore, to name a few, have developed proprietary software to satisfy their own needs.

Color matching is a science that is best left to formulators and quality control experts. For the applicator, it can be expressed in more general terms. There are many pigments that can be used to obtain a color. All these pigments vary from one another within the same color group. For example, there are red pigments with yellow cast, orange cast, blue cast, and so forth. The pigments can be clean, meaning that they reflect light in a very limited wavelength range, or they can be dirty and reflect light in a much wider wavelength range.

For a new color, the paint supplier should be consulted to see if the color is feasible. They can usually determine how easy it will formulate a color match. If a color standard is already available, the paint supplier can match the established standard and submit a sample. When the sample is approved a master standard is prepared. The painted panels can be used as the master standard and working standard. Color tolerance charts can be prepared to monitor differences.

Since color standards change with time and use, the Macbeth color program ensures a consistent standard. They can also generate color tolerance charts that establish a range of color variation that is acceptable for production color matching and establish the visible limits of an acceptable color match.

For visible color match system, the following parameters should be included in color control:

1. Light source to be used for viewing, that is, daylight or CWF. By using only one light source one can eliminate metamerism from one light source to another. The light should be the one that the product will normally be seen in.

2. Angle at which panels will be viewed.
3. Viewing distance from panel. At 50 ft, all colors look good. At 6 in, color matches are very difficult.

Remember, when comparing colors, that the appearance of the panels must be the same, that is, both smooth or both textured. In addition, the gloss should be the same within limits, generally ±5 gloss units.

11 Liquid Paint Application Systems

Liquid paint can be applied in many different ways. The three major components of a paint line are pretreatment, application, and cure. The type of industrial application process will be determined by the product, the volume, the quality requirements, and other factors. It may be as simple as a dip tank and a method of hanging parts, or it may be a fully enclosed automated system that allows continuous pretreatment, application, and curing. A basic conveyorized paint line layout is illustrated in Figure .1.

The actual layout and specific equipment vary widely, but all paint lines have the same general stages: a loading area, pretreatment or cleaning stage, application stage, curing stage, and unloading area.

The loading/unloading stages are where parts are put on and taken off the paint line. The pretreatment stage is where the parts are prepared for the coating application stage. Cleaning can be done mechanically, such as blasting, or chemically by solvent washing, solvent wiping, power washing, or dipping. After cleaning, the parts are dried by heat or forced air or a combination of the two so that they are ready to be painted.

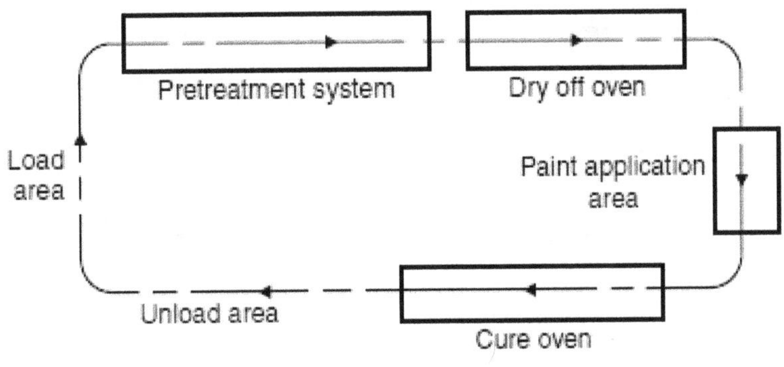

Figure .1 — Basic paint system layout.

Paint can be applied by spray method using conventional handheld spray guns, high-volume-low-pressure (HVLP) guns, automatic guns, automatic rotary bells, or disks. The spray equipment can be either nonelectrostatic or electrostatic. Dip systems, flow coaters, curtain coating, or roll coating are other application methods.

The curing stage is where the paint film is formed. Paints can be cured by air drying or heating. Ovens for heating are conventional gas-fired, high air velocity gasfired, electric, infra red, UV energy, electron beam, or induction. The ovens can be either batch type or continuous. Parts are moved from one stage to another by manual or automated conveyor systems.

PART PREPARATION

As discussed in Chapter 1 on pretreatment, a properly prepared surface is critical to the success of the coating. Paint does not hide defects or clean a substrate. If the substrate is dirty or has a film on it, the paint will adhere to the dirt film, since it is the first surface it comes in contact with. If that film adheres poorly to the substrate, then the paint will lift off.

In many cases, the dirt or film may not be compatible with the paint. This will cause the paint to pull away from that area and cause fish eyes or crawling on the surface of the substrate. With the need to use high-solids, solvent-based paints, or water-based paints, the cleaning step becomes more critical. Prior to the regulation of organic solvent emissions, paints were commonly 20–30% solids—the remainder being solvent. In many cases, this excess of solvent could attack the surface film and allow the paint to adhere through it to the substrate. Today this is not the case.

With regard to surface defects such as scratches, welding marks, porosity, and so forth paint will only flow over a surface in a uniform manner and will not hide a defect as shown in Figure .2. The only way to hide the defect at this point is to sand the paint until smooth and then recoat.

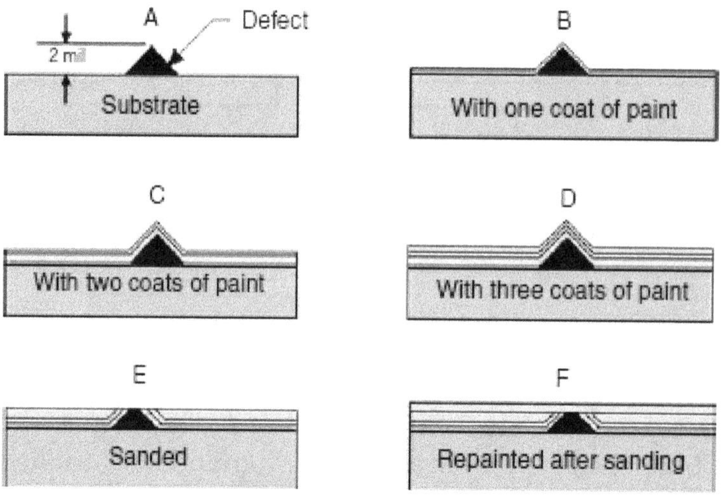

Figure .2 — Repair of a paint defect caused by particulate on the surface.

Surfaces to be painted must be clean and dry to ensure paint adhesion. They must also be nearly as smooth as the desired final finish. Generally, paint cannot be used to hide surface defects. Paint will often highlight surface defects by reflection and make them more noticeable.

In addition to cleaning and smoothing, surfaces are often converted to a material different from the original before painting. Phosphated steel and flamed plastic surfaces are examples. Surface conversion is generally used to improve paint adhesion, reduce corrosion, or both.

Smoothing

Some surfaces require grinding, filing, or sanding to make them smoother before the paint is applied. Grit blasting or heavy grinding can be used to remove burrs or scale. In some cases, the substrate will require a combination of these processes. Large pits, holes, seams, or scratches in a surface can be filled with a liquid or paste material and then sanded after they harden.

Typical grinding papers use aluminum oxide or silicon carbide (carborundum) abrasives. Blasting abrasives may be sand, glass, metal grit, or walnut hulls. These media are available in a variety of grit sizes. Abrasives must be replaced or cleaned regularly so that they do not become oily and contaminate the surface.

SPRAY APPLICATION OF PAINT

Paint can be applied to a substrate in many different ways. Each method has advantages and limitations. The needs of the product and the manufacturing process will help determine the correct application method. An understanding of the different application processes is necessary in order to evaluate them for a specific task.

Most industrial paint systems apply liquid coatings with spray equipment. The liquid coating is delivered from a container to a spray device that uses pressure to break the liquid stream into fine droplets. The even distribution of the liquid paint into a finely divided mist of droplets is called atomization. There are several different types of spray guns and different methods of atomization and pattern control that can be used to control quality and efficiency.

METHODS OF ATOMIZATION

- Air-spray is one of the earliest methods of atomization. An air-spray device mixes a high-velocity air stream with a low-velocity fluid stream and the resulting friction creates the paint droplets.
- Hydraulic atomization occurs when a high-velocity fluid sheet that is formed by a small orifice at the spray gun tip exits into a stationary atmosphere. The resulting friction causes atomization into droplets.

- Centrifugal atomization occurs when a fluid stream is introduced at the back of a disk that is rotating at high speed. The centrifugal force of the rotating disk delivers the paint to the edge of the disk and breaks it into a fine mist.

There are certain features that all spray applicators have in common, but each different spray device has particular differences that make it fit with a certain application. An understanding of the different applicators and their strengths is useful in the selection of the correct device for a particular application.

Manual Spray Guns

There are a number of different spray gun technologies used for manual spray application of industrial coatings. Most coating systems use manual spray guns, alone or as reinforcement for automated application. Manual spray equipment provides the freedom to spray in hard to reach areas at different angles that would be difficult to reach with a stationary automatic gun.

Manual spray technology is available with or without electrostatic capability for better transfer efficiency. Handheld spray guns are typically identified by the following names:

- Conventional air-spray
- HVLP
- Airless
- Air-assisted airless (AAA)
- Electrostatic HVLP
- Electrostatic air-spray
- Electrostatic AAA

Each of these devices has features that make it work well in a given situation. Selection often depends on the cost of the device, the coating(s) to be used, and the difference in application efficiency. Other factors may also influence the specific gun used in a certain application.

Automatic Spray Equipment

Automatic spray equipment is used when the volume of work exceeds the capabilities of manual operators, the product is too large for men to reach, there is an interest in reduced labor, or there is an interest in automation for more precision. It is available in all standard forms of air atomization and also uses centrifugal or rotary devices for atomization.

Most automatic spray systems use a device that moves the spray guns and controls application. Programmable logic controllers (PLCs) use part recognition systems to trigger guns on, initiate color change, and adjust the equipment for optimum efficiency. In addition to control of paint application, PLCs are used for paint mixing, color changes, and other control tasks.

Spray Systems

Conventional air-spray guns are common spray tools that can be used with and without electrostatics. Aconventional air-spray gun is a precision tool that uses energy in the form of compressed air to atomize the material. Air and paint enter the gun at different points and meet at the head of the gun. The high-velocity air stream from the annular hole around the nozzle passes the low-velocity fluid stream and the resulting friction creates a pattern of fine droplets. The pattern of fine paint particles provides a uniform thin film of coating on the part surface.

When the gun trigger is pulled back slightly, an air valve, usually located in the upper portion of the handle, will open and allow air only to reach the gun tip. When the trigger is pulled back further, the fluid needle is retracted and paint begins to flow from the fluid nozzle. Immediately upon leaving the nozzle, it is mixed with air from the air nozzle, and the spray pattern is formed. Figure .3 shows the inside of a typical air-spray gun.

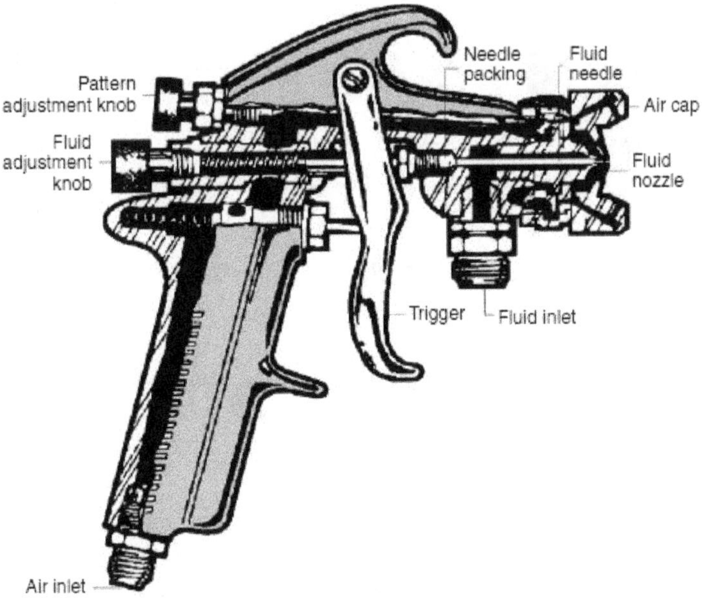

Figure .3 — Spray gun components.

Suction or Siphon Systems

A suction or siphon feed gun uses a stream of compressed air to create a vacuum, allowing atmospheric pressure to force the fluid material from an attached container to the spray head. A siphon cup is a simple system for testing, for very short runs, for shading operations, or similar situations that require a small amount of paint and a short run. These systems typically use a one-quart container, and they are not very effective for high-viscosity materials. A siphon cup system is shown in Figure .4.

Another system that works well for small volume applications uses gravity to feed the paint to the gun tip. It is lightweight and uses virtually every drop of paint in the cup so that almost no paint is wasted in the cup.Agravity cup is shown in Figure .5.

Pressure Feed Systems

A pressure feed system applies paint that has been forced to the gun by pressure, either from compressed air or a pump. Pressure-feed systems are used in situations requiring more volume, delivered from a larger vessel, rather than the small cups used in gravity and siphon systems. It is also suitable for paints with higher viscosity.

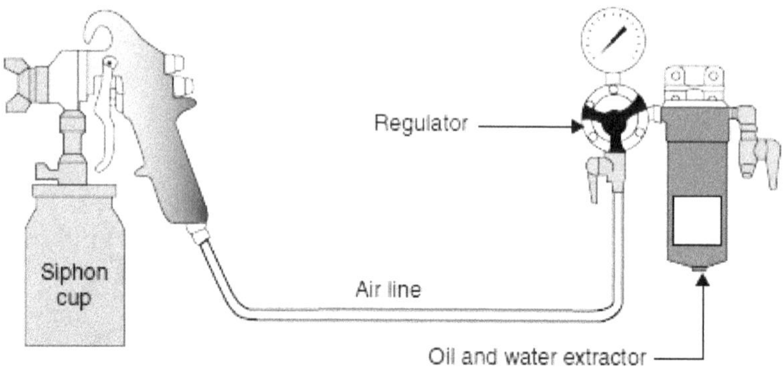

Figure .4 — Siphon feed system.

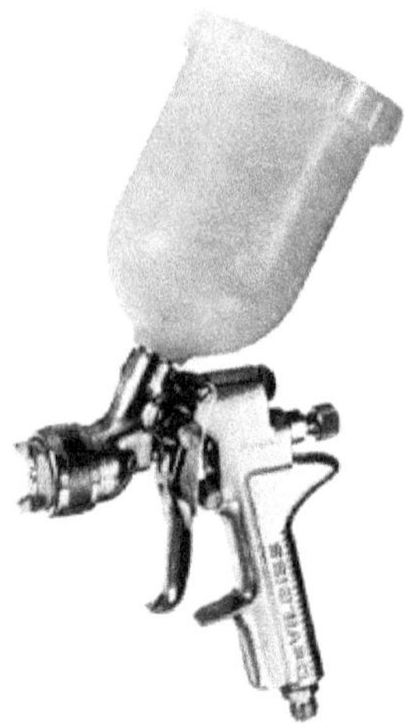

Figure .5 — Gravity feed system.

Pressure pots use compressed air to create fluid pressure and force the paint through the fluid hose to the spray gun. There are many sizes of pressure vesselsthat can be used depending on the size of the operation. Five-gallon pots are fairly common. A system with a 5-gal pot is shown in Figure .6. A pump is often used to supply coating material from a larger container such as a 55-gal drum.

Circulating Systems

Some systems deliver paint to the booth but do not circulate it back to the container. Circulation systems are more effective for maintaining a stable paint material and limiting waste materials.

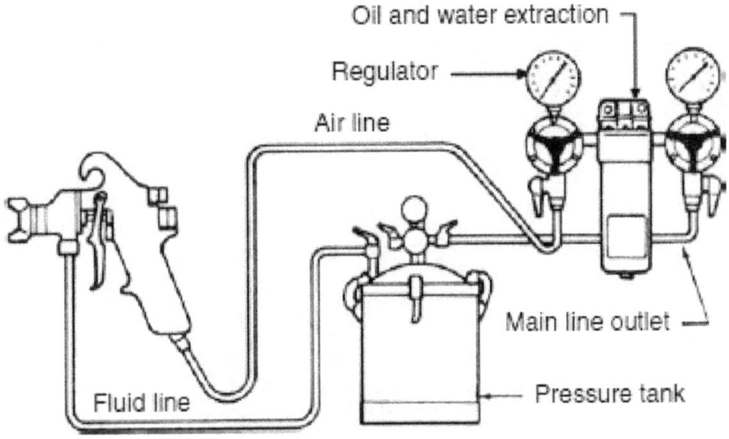

Figure .6 — Air-spray pressure feed system.

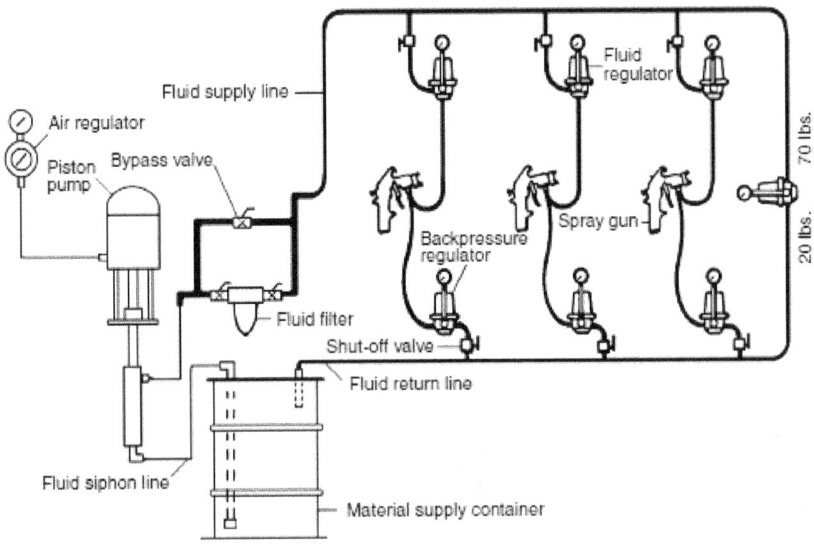

Figure .7 — Paint circulation system.

Many paint lines are installed with paint supply systems that bring fluid to the booth and circulate it back to a supply container in a "paint kitchen." A typical circulating system is shown in Figure .7.

Plural Component Mixing Equipment

Two component materials must be mixed prior to spray application. Mixing can be as simple as mixing the two materials at the desired ratio into a siphon cup or pressure feed tank in an appropriate batch size. For high-volume mixing requirements a mechanical proportioner, gear driven mixer, or an electronic mixer may be the better choice. In all cases, the objective is to consistently mix two or sometimes three materials at the recommended mix ratio while always generating little or no waste. Mix ratios may range from 1:1 to 50:1 or greater. It all depends on the coating and the application requirements.

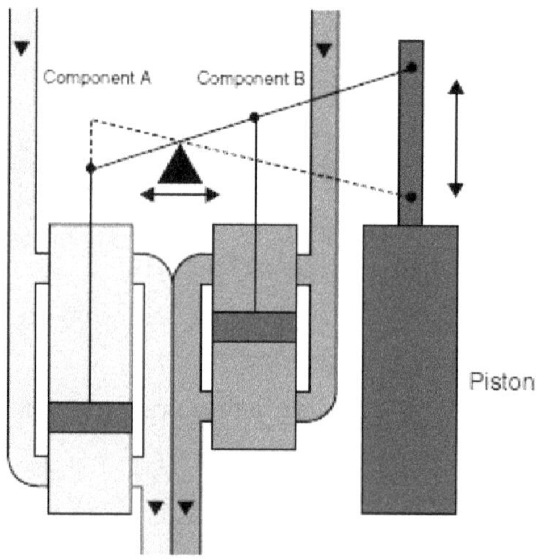

Figure .8 — Dual piston pump mixing system.

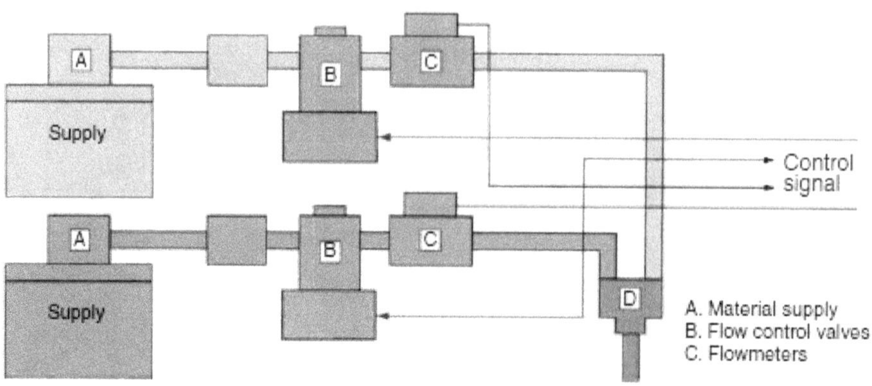

Figure .9 — Two component proportioning system.

Mechanical systems, like the dual piston pump system shown in Figure .8, cost less money than the cost of more sophisticated systems, but they are maintenance prone and flushing is difficult and expensive. Ratio is adjusted by move the fulcrum position by trial and error.

Electronic systems provide precision mixing and information such as notification if the material supply is low. A well-designed system can supply reliable mix ratios and provide alarm conditions so that problems are quickly identified and corrected. A typical system is shown in Figure .9.

Air Nozzles

The air nozzle directs the flow of compressed air through a series of holes into the paint stream at the gun tip. The air atomizes the paint and directs the flow of atomized paint into the pattern. Aprofile is shown in Figure .10 with fluid flowing around the fluid needle and air introduced by the air horns at the gun tip. Figure .11 shows the air nozzle.

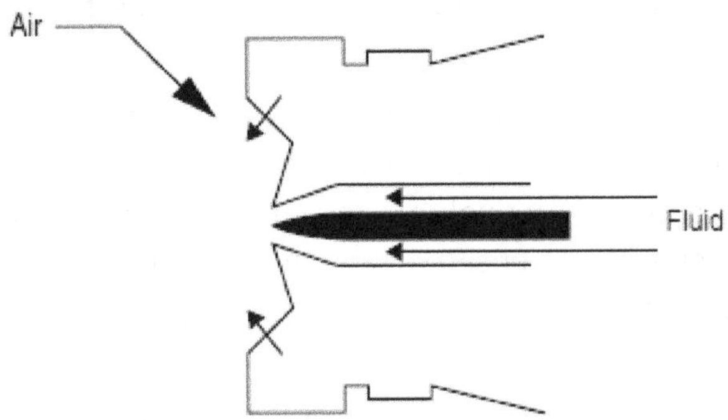

Figure .10 — Spray gun air nozzle.

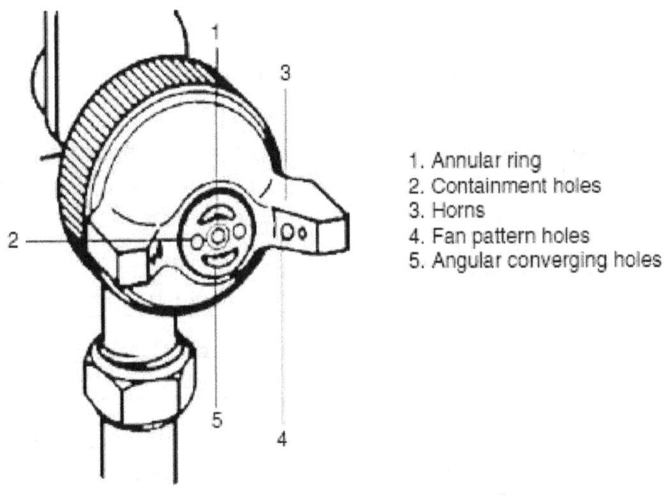

Figure .11 — Air-spray air nozzle.

Some air nozzles are referred to as internal-mix.With an internal-mix nozzle, the air and paint are mixed inside the nozzle and the pattern is formed as the mixture exits through a hole in the nozzle tip. An air nozzle introduces the compressed air to the fluid stream at the gun tip.

Fluid Nozzles

The fluid nozzle and needle combination is used to control the volume of coating material delivered at the spray tip. The fluid needle is pulled back when the gun is triggered and fluid flows through the nozzle. When the trigger is released, the needle slides back into place and shuts off the flow. Atomization occurs at the gun tip when compressed air is directed into the paint stream as it exits the fluid nozzle. The fluid orifice in the nozzle assembly provides the lowest possible velocity fluid stream.

Selection of a particular fluid nozzle depends on the viscosity of the coating material and the required level of atomization. Different sized fluid nozzles are available for materials including paint, adhesives, mastics, flocking materials, and so forth.

Usually, a nozzle with a smaller orifice will work fine with a lower viscosity material while a higher viscosity material may require a larger nozzle. If the nozzle is too small to deliver the volume of paint needed, the fluid pressure must be increased in order to compensate. Excessive pressure can create application problems so it is important to select the correct nozzle for the material being sprayed.

Nozzle size, fluid flow, and pressures are related. Selection of a fluid nozzle should consider the viscosity of the material, the amount of material required to achieve the desired film thickness, and the maximum desired fluid flow rate.

Table .1 shows some examples of different fluid tip orifice sizes and the types of products that they are likely to be used with them.

Table .1 — Fluid Tip Orifice Sizes

Orifice Rang	Viscosity Range	Example
0.022-0.028	Very thin	Solvents, stains
0.040-0.052	Thin	Lacquers, primers
0.059-0.070	Medium	Varnish, urethanes
0.086-0.110	Heavy	Expoies, vinyls
0.125-0.500	Very heavy	Block fllers

Table .2 shows the difference in fluid flow at different pressures for a material with a viscosity of 21 s in a #2 Zahn viscosity cup. Control of fluid flow can be adjusted by changing the fluid nozzle, by turning the fluid adjustment knob located on the rear of the gun handle, or by changing the air pressure through the fluid regulator.

Table .2 — Delivery in Ounces per Minute at Different Pressures

Orifice Size	At 8 psi	At 12 psi	At 18 psi
0.040	14	18	24
0.046	18	22	30
0.052	22	28	36
0.059	31	36	48
0.070	40	48	60

The fluid control knob on the gun is provided so that the gun can be adjusted when used with a siphon feed system. A pressure feed system should typically be adjusted by air pressure regulation with the fluid control knob left in the full-open position. Fluid control by adjustment of the needle position is not recommended.

Triggering a spray gun with no atomizing air can be useful for setting the proper fluid pressure. The resulting paint stream should flow in a solid stream for approximately 1′0″ to 2′0″ before falling off. Anything in excess of 2 ft indicates that the fluid pressure is too high.

Proper Fluid Pressure

The fluid needle fits into the orifice of the fluid nozzle. When the gun trigger is pulled, the fluid needle is pulled back from the orifice and paint material is allowed to flow from the gun. When the trigger is released, the fluid needle moves forward and stops fluid flow.

Fluid needles are machined from hardened steel or stainless steel with a tapered end. Fluid needles are also made of engineered plastics. The tapered end is designed to match the orifice of a family of fluid nozzles.

The fluid nozzle and needle is a pair. If a needle does not match a particular fluid nozzle, the taper will not match and fluid will leak from the nozzle tip, the gun will spit and drip. The nozzle and needle should be replaced. The manufacturers catalog will list the correct needle and nozzle families.

Sometimes it is necessary to make a needle adjustment to compensate for wear. Every time the gun is triggered, the needle slides back across the inner surface of the fluid nozzle and cause a small amount of wear. Eventually, there is enough wear to create a gap between the needle and the seat of the orifice and the gun will leak.

AIRLESS SPRAY GUNS

Airless spray systems use hydraulic pressure to force the fluid through a spray tip or orifice at high pressure. The combination of pressure and orifice size creates a pattern. Orifice sizes are catalogued by flow capacity and degree of spray angle.

An airless spray gun does not use compressed air for atomization, so there is no air nozzle with atomizing ports like those found on an air-spray gun. There is only one trigger position, full open.

An airless system works with a high-pressure pump that delivers fluid to the gun tip at 500–6000 psi and forces the fluid through the orifice.

Airless systems can be used with heaters to reduce the spray viscosity and improve finish quality. The heater helps to control spray viscosity for consistency and may be particularly helpful with thicker materials. Figure .12 shows an airless spray system.

FLUID NEEDLES AND TIPS FOR AIRLESS SYSTEMS

Usually an airless needle will be a permanent assembly having a ball, approximately 1/8″ diameter, silver soldered to the end. This ball will “seat” or act as a seal to prevent fluid from leaking through the fluid tip when the gun is not in use.

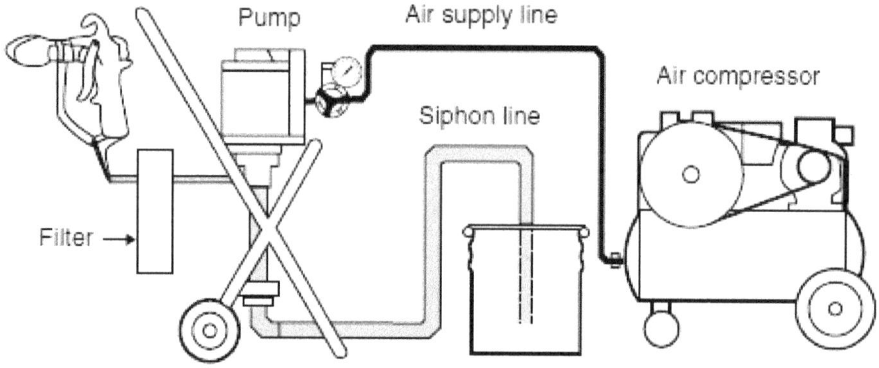

Figure .12 Airless spray system.

Table .3 : Orifice Sizes for Airless Systems

Orifice Size	Viscosity Range	Material
0.009-0.010	Very thin	Acetone
0.011-0.013	Thin	Water
0.015-0.021	Medium	SAE #10 oil
0.026-0.036	Heavy	SAE #50 oil
0.043-0.072	Very heavy	Vaseline
0.072-0.081	Extra heavy	Mastics

Table .3 — will give a general idea of orifice size for typical materials.

AIR-ASSISTED AIRLESS SPRAY GUNS

TheAAAgun uses a lower fluid pressure (typically 400–800 psi) with an airless tip and an air nozzle for atomization. This combination of elements provides faster application of material than air atomized spray provides, reduces overspray and rebound, and provides enough forward velocity to penetrate some recesses and cavities. The pressure range used with AAA spray guns is much higher today for high solids and ultra high solids materials. A picture of an AAA gun is shown in Figure .13. Note the "duck bills" at the gun tip that provide protection against accidental exposure to the high-pressure spray.

AAA does not provide the fine atomization necessary for extremely high-quality finishes. It is very popular in the wood finishing industry where stains, sealers, fillers, glazes, lacquers, and polyurethane materials are common. It is also very effective for applications that require high-TE, low overspray, fast application, and modest appearance standards.

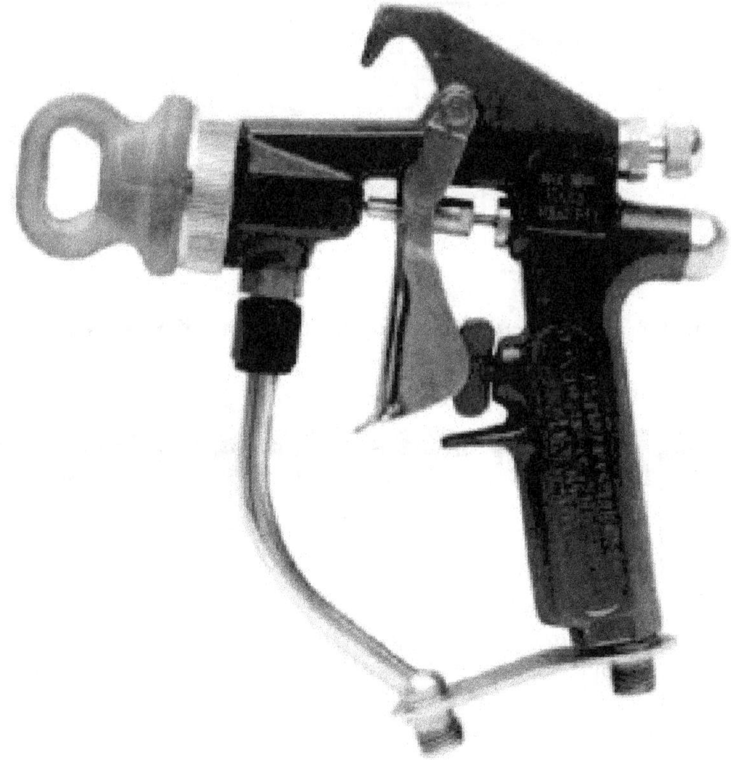

Figure .13 — Air-assisted airless spray gun

Airless systems are often used for field painting or shop painting that does not require fine atomization.AAAsystems is suitable for a wider variety of applications. Like an airless system, atomization is achieved by the choice of fluid tips and the pressure used to propel the coating through it. The added feature of compressed air has very little effect on atomization. Its main function is to fill the pattern so that the coating is uniformly deposited on the surface. Figure .14 shows a typical AAA system with a heater.

A comparison of the three gun types that have been discussed is outlined in Table .4.

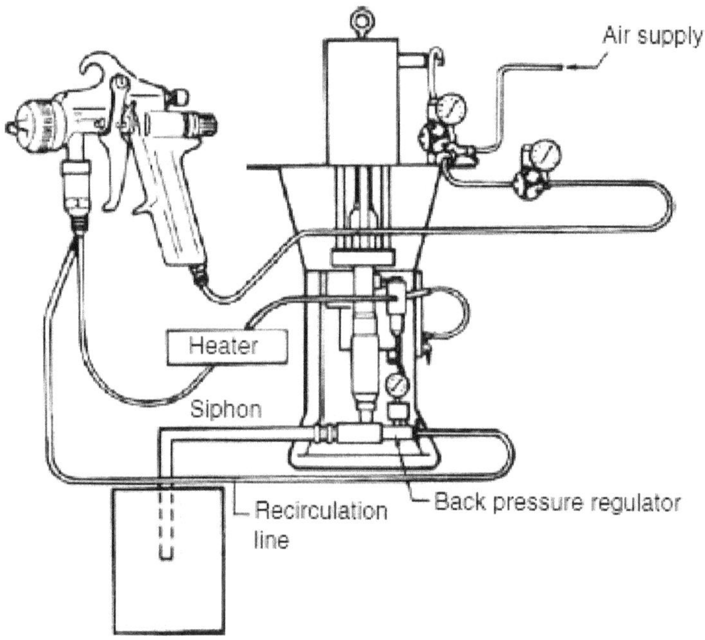

Figure 14 — Air-assisted airless system with heater.

Table .4 — Comparison of Different Spray Guns

	Air Atomized Spray	Airless Spary	Air Assisted Airless
Finish appearance	Excellent	Coarse	Very good
Transfer efficiency	Low	Good	Good
Fluid flow rates	Very low to medium, usually less than 30 fluid oz/min	High or very high flow rates, over 80 fluid oz/min	Wide range, 7 fluid oz/min to 78 fluid oz/min
Booth maintenance	Lot of rebound and overspray, booth maintenance is high	Less rebound and overspray, less maintenance	Least rebound and overspray, maintenance is low
Pattern adjustment	Yes	Tip selection	Yes (limited)
Tip plugging	Virtually none	Yes	Yes
Equipment life	Very long due to very low fluid pressure	Very high fluid pressure, high tip wear and shorter pump life[a]	Less tip wear, lower fluid pressure
Noise	Noisy	Quiet	Quiet
Emission level	High	Lower	Low

[a] if the pump is sized properly to reduce cycles, the wear is not different than the AAA.

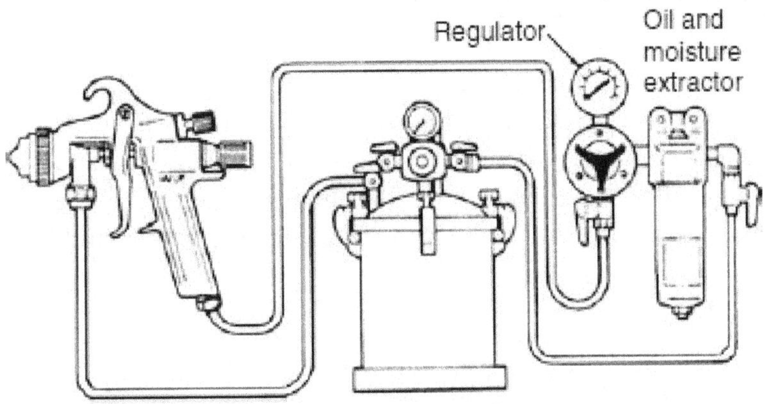

Figure .15 — High-volume, low-pressure spray system.

HIGH-VOLUME LOW-PRESSURE (HVLP)

HVLPgun technology is a form of air-spray that uses a restrictor to reduce the velocity of the coating material for increased transfer efficiency. It has become a favored technology by environmental agencies, because it uses about the same volume of air as a conventional air atomized gun, but it operates at much lower air cap pressures. An HVLP system is shown in Figure .15. HVLP can provide higher transfer efficiency compared to other conventional spray guns without electrostatic charging. This makes it particularly useful for nonconductive surfaces such as plastic and wood.

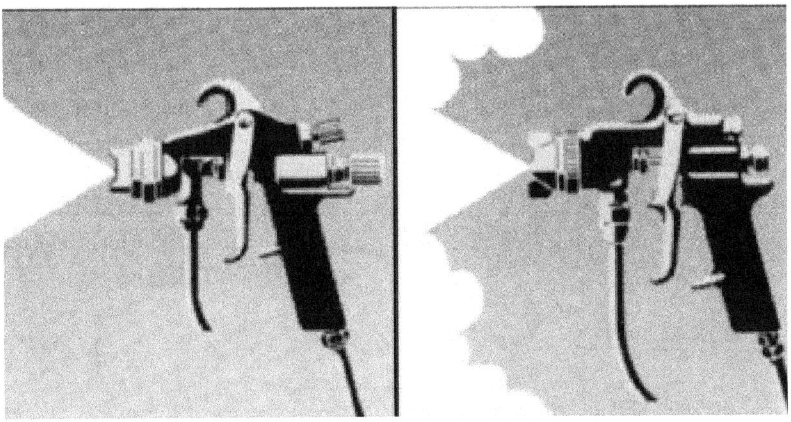

Figure .16 — HVLP versus conventional air-spray gun.

It takes a specific velocity of air to effectively overcome the cohesiveness of a paint material and atomize a coating. With air-atomized systems, more compressed air volume will increase the breakup of the material and produce finer atomization. The illustration in Figure .16 shows the difference in the spray pattern between an HVLP and a conventional air-spray gun.

ELECTROSTATIC APPLICATION

Electrostatic application is designed to enhance the attraction of the paint material to an earth-grounded surface. The basic features of the spray gun are the same, but a power supply is used to provide voltage and charge the sprayed coating material. Electrostatic attraction uses the elementary science where two surfaces that have the same charge will repel each other and unlike charges attract. Figure .17 shows a hand-held electrostatic spray system with an air-spray gun and Figure .18 is a system with an airless gun.

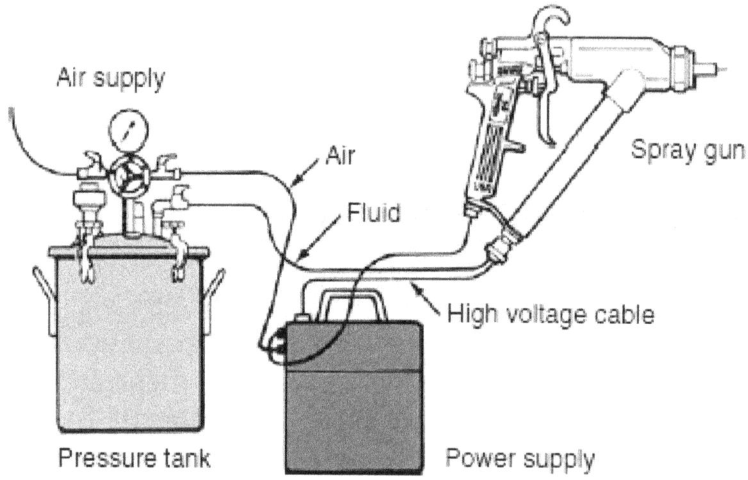

Figure .17 — Electrostatic handheld air-spray system.

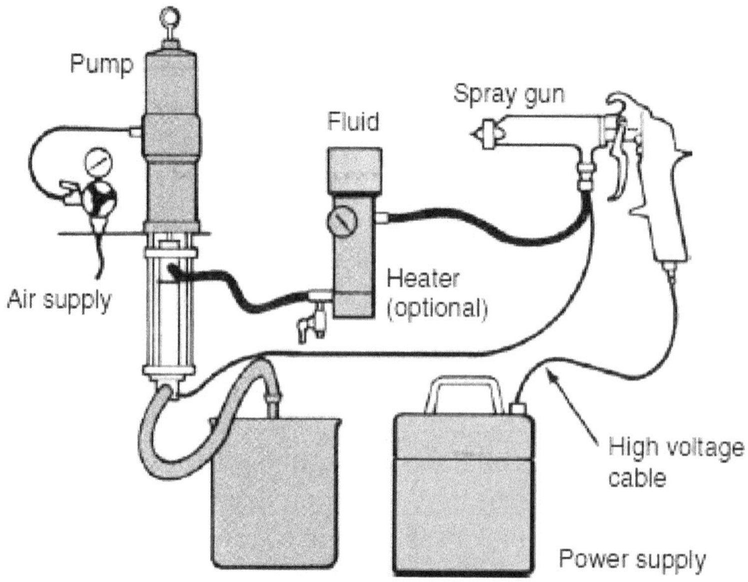

Figure .18 — Airless electrostatic system.

High-voltage electric charge is added to the atomized paint particles as they leave the gun. A voltage generator is connected to an electrode at the gun tip. The voltage source can be external or internal. When the gun is triggered, the electrode discharges a concentrated field of negative ions. Paint particles travel through this ionized area and become charged with negative ions.

The charged particles of atomized paint are directed toward the earth grounded part from the spray gun. As they approach the target surface, the electrostatic charge creates a strong attraction to the part surface. The charged particles are strongly attracted to all bare surfaces of the grounded part as shown in Figure .19.

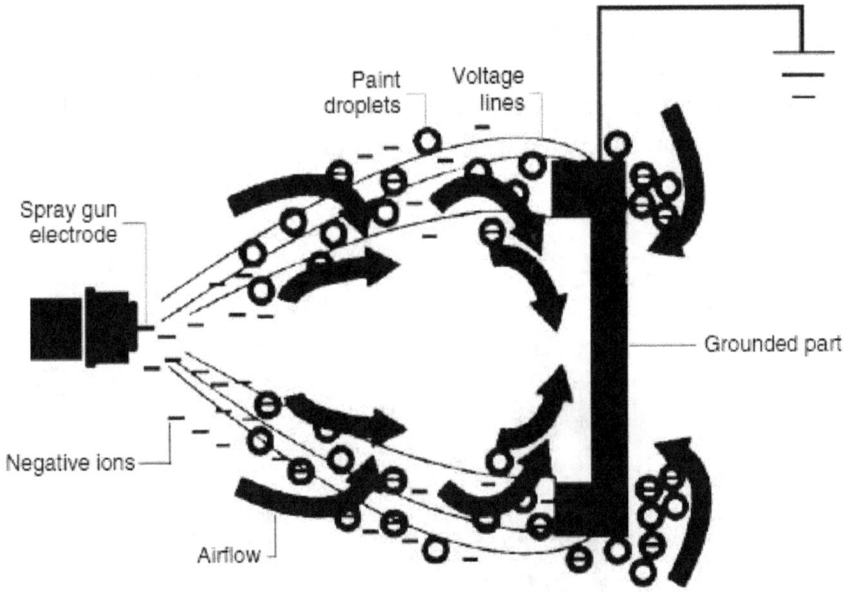

Figure .19 — Electrostatic attraction.

Effective use of electrostatic systems requires certain conditions to be met.

1. The coating must be in the proper conductivity range. The majority of solvent-borne coatings will work with electrostatic systems.
2. The atomized droplets should be round and remain as wet as possible. Sometimes a slow evaporating solvent is added to maintain this wetness.
3. An electrical field must be established and maintained in the proper position to direct the charged particles toward the part.
4. Forces such as excessive spray booth velocity and abnormally dry air will interfere with the electrostatic attraction process.
5. The parts to be coated must be electrically conductive to earth ground.

Manual Electrostatic Spray Equipment

There are several types of handheld, manually triggered electrostatic spray guns used for industrial application.

- Air-spray electrostatic guns
- Airless electrostatic spray guns
- AAA electrostatic spray guns
- HVLP electrostatic spray guns
- Electrostatic rotary atomizers

Isolation systems are used for safety reasons and for optimum system performance.

Isolation stands were first used to protect workers from the "hot" container. An isolation stand is constructed of a nonconductive material that keeps the container and the fluid supply hose away from ground and typically, they are enclosed in a cage where operators cannot access the container without automatically grounding the system. An isolation cage used with a waterborne coating is shown in Figure .20.

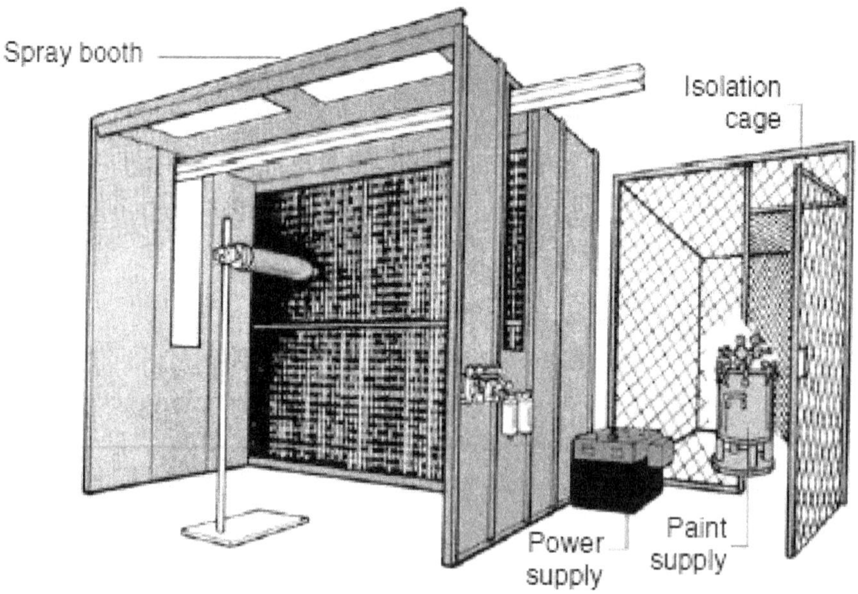

Figure .20 — Isolation cage for waterborne electrostatic spray application.

Table .5 — Solvent Polarity

Non or Low Polarity	Mineral spirits	High Polarity
Medium Polarity	Ethyl acetate	MEK, MIK, acetone
VM and P naphtha	Butyl carbitol	Isopropyl alcohol
SC-100, SC-150	Ethyl alcohol	
Xylol, toluol	Methyl alcohol	
Butyl acetate	Methyl acetate	

Coating Formulation for Electrostatic Application

In handheld systems, the polarity of the solvent should provide a coating that is stable under voltage at above 10 meg-ohm cm3 (typical). For efficient application, it is usually recommended not go beyond 200 meg-ohm cm3 (see Table .5).

Addition of polar solvents may be necessary when the coating is so dead, or so nonconductive that without the solvent it will not accept a charge. Electrostatic systems require special care and maintenance practices. All parts of the system must be kept clean, including the outside surfaces of the fluid hose and spray gun. Any possibility of a direct path to ground, resulting in a system shut down, must be eliminated. The exterior of the gun must be kept clean of overspray that will allow the voltage to "track back" to the operator who provides a great ground path. Automatic systems are even more demanding with regard to system cleanliness, because they operate at even higher voltages. Advantages and disadvantages of electrostatic spray.

Advantages

- Material savings due to electrostatic attraction and wrap; more material on the parts, less on the filters
- Reduced VOC emissions
- Reduced booth cleaning and frequent filter changes
- Increases in production rates
- Reduced equipment wear because of lower fluid and atomizing air pressures

Disadvantages

- Minor coating adjustment may be necessary.
- The part must be conductive to ground or must be prepped with a solution that will provide a ground path.
- Conveyors and parts carriers must be well grounded and kept clean.
- The relatively high cost of an electrostatic gun compared to that of a conventional air atomized gun or and HVLP gun. Efficiency should still provide the necessary return on investment (ROI).

An electrostatic spray system can be a great advantage in increased transfer efficiency, reduced VOC emission, reduction of labor, and reduced coating usage. On the other hand, if it is not fully understood, not properly used, or not maintained, it can become an expensive air-spray gun.

There are a few simple rules to remember in order to use an electrostatic system effectively and safely.

- The parts being coated must be electrically conductive to ground.
- Part carriers, racks, fixtures, carts, and so forth must be designed to provide a consistent ground path, and must be kept clean in order to maintain a good ground path.
- All items in the vicinity of the spray booth must be grounded. They can develop a static charge and discharge it to ground. A spark may ignite volatile vapors.
- The spray operator should never wear gloves that isolate them from the grounded gun handle. If gloves are worn, they should be designed with a conductive palm or a hole should be cut that allows skin to contact the gun handle. In addition, metallic objects, coins, watches, keys, and so forth that are normally kept in pockets should be removed as they, also, can build up a static charge. Rubber boots or insulated shoes should not be worn; leather soles will dissipate any static charge.
- Flooring in and near the spray booth should be conductive. In older buildings where wooden floors are still found, sheet metal should be used as a covering, making sure all pieces are mechanically connected to one another and the booth walls.
- It is very important that only items designed for electrostatic spray be used. Air hoses should contain a ground wire built into the jacket. Air or fluid moving through the hose can create a static charge. Be sure to use the proper high or low-voltage cable for your gun. Does not mix brand "A" with brand "B".
- The voltage should be turned off before servicing or cleaning the electrostatic device. If the gun has a built in cascade type generator or multiplier, the atomizing air must also be turned off. Be sure the gun is wiped clean and that no solvent remains in the fluid line before turning on the voltage.

Automatic Spray Equipment

Automatic spray equipment can add substantial capability to a system by increasing the amount of coating that can be delivered in a given time and controlling the volume of output. Automatic application reduces the labor required for application and it contributes to the consistency and quality of the product. Many factors enter into the design of a paint application system and contribute to the decision to apply paint with manual equipment only or add automation to the process.

Depending on the level of automation and electronically controlled functions, the operator will need to at least monitor and maintain:

- Applicator operation

- Fluid flow controls and components
- Equipment programming
- Applicator movers
- Material supply system
- Safety interlocks

Once they have been properly set, automatic gun settings should not be continually adjusted, but they should be routinely monitored. In spite of the added responsibility and skills required for their operation, automatic application equipment can provide needed volume and superior control. An automatic application system can cover substantially more area than a single manual operator.

In order to take advantage of the additional capacity, the application equipment must be installed with complimenting control equipment and it must be well maintained, workers must be well trained, and good scheduling practices must be in place to ensure that the system capacity is used. Racking arrangements must be more precise and consistent to get the best transfer efficiency. Part recognition systems must be well designed and maintained to make sure that the guns are triggered accurately. There are many different models and types of automatic application equipment from many different manufacturers, but they break down into three main categories; air-spray guns, bells, and disks.

Automatic Air-Spray Guns

An automatic air-spray gun closely resembles a manual spray gun without a handle and manual trigger. It may be mounted in a stationary position or designed for use with a reciprocator or robot. They have one or dual spray heads and different spray angles to best fit the application. Automatic spray guns are triggered by a part identification system that "reads" the position and size of the part as it approaches and turns the guns on and off at the appropriate times. A more detailed triggering system uses a manually entered part number or style code at a manual input station (MIS) that utilizes a detailed application program that is specific to that part style and color.

Fluid tips and pattern shapes can be changed for different applications, making these guns flexible for a variety of products and paint materials. Air-spray guns will typically provide better penetration into recesses than a bell or disk. Typical transfer efficiency can be very good, if parts are properly racked and the triggering programs are well defined. However, the air pressure required for atomization can result in more overspray than a bell. The bell atomizes the paint in a finer droplet and more narrow range, so the grounded part is able to attract a higher percentage of the paint sprayed. The bell also has a lower forward velocity, so the paint is less likely to be carried past the target by the force of the spray.

Rotary Atomizers

A system that uses a "bell" (rotary atomizer) or a disk is sometimes referred to as a true electrostatic system. The rotating bell or disk provides the atomization, and the high voltage causes the coating to go to the part. The rotation of the bell or disk causes the paint to move parallel to the target with the voltage off and the high-voltage field redirects the atomized droplets toward the target. Shaping air is also used with bells to control paint pattern size and aid penetration into recessed areas. Shaping air is used to direct the droplets toward the part where electrostatic attraction helps to draw the paint droplets onto the target. Figure .21 shows atomized paint at the edge of a bell.

Bells range in size from about 15–70 mm in diameter. Rotational speeds are anywhere from 10,000 to 60,000 rpm. Usually, a smaller bell diameter will have a faster rotational speed. The bell can be fed from a pressure pot as shown in Figure .22 or from a larger container.

Typically, a faster rotational speed will provide more finely atomized paint. High solids and waterborne coatings may be difficult to atomize at rotational speeds below 15,000 rpm.

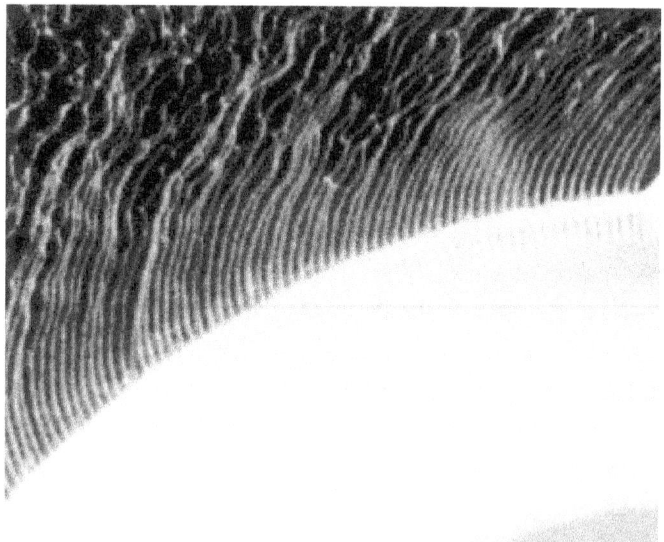

Figure .21 — Paint filaments from the serrated edge of a rotating bell.

On a bell, the fluid is normally fed through a tube down the center of the turbine shaft where it flows evenly over a cone at the back of the cup. The centrifugal force of the rotating bell evenly disperses the fluid to the edge of the rotating cup. The fluid passes through small holes on the outside of the bell; it is centrifugally atomized and attracted to the grounded work surface by electrostatic attraction. The shaping air is directed from an opening behind the bell to provide pattern control from 15″to 48″in diameter. The inside of the bell is shown in Figure .23.

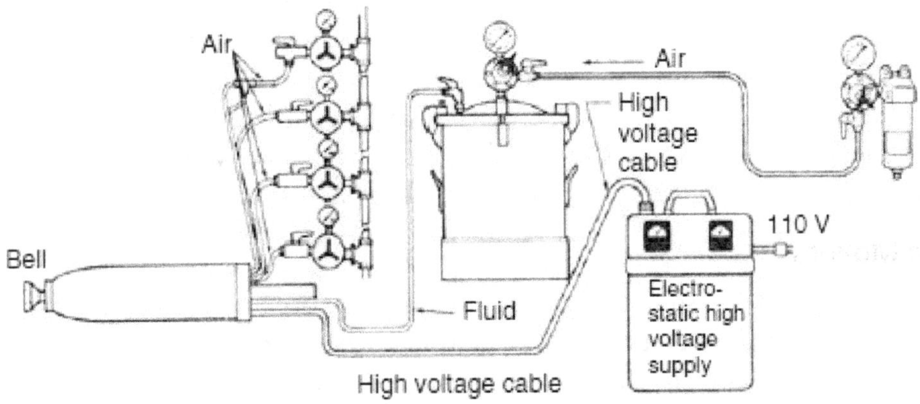

Figure .22 — Electrostatic bell system

As mentioned previously, an automatic spray gun employs compressed air to deliver the paint to the part. It will provide somewhat better penetration of the Faraday areas and deep recesses but has the potential to generate more overspray. It is important to consider this and compare the efficiency of a bell or disk and the related touch-up required to that of an automatic electrostatic spray gun when planning a system.

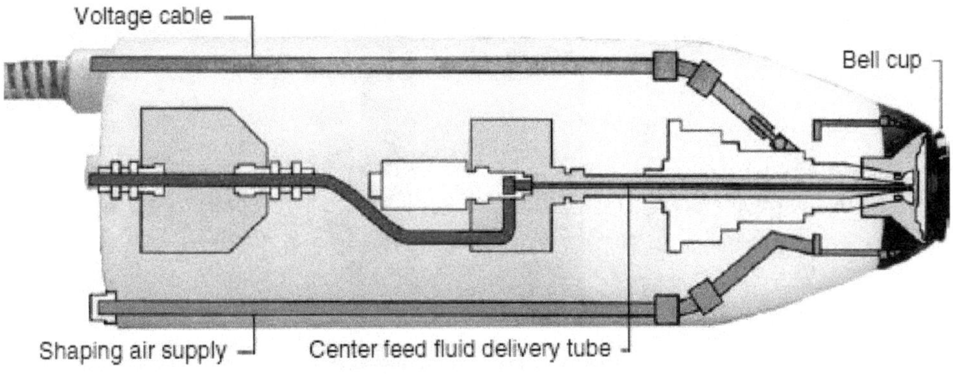

Figure .23 — Inside of a rotary bell

Testing is important to see just what device works best. Typically, a bell works well for simpler geometries requiring high transfer efficiency and even film or for the first application in a multiple application process. An air-spray gun may be the best application device for more complex geometries or for cut-in and touch-up.

Automatic Spray Supporting Equipment

Automation requires much more than just the correct selection of application equipment. There are many small but important application equipment and fluid flow control components. Multiple gun stations must be designed for accurate and controllable flow of air, fluid, voltage, and movement. Hoses must be kept as small and as short as possible for quick color change and low material waste. Color change manifolds should be located near the spray station so that the purge of material from the supply line is quick and there is very little material in the line. Color change programs should be written to fully utilize the material from the color change manifolds to the applicators to further reduce waste and color change time.

Gun Movers

Automatic guns are sometimes mounted in a stationary position. More often, they are mounted on a reciprocator or robot. There are long-stroke (Figure .24) and short-stroke reciprocators with vertical or horizontal movement. There are multiaxis gun movers that can move the guns in three dimensions: in/out, up/down, and in the direction of the conveyor. Moreover, there are robots (Figure .25) that can move the gun in multiaxis patterns similar to a manual operator but with much more precision.

Figure .24 — Long stroke reciprocator with bells.

The use of a gun mover is product dependent. They can help the guns to cover more surface of product and get into deep recesses, such as cars, office furniture, or appliance shells.

One important thing to consider is that the machine selected is a durable piece of industrial equipment. Examine the construction and motion design to make sure that it is solid and will last. These machines will be expensive, so the testing must be thorough.

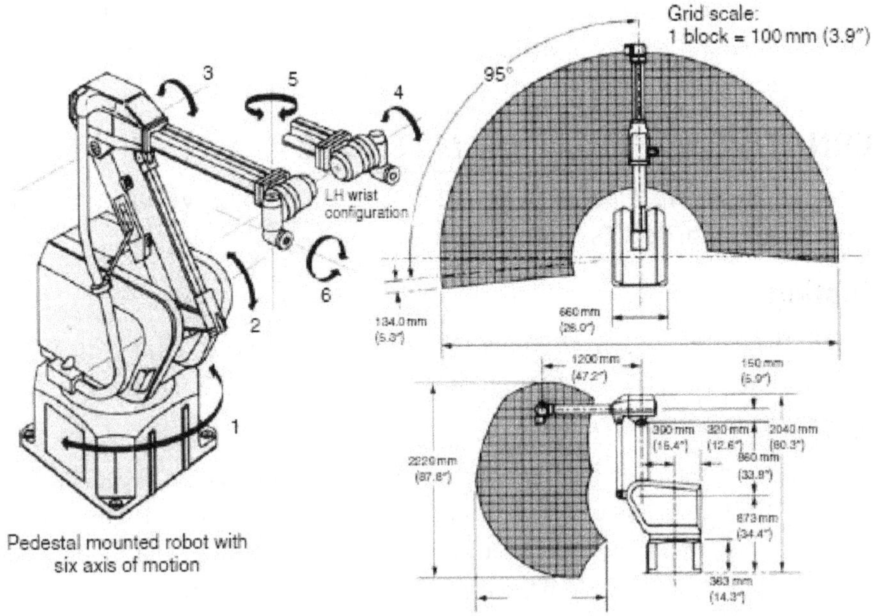

Figure .25 — Movement of robotic arm.

The stroke of the machine must be electronically integrated with the product and the line speed and controlled accordingly to avoid irregular patterns on the part surface. Establishing the correct stroke distance and stoke speed based on the line speed, surface to be covered, and desired film thickness will provide uniform coating application.

Reciprocator mounting bars can be configured with a variety of patterns for different products. They can be mounted over the top of product on a floor-mounted conveyor or in vertical positions to coat from the side. One short stroke reciprocator can hold a double bar of three or four spray guns or bells each. This kind of versatility is one of the advantages of automation. Several guns can be brought to bear on the parts in a very short time span. The number and position of the guns can be designed to provide excellent coverage with low overspray.

Robotic Painting

A spray gun or bell mounted on a robotic arm provides the advantages of manual application (focused application and multiaxis movement) with much more accuracy and consistency. This provides an opportunity for improved quality, productivity, and reduced labor.

Robotic arms come in many different sizes with different control options. Atypical software driven controller can manage color change, provide analog fluid and air control, part queuing, arm homing, conveyor synchronization, path and function editing and simple lead through, and Cartesian teaching methods.

With a slim wrist design, the robotic arm can provide 360° of roll and 90° of pitch and yaw with flexible spatial coverage. Large, complex parts can be painted very effectively with exceptional precision using a robotic arm.

ALTERNATIVE APPLICATION METHODS

In addition to spray application, there are several alternative ways to apply liquid paint to industrial products.

Dip Coating

In a dip coating application, the part is simply immersed in a tank of paint. Excess paint drains back into the tank once the part is removed. Parts may be handled in batches or with a conveyor.

Dip tanks can be built in any size needed for small or large parts. Provisions must be made for circulation, filtering, and control of temperatures and viscosity. Because of the fire hazard involved, dip operations using flammable paints must be equipped with CO2 fire extinguishing systems and dump tanks.

Advantages of Dip Coating

- Simple equipment
- Low-skilled labor requirement
- Easily automated
- Close racking of parts
- Various part shapes can be mixed
- Little paint is wasted if the part is allowed to drip over the tank

Limitations of Dip Coating

- Viscosity control is critical. If the paint is too viscous, too much paint is used (wasteful) and film thickness is too high. Low viscosity results in paint films that are too thin. Consequently, corrosion resistance may be too low for primers and hiding power may be inadequate for topcoats.
- Possible settling of the paint in the circulation system.
- Gradual tank contamination is possible. Parts entering the tank must be clean and dry to avoid carrying in contaminants.
- Some parts are hard to immerse, because trapped air keeps them afloat.
- Tanks must be large enough to immerse the entire part.
- There is a fire and toxicity hazard associated with large volumes of paint solvents.

- Film thickness control is a problem. Paint may become trapped in recesses and drain slowly or not at all. On vertical surfaces, the paint tends to be thicker at the bottom of a panel than at the top. This is caused by an increase in viscosity as the paint loses solvent while draining. It often results in fatty edges that can later break off under thermal or mechanical shock.

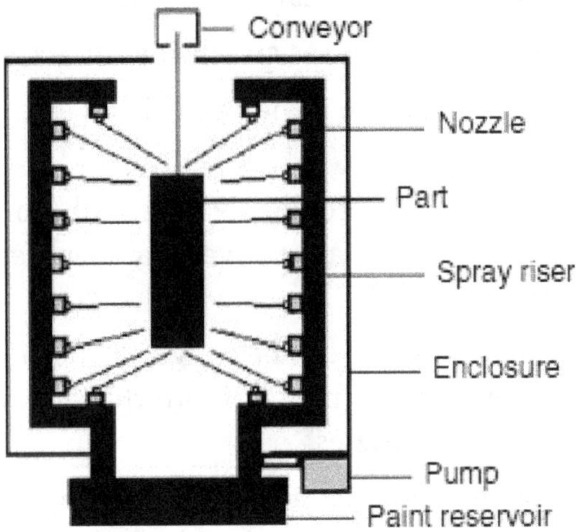

Figure .26 — Flow coating

- The drain-off area must be cleaned regularly. Since this is usually done manually, some dried paint may fall into the circulation system and clog the filters, pumps, and nozzles.

Flow Coating

In a flow coating operation, the part is held over a tank and a stream of paint is pumped over the part through a nozzle (Figure .26). The excess paint drains into the tank and is recirculated over the parts. The variables that are important for flow coating include pumping, filtering, viscosity, temperature, ventilation, and fire control.

Advantages of Flow Coating

- Simple equipment
- Little labor required
- Easily automated
- Dead air spaces can be coated if a stream of paint can be directed upward into the recess
- Smaller volumes of paint are required, because the parts need not be immersed
- Little paint is wasted if the parts are allowed to drain over the tank

Limitations of Flow Coating

- High solvent loss by evaporation. Solvent must be replenished on a regular basis.
- Possible settling of the paint in the circulation systems.
- Fire and toxicity hazards associated with the high solvent loss.
- Inability to coat certain recesses and inside surfaces.
- Gradual paint contamination if the paint rinses dirt, salts, oils, and so forth from the parts.
- Film thickness control is a problem. Paint may become trapped in recesses and drain slowly or not at all. On vertical surfaces, wedging may occur resulting in fatty edges.
- The drain-off area must be cleaned regularly. Since this is usually done manually, some dried paint may fall into the circulation system and clog the filters, pumps, and nozzles.

Dip andflowcoating methods should be considered when high production painting of relatively simple shapes is required and quality appearance and corrosion resistance is not important. Viscosity and film thickness control are the major problems to be expected. Efficient use of paint and low labor rates are the primary advantages. These processes are limited to waterborne coatings because of environmental considerations.

High-solids coatings do not lend themselves to these types of application because of the higher viscosities that cause poorer flow characteristics and high film builds. Paints with high solvent content produce too much VOC for most installations and they create an unsafe working environment.

ROLLER COATING

A roller coating operation is capable of coating large areas in the form of sheet or strip at high speed and with a continuous and even thickness of paint or varnish. The rollers can also be used to laminate plastic film to metal substrates.

There are two methods of liquid transfer for roll-coating systems on flat sheets. One system uses three rollers and the second uses four rollers. In the first process, the liquid is fed between the feed and applicator rollers rotating in opposite directions. The stream of paint is then fed through the rollers onto the sheet. The third roller acts as a pressure roller to ensure that the sheet comes into correct and even contact with the wet applicator rolls (Figure .27).

In the four-roller process, a fountain roller picks up liquid from a trough, transfers it to the feed roller, which in turn presses it to the applicator roller. The fourth roller acts as the pressure roller, as in the three-roll system. In coil coating applications, where both sides of the sheet are coated, the four-roller system is used. The applicator rollers act as pressure rollers.

Fountain and feed rollers are normally made of stainless steel, while the applicator roller has a steel core covered with a rubber, gelatine composition, or polyurethane face. These surfaces become deformed during use and need to be remachined to offer a continuous smooth surface. Gelatine rollers are resistant to many organic coatings but are not resistant to water-based coatings. Polyurethane rollers produce a solvent and abrasion resistant surface. They are normally found as applicator rollers on fast running coil coat lines.

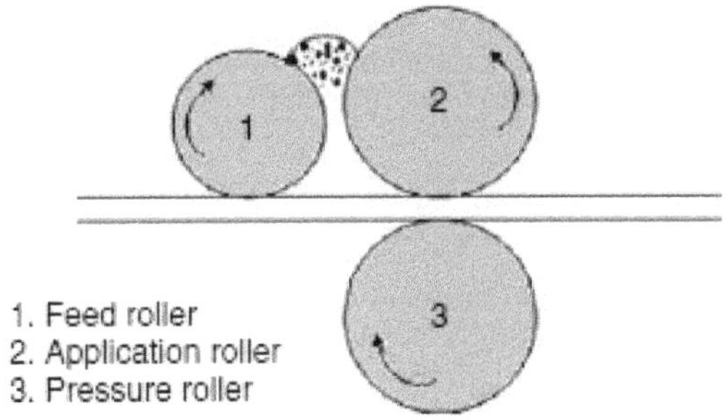

Fugure .27 — Roll coating

Advantages of Roller Coating

- Smooth, even, dust and “fat edge” free films can be produced
- Films dry relatively quickly
- No excessive solvent or paint losses due to evaporation occur
- There is no fire hazard

Disadvantages of Roller Coating

- Limited to thin sheet stock
- Not able to apply the variety of finishes used in spray coating

CURTAIN COATING

Curtain coating is a sophisticated method of pouring a continuous film of lacquer or paint over flat sheets. The sheets are conveyed under a trough with an adjustable slot. The trough acts as a pouring head and control of the gap regulates the volume of material, controlling the film thickness deposited on the sheets. Curtain coating works well for coating flat sheets of metal or wood.

A second conveyor moves the coated panel away from the gap. Initial application efficiency is excellent and any material that does not adhere to the product is recirculated to the trough by means of a pump.

In some instances, moisture-containing substrates are prewarmed in a tunnel prior to coating. The warm panel aids the set-up characteristics of the paint or lacquer. An alternative is to heat the lacquer with a heat exchanger located between the reservoir and trough.

The most important part of curtain coater is the adjustment slot (or gap) in the trough. The gap can be accurately controlled to develop predetermined film weights. The trough and slot must be kept scrupulously clean, because any dirt or solid material will break the curtain and cause a surface defect.

12 The Painting of Ships

The painting of ships and structural steel work, such as oil and gas rigs, poses some of the most formidable problems in respect of corrosion protection that the paint technologist has to face. Salt-water immersion or partial immersion, spray followed by drying winds, are extemely aggressive environments. They call for the development of paints which protect the steel from which the ship's hull is constructed, by means of a number of different mechanisms which we shall discuss briefly. In addition to problems of corrosion, the fouling of ships' bottoms with marine organisms leads to increased drag on the vessel and hence to increased fuel consumption.

CORROSION

Corrosion is an electrolytic process involving the reaction between iron metal, oxygen, and water resulting in the formation of hydrated ferric oxide or rust. The overall reaction is:

$$4Fe + 3O2 + 2H2O \rightarrow 2Fe2O3 \cdot H2O$$

The generally accepted reaction mechanism is one which involves anodic and cathodic processes. The surface of the iron or steel in contact with water develops localized anodes and cathodes at which these processes take place. Electron flow (constituting'corrosion current') occurs through the metal, and this is complemented by an equivalent transport of charge through the water or electrolyte by hydroxyl ions. The process is shown diagrammatically in Fig. 1. The individual electrode processes are as follows.

At the anode: the formation of ferrous ions by the loss of electrons,

$$4Fe \rightarrow 4Fe^{++} + 8e$$

and at the cathode: the formation of hydroxyl ions,

$$4H_2O + 2O_2 + 8e \rightarrow 8OH^-$$

The initial product of oxidation is thus ferrous hydroxide,

$$4Fe^{++} + 8OH^- \rightarrow 4Fe(OH)_2$$

In the presence of excess oxygen the ferrous hydroxide is oxidized to hydrated ferric oxide, the all too familiar red oxide which is rust.

The anodic and cathode regions at the surface of the metal arise from compositional heterogeneity of the surface. This may be due to a number of factors including grain boundaries, stresses, and microscopic faults that cause local concentration gradients of electrolyte or oxygen in solution. Any of these giving rise to a potential difference between adjacent areas in the surface is sufficient to cause galvanic action.

In dealing with ships under construction one is faced with painting steel that has at its surface a layer of oxides up to 60mm in thickness that is formed when the steel is manufactured. The steel is subjected to hot rolling to the required thickness, and this is carried out in the temperature range 800–900°C. Oxidation of the steel takes place during cooling. This oxide layer is known as 'millscale'. Millscale, unless removed, can cause corrosion because there is a significant potential difference between the millscale and the bare steel (about 300 mV) when the steel is immersed in electrolytes such as sea-water.The millscale is cathodic, and the bare steel anodic.

Thus where there is a crack or fissure extending through the millscale layer to the metal a galvanic cell is formed and the corrosion process begins.The rust is formed

in the vicinity of the anode which is in contact with the salt-water, but not directly on the anodic site.Thus the anode is slowly dissolved, and pitting occurs. The cathode site undergoes no such dissolution. All millscale is now removed by blast cleaning, and the steel is 'shop primed' before fabrication. However, this does not mean that corrosion is entirely eliminated.

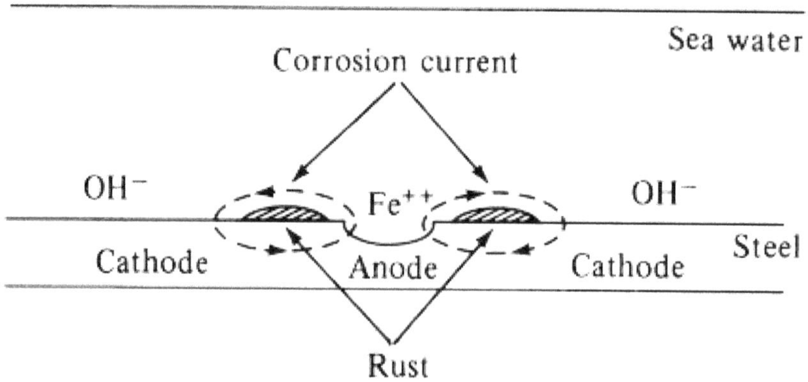

Figure 1 — The corrosion process.

How, then, can corrosion be reduced or eliminated? Two approaches are possible:

- The elimination or inhibition of the electrolytic processes.
- The exclusion of water or oxygen from the potential corrosion site.

The cathodic process requires the presence of water and oxygen at the metal surface, so that it can be seen that the two factors indicated above are interrelated. In practice it is very difficult to prevent oxygen and water reaching the metal surface because most paint binders have quite high permeability to these agents. Materials that are low in permeability to water and oxygen are, for example, waxes, because of their crystalline nature. Similarly, crystalline polymers exhibit low permeability, but these are difficult to formulate into air-drying compositions. In general, conventional marine paints do not confer cathodic protection. At the anodic surface, for corrosion to take place ferrous ions are required to pass into the electrolyte.This may be suppressed by the use of paint systems of two kinds. The first uses metallic zinc as pigment at a very high pigment concentration, such that contact between the zinc particles and the iron surface takes place. In this case, because zinc is more electronegative than iron, a galvanic couple exists between the two metals, the iron being cathodic and the zinc anodic. The net result is the suppression of the dissolution of ferrous ions and the consequent corrosion of the zinc. The second kind of cathodic suppression can be achieved by using paints containing inhibiting pigments such as red lead, white lead, or zinc oxide. These pigments are used in conjunction with binders based upon drying oils that undergo autoxication. In addition to the formation of crosslinked films some scission of the drying oil triglycerides occurs with the formation of oxidation products such as azelatic acid.These acids react with the pigment to form lead or zinc soaps as the case might be. The soaps react with the surface oxide film on the metal and prevent the ingress of water which is necessary for the formation of rust. A second group of pigments that act in a similar way are those that exhibit limited water solubility. These include the chromates of zinc, barium, strontium, and lead. In these cases the oxide film is protected by the formation of an iron/chromium oxide complex.

Two other ways of suppressing the electrolytic processes that are responsible for corrosion have been used on ships:

- The use of 'sacrificial anodes', in which anodes are built onto the hull, using alloys that are electropositive with respect to the steel of the hull. These anodes are unpainted and their purpose is to corrode away, being replaced when necessary.
- The application of a controlled external current to the hull to suppress the dissolution of ferrous ions. This is called the 'impressed current method'.

Neither method eliminates the need for painting the hull for other obvious reasons.

The foregoing is undoubtedly an oversimplification of the processes involved in the corrosion of steel when immersed in natural sea-water. For example, bacteria and other microorganisms can cause corrosion to take place even under anaerobic conditions. The reason that a film of microorganisms can influence an electro-chemical process such as corrosion is because of its ability to change the chemistry of the environment at the locus of corrosion. Thus, a biofilm can act as physical barrier to the diffusion of oxygen through the film and also act as an active oxygen scavenger during respiration; it can also have a significant effect on the local pH. The influence of biofilms formed on metal surfaces in natural sea-water is thought to account for differences in rates of corrosion observed between metal panels tested in natural sea-water and those tested in artificial sea-water used in laboratory testing. It is not possible within the scope of this book to pursue this topic further

Surface Preparation

To obtain satisfactory performance from a paint system it is essential for millscale to be removed from the surface of the steel. Not only does it contribute directly to corrosion, but it will eventually become detached from the steel surface and is therefore an unsatisfactory substrate for painting. Surface preparation is vitally important when one is seeking to obtain the best performance from a paint system.

Blast Primers

Blast primers are applied by spray immediately after blast cleaning. They must dry quickly (within a few minutes of application) to enable the steel to be handled without significant damage to the coating.The primer must not adversely affect the subsequent fabrication process. For example, it should not prevent successful welding. It must also provide a suitable surface for subsequent painting.

Four types of blast primer are in common use:

- Phenolic/polyvinyl butyral (PVB), acid cured, one- and two-pack compositions;
- Two pack cold curing epoxy compositions, pigmented with red iron oxide (and optionally an inhibitive pigment such as chromate);
- Two-pack cold curing epoxy compositions pigmented with zinc dust; and
- Zinc silicate one- and two-pack compositions.

Typical formulations of these types of prime are shown in Table .1.

Paint Systems for Ships

The protection of steel plate after manufacture has been described briefly above. The subsequent painting of ships during and after construction is a highly complex technology. As has been indicated, different surfaces require different treatment. Each surface requires a system, since it is impossible to combine all the requirements for any one surface within one paint. Thus multi-coat systems are essential, and for any one type of paint it has become the convention to apply several coats to build up film thickness to minimize corrosion occurring, for example, through pinholing. Typical systems for topside and superstructure paints, boottopping, and bottoms are summarized for what are regarded as traditional or conventional marine paints in Table .2.

Table .1 — Formulations for blast primers

One-pack phenolic/PVB blast primer	
Paint base	% by
PVB (e.g. Mowital B6OH)	4.7
Phenodur PR263 (phenolic resin)	4.7
Red oxide (synthetic)	8.6
Asbestine (magnesium silicate)	1.4
Phosphoric acid, SG 1.7	1.0
Methyl ethyl ketone	30.0'
n-Butanol	19.6
iso-Propanol	30.0
	100
Two-pack epoxy blast primer	
Paint base	% by weight
Epikote 1004	12.0
Zinc phosphate (to BS 5193:1975)	8.0
Red oxide (synthetic)	10.0
Asbestine	1.5
Talc (5μm)	15.3
Bentone 27 (anti-settling agent)	0.4
Toluene	23.0
iso-Propanol	11.5
Crosslinking agent	
Versamid 140 (polyamide)	1.8
Toluene	11.0
iso-Propanol	5.5
	100.0
Two-pack zinc-rich epoxy blast primer	
Paint base	% by weight
Zinc dust (Zincoli 620 or Durham Chemicals 'Ultrafine')	80.0
Bentone 27	1.0
Calcium oxide	0.3
Epikote 1001	4.0
Toluene	6.6
iso-Propanol	3.3
Crosslinking agent	
Versamid 115	1.8
Toluene	2.0
iso-Propanol	1.0
	100

Note: The calcium oxide acts as a water scavenger, reacting with moisture in the solvents. The moisture would otherwise react with the zinc to form hydrogen in the can.	
Zinc/ethyl silicate blast primer	
Paint base	% by weight
Ethyl silicate (e.g. Dynasil H500)	20.0
Bentone 38, anti-settling agent	4.0
Talc (5μm)	5.3
Toluene	5.3
Cellosolve	4.0
Zinc dust (e.g. Zincoli 615)	60.0
The zinc dust is stirred into the base at the time of use. The product then has a useful life of about 8 hours.	

Thus a new generation of paints has been developed that enable the periods between dry docking to be increased from about 9–12 months to 24–30 months.

Table .2 — Summary of 'conventional' marine paint types

Surface	Paint System	Binder type	Coats	Dry film thickness (μm)	Total average film thickness (μm)
Bottom	Printer	Oil-modified phenolic; bitumen	3-4	50-60	
	Barrier coat No. 1 anti-corrosive	Bitumen/limed rosin	1	50-60	
	Anti-fouling	'Soluble matrix' based on boiled oil/limited rosin; Cu or tributyltin oxide (TBTO) toxicant	1	50-75	~265
Boottopping	Primer 655	Tung/linseed-modified phenolic	2-3	40-60	
	Finish	Tung-modified phenolic	2	40-60	~200
	Primer				
Topside and superstructure	Primber (quick-drying) Undercoat	medium oil length (a) linseed alkyd (55-60% OL)	1	50	
	Enamel	Long oil length soya alkyd (60% OL)	1-2	35-50	~200

Topside and superstructure paints

Conventional paints meeting the requirements of topsides and superstructures come closest to those used in the trade paint market, requiring primers, under coats, and gloss coats that dry at ambient temperature. These have for many years been based upon oleoresinous and alkyd binders. Thus, a primer would use an oleoresinous vehicle (e.g. an oil-modified phenolic) with an inhibitive pigment, usually red lead. An undercoat would be formulated on a medium oil length linseed alkyd.

The topcoat might be based on a long oil alkyd or a vinyl-toluenated alkyd. A typical epoxy system for topsides would, for example, involve applying two coats of a high-build primer/undercoat followed by one coat of epoxy enamel, giving a total dry film thickness of about 300mm. An alternative for the superstructure, where gloss retention is more important, would be to use one coat of primer, followed by an epoxy primer/undercoat and a polyurethane topcoat.The dry film thickness would be approximately 200mm in this case. Examples of a red oxide epoxy thick coating and white epoxy enamel are given in Table .3.

Epoxies can be crosslinked with amines, diethylene triamine and triethylene tetramine commonly being used. Amines are often pre-reacted with epoxides to

Table .3 — Formulation of a red oxide thick coating and a white epoxy enamel

Paint base	*% by weight*
Synthetic red oxide	5.0
Asbestine (magnesium silicate)	22.0
β Crystoballite (silica) to pass through a 325 mesh sieve	18.0
Epikote 1001	16.0
Bentone 27	2.0
Xylene	14.0
n-Butanol	5.0
Curing agent	
Versamid 115 (polyamide)	8.0
Cellosolve	10.0
	100.0
A typical white epoxy enamel formulation:	
Paint base	*% by weight*
Rutile titanium dioxide (e.g. Tioxide R-CR2)	30.0
Aerosil 380 (anti-settling agent)	0.5
Epikote 1001	23.0
Solvesso 100 (naphtha)	16.5
Methyl isobutyl ketone	8.0
n-Butanol	3.0
Curing agent	
Versamid (polyamide)	11.0
Accelerator (e.g. Anchor K54)	1.0
Xylene	7.0
	100

Typical formulae for a primer, undercoat, and topcoat all based upon chlorinated rubber are given in Tables .4–.6.

The composition given in Table .4 calls for careful formulation and manufacture.

The plasticizer is a preformed blend of solid (waxy) chlorinated paraffin with liquid chlorinated paraffin. The thixotropic agent is based on hydrogenated castor oil. This develops structure only when it is incorporated during the milling stage, when the temperature rises, depending on the scale of manufacture, to over 40°C. The temperature should not exceed 55°C, otherwise the thixotropic

Table .4 — Chlorinated rubber primer

Aluminium paste, non-leafing (65% in naphtha)	% by weight 20.0
Zinc oxide	1.0 10.0
Blanc fixe	0.5
Thixatrol ST (thixotropic agent)	15.0
Alloprene R10 (chlorinated rubber)	7.5
Cerechlor 70 (chlorinated wax plasticizer)	7.5
Cerechlor 42 (chlorinated wax plasticizer)	0.1
Propylene oxide (stabilizer)	28.4
Xylene	10.0
Solvesso 100 (Naphtha)	100.0

Table .5 — Chlorinated rubber undercoat: white

	% by weight
Titanium dioxide (e.g. Tioxode R-CR2)	16.0
Blanc fixe	14.0
Zinc oxide	1.0
Thixatrol ST (thixotropic agent)	1.0
Alloprene R10 (chlorinated rubber)	14.0
Cerechlor 70 (chlorinated plasticizer)	7.0
Cerechlor 42 (chlorinated plasticizer)	7.0
Propylene oxide (stabilizer)	0.1
Xylene	29.9
Solvesso 100	10.0
	100.0

Table .6 — Chlorinated rubber topcoat: white

	% by weight
Titanium dioxide (e.g. Tioxode R-CR2)	18.0
Blanc fixe	2.0
Zinc oxide	1.0
Bentone 38 (thixotropic agent)	1.0
n-Butanol (swelling agent)	0.5
Alloprene R10	16.0
Cerechlor 70	12.0
Cerechlor 42	4.0
Propylene oxide	0.1
Xylene	30.4
Solvesso 100	15.0
	100.0

In the formulation given in Table .6, the reduced amounts of filler (cf. undercoat), the use of Bentone 38 instead of Thixatrol ST, and the higher proportion of Cerechlor 70, are all measures designed to improve the gloss of the topcoat. Even so, the gloss is low compared to epoxide or polyurethane paints. These paints are all applied by airless spray.A typical system would be:

	Dry film thickness (μm)
Primer, one coat	50
Undercoat, two coats	200
Topcoat, one coat	50
Total	300

Boottoppings

As distinct from the traditional boottoppings mentioned previously (Table .1), high performance compositions for this area are based essentially on the same binders as the topside paints. In this category the chlorinated rubbers perform very well because of their excellent intercoat adhesion. This is particularly important because of the damage to the paint system from abrasion and impact with fenders and quays, and the consequent need for frequent repainting.With VLCCs, fouling can also be a major problem in the boottopping area, and it is common practice to use an anti-fouling composition as the finishing paint.Thus it is usual to employ the same anti-fouling as is used on the bottom.

Bottom paints

The high performance/high build systems described for topsides and boottoppings are also used for bottoms. However, since asethetic reasons are not predominant it is possible to formulate such paints at lower cost, without sacrificing performance, by the incorporation of coal tar. The paints are thus usually chocolate brown or black, but this is of no real significance.The proportions of coal tar in an epoxy coal tar thick coating can be varied over a wide range. The higher the coal tar content, the poorer the oil or chemical resistance, but this is less important in underwater compositions. The coal tar pitch in a formula complying with the UK Ministry of Defence (Ship Department) Specification DGS 5051 would represent about 60–65% of the total binder. It would be selected according to its compatibility with the resin; and, although regarded as essentially unreactive, it can react with the epoxy groups because it contains some phenolic hydroxyl groups.

For this reason it is often incorporated into the curing agent part of the two-pack composition, otherwise the storage life of the base would be inadequate.A typical formulation is given in Table .7.

After mixing, the paint is applied by airless spray to give a dry coat thickness of 125 mm per coat. The pot life is about 4 hours at 15°C. The reaction rate doubles, approximately, with a 10°C rise in temperature, or is halved with a 10°C drop in temperature. Thus the paint can have a pot life varying from about 8 hours at 5°C to only 2 hours at 25°C. The rate of cure also affects recoatability. To achieve satisfactory intercoat adhesion and to avoid lifting, maximum and minimum recoating times are specified, according to the temperature of application.

Table .7— Typical formulation for two-pack bottom paint

Base	% by weight
Barytes	20.0
Asbestine	20.0
Epikote 1001	11.0
Bentone 27	1.0
Xylene	10.0
Solvesso 100	5.0
n-Butanol	3.0
Curing agent	
BSC Norsip 5 (75% aromatic coal tar in xylene)	25.0
Versamid 140	2.5
Synolide 968	0.3
Xylene	2.2
	100.0

Table .8 —Aluminium tie coat

	% weight
Aluminium power, non-leading	14.0
Basic lead sulphate	14.0
Gilsonite (bitumin)	22.4
Linseed stand oil, 50 poise	5.6
Coal tar naphtha, 90/160	44.0
	100.0

This type of composition may be used both on ships under construction and those undergoing 'in service' repainting. The former poses a problem in that the application of an anti-fouling will not take place until shortly before the ship goes into service, and this may well be long after the optimum time for overcoating the coal tar epoxy paint. In this case a 'tie coat' is used. A composition that meets this requirement is given in Table .8.

Anti-Fouling Coatings

The Problem of Fouling

Marine organisms, both plant and animal, are able to attach themselves to the hulls of ships. Plant growth requires some daylight to sustain it, so attachment occurs to the sides of the hull in the upper regions of the underwater area.The most common of these are of the genus Enteromorpha, seaweeds that have long green tubular filaments that grow very rapidly under favourable conditions. The brown and red seaweeds require less light, and these can grow at lower levels or even on the bottom of the ship. Firm attachment of these plants can take place in a few hours unless they are prevented from doing so.

Aquatic animals that are capable of attaching themselves to ships include barnacles, mussels, polyzoa, anthozoa; hydroids, ascidians, and sponges.The best known (and most troublesome) are the barnacles. These animals do not require light to sustain growth, so they are to be found on all immersed parts of the hull. The larvae require about 48 hours to become firmly attached.The consequences of fouling have already been mentioned, but it is worth repeating that the most important consequence of fouling with large modern ships is the increased cost of propelling the vessel through the water as a result of the increased drag that fouling imposes.

The first 'anti-fouling' paints based upon the use of toxicants were introduced in the middle of the 19th century. These were oleoresinous compositions containing mercuric and arsenious oxides. Subsequently, because of the high human toxicity and the increasing cost of mercuric oxide, alternative paints using cuprous oxide were developed. Cuprous oxide remains the most commonly used toxicant in the 'conventional' types of composition today.

Conventional Anti-Foulings

There are two types of conventional anti-fouling paints, classified according to their mode of action. These are generally known as 'soluble matrix' and 'contact' types.

Soluble Matrix Type

In the soluble matrix type the toxicant is dispersed in a binder that is slightly soluble in sea-water.The toxicant is slowly released as the binder dissolves.The rate of dissolution has to be controlled very carefully since an inadequate concentration of the poison at the surface will not prevent the attachment of the marine organisms, whereas too rapid dissolution will mean that the effectiveness of the anti-fouling will be too short-lived. The binder used in this type of anti-fouling might be a 3: 1 limed rosin/boiled linseed oil mixture.

A better 'supertropical' grade might have the following composition given in Table .9. 'Atlantic' or 'tropical' grades would not necessarily call for the use of tributyltin oxide.

Table .9 — Formulation for supertropical anti-fouling composition % by weight

Cuprous oxide	24.0
Zine oxide	5.0
Tributyltin oxide	2.0
Rea iron oxide	8.0
Paris white	9.0
Talc (10 μm)	6.0
Bentone 38	0.3
n-Butanol	0.1
Lied rosi/boiled linseed oil, 3:1, 60% in naphtha	41.0
White spirit	4.6
	100.0

Contact Type

The contact type of anti-fouling composition is so named because it is formulated at a very high pigment volume, such that the pigment (toxicant) particles in the disperse phase are in contact with each other in the dry film.The binder is largely insoluble in sea-water, so that when dissolution of the toxicant takes place a porous film of the paint binder remains on the surface.The binder is not entirely insoluble, and the balance between binder solubility/insolubility plays a part in controlling the rate of leaching of the toxicant. Although these paints are formulated to give a closepacked pigmentary structure in the dry film, it is doubtful if this is achieved or is desirable in practice. Any significant flocculation of the particles of toxicant will lead to a more open structure, and the paint may in fact be underbound. This does not appear to be a major problem in practice.A typical formulation for a contact type anti-fouling is given in Table .10.

Modern Anti-Foulings

Just as the development of larger vessels (e.g. supertankers) has required paint systems that call for less frequent maintenance occasioned by paint breakdown and corrosion, the prevention of fouling over a longer period is similarly required. Subsequently, tributyltin oxide was used in combination with copper oxide. However, with the expectation that the effective life of an anti-fouling coating should exceed 18 months, it became clear that this could not be achieved with cuprous oxide at practical film thicknesses. Thus, the use of organo-tin derivatives has increased, and a 'family' of organo-tin derivatives is available to the paint formulator will not be any more effective than copper. Thus the paint requires very careful formulation in order to make the maximum effective use of the toxicant. Several organotin derivatives have been used successfully as toxicants in the marine paint market, the principal ones being tributyltin oxide, tributyltin fluoride, and triphenyltin fluoride.

Table .10 — Typical formulation for contact type anti-fouling % by weight

Cuprous oxide	57.4
Asbestine	2.4
Rosin	16.1
Modified phenolic/linseed stand oil, 1:2, 60% in 90/190 solvent naphtha	9.0
Chlorinated plasticizer	5.4
Solvent naphtha, 90/190	9.7
	100.0

Tributyltin oxide is a liquid, miscible with common paint solvents. It has a relatively high salt-water solubility, 25ppm that enables it to be used when a high leaching rate is required. It has a plasticizing action on the film in which it is incorporated, and this limits the amount that can be used to about 13% by weight in a typical vinyl system. In current commercial use it is usually a cotoxicant with cuprous oxide at about 2–5% by weight of the dry film. It is apparently biodegradable. Photolysis and biodegradation processes convert it ultimately to tin oxide which is quite low in toxicity. Moreover, although tributyltin oxide is soluble in sea-water to a limited extent, it is readily adsorbed onto suspended matter and tends to become associated with the silt after release where it undergoes degradation.

Tributyltin fluoride is a white, high melting waxy solid that is insoluble in common paint solvents. It is therefore incorporated as a pigment up to about 30% by weight of the dry film. It is less soluble in sea-water than the oxide, 10ppm, and is often used as the sole toxicant in vinyl/rosin or chlorinated rubber/rosin-based paints.Tributyltin fluoride is available either in powder form or as a paste. The paste offers more convenient handling and can be readily incorporated by high-speed mixing.

Triphenyltin derivatives were also found to be useful toxicants.Triphenyltin fluoride (TPTF) has been used as an agricultural fungicide. It is a white powder with a sea-water solubility of less than 1ppm. Paints containing it can give effective protection against marine growth for up to two years. The hydrolysis products of triphenyltin fluoride in sea-water are triphenyltin chloride and triphenyltin hydroxide, both of which are solids that will not diffuse through the film. TPTF is therefore most effective when used in compositions that dissolve or erode.

Polymers that incorporate organo-tin moieties are simple random color ter polymers having, typically, the following structure derived from tributyltin acrylate and methyl methacrylate:

Binding the toxicant into a polymer offers many advantages over the use of nonpolymeric derivatives that generally have to be incorporated into the paint in a pigmentary form. The toxicant content of the polymer can be varied at will between wide limits, and the Tg of the polymer can be varied to meet the requirements of use. Perhaps the biggest single advantage is the opportunity to provide controlled release of toxicant through the chemical process of hydrolysis rather than by physical means alone.

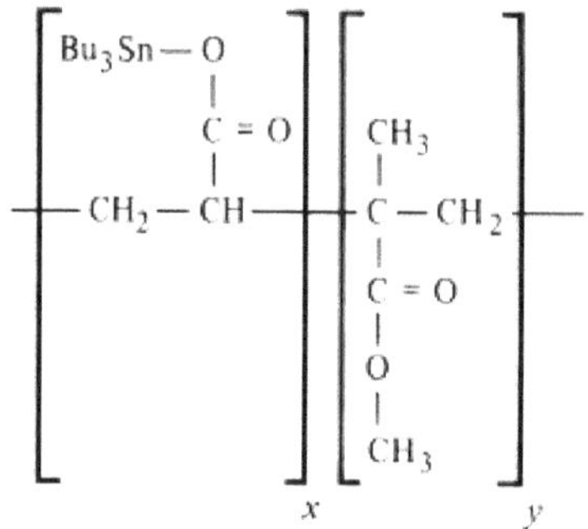

The advantages claimed for the newer organo-tin polymer anti-foulings include:

- Constant rate of release of toxicant;
- Controllable rate of erosion and toxicant release;
- Self-cleaning (and at high erosion rates, self-polishing);
- High utilization of toxicant is achieved, and as a result of the erosion there are no significant problems due to residues when repainting is necessary.

The formulation of organo-tin polymer-based anti-fouling paints did, however, present very special problems. The use of organo-tin biocides in rapidly eroding systems increases the initial cost to the ship owner, but this may be more than offset by the reduced requirement for dry docking. The main protagonists of organo-tin polymer anti-fouling paints, International Paint (in the UK) and the Nippon Oils and Fats Company in Japan, developed their products in different directions.

THE PAINTING OF OFF-SHORE STRUCTURES

Drilling rigs, off-shore platforms, and underwater pipelines all require protective coatings similar to those used in the painting of ships. They are commonly referred to as 'heavy duty coatings'. Drilling rigs are of three types: semi-submergible, 'jack up', and vessel (i.e. floating). All can be considered to be static during drilling operations. Platforms at the junction of the pipeline from a gas or oil field and the main connecting pipeline to the mainland are static structures which resemble drilling rigs, but do not have drilling equipment; nevertheless because they are located permanently in hostile environments they also require effective protective coating systems that call for the minimum maintenance. Corrosion protection is to some extent more important than antifouling, except where marine organisms contribute to the onset of corrosion; thus, unlike marine paints the energy savings obtainable by the reduction of drag (applicable to fast moving ships) do not arise.

Table .11

1. Main exterior surfaces (hull, tanks and ladders, etc.) **Primer:** Zinc silicate primer (one coat, 70µm dry film thickness) **1st build coat:** Micaceous iron oxide-pigmented polyamide-cured-epoxy paint. (1 coat, 120 µm dry film thickness (dft)
2nd build coat: Finish: High build vinyl paint (one coat, 100 µm dft) **Finish:** Vinyl acrylic paint (one coat, 50 µm dft)
2. Main exterior decks (including heliport deck) **Primer :** Zine silicate paint (one coat, 70 µm dft) **Build coat:** Two-pack, high build polyamide-cured epoxy paint (one coat, 120 µm dft) **Finish:** Two-pack, high build polyamide-cured epoxy paint (one coat, 120 µm dft) (incorporating flame dried silica sand 300- 500 µm).
3.1 Submerged zone - Legs, topsides of spud tanks, jack-up rig **Primer:** Two-pack, zinc phosphate-pigmented polyamide-cured epoxy primer (one coat, 50 µm dft) **Build coat:** Two-pack polyamide-cured epoxy paint (three coats, 360 µm dft)
3.2 Below drilling load line (pontoons, columns, bracings, etc.) **Primer:** Vinyl wash primer (one coat, 5 um dft) **Build coat:** Two-pack high build polyamide-cured epoxy paint (three coats, 360 µm dft) **Intermediate coat:** Aluminium pigmented chlorinated rubber paint (one coat, 40 µm dft) Finish:High build, self-polishing, antifouling paint (two coats, 200 µm dft)
4. Mud zone **4.1 Mud pits room and mud pump room; mud pits, mud tanks and mud pumps (exterior)** **Primer:** Two-pack zinc phosphate pigmented polyamide-cured epoxy (one coat, 50 µm dft) **Build coat:** Two-pack, high build, polyamide-cured epoxy paint (two coats, 240 µm dft)
4.2 Mud pits and mud tanks (interior) **Primer:** Vinyl wash primer (one coat, 5um dft) **Build coat :** Two-pack, high build polyamide-cured coal tar epoxy paint (two coats, 240µm dft)
4.3 Drilling floor **Primer:** Zinc silicate paint (one coat, 70µm dft) **Build coat:** Two-pack, high build polyamide-cured epoxy paint (three coats, 360 µm dft)

Like ships, off-shore structures require multicoat systems, which together have a passivating effect on the steel substrate, and a barrier effect. The initial painting of the structure is carried out on shore in a more desirable microclimate. In service, painting is confined largely to the retouching or repair of the original coatings as a result of damage to the coating or its eventual breakdown.Three main zones, which are exposed to different environmental conditions, are recognized. These are the 'atmospheric zone' (akin to the topside and superstructure of a ship), the 'waterline zone' (akin to the boottopping of a ship), and the 'submerged' zone (akin to the bottom of a ship). Giudice identifies some 12 types of surface, both interior and exterior, requiring different paint finishing systems.The main exterior zones employ systems which are exemplified in Table .11. Other surfaces include those of ballast and freshwater tanks, machinery, etc.

13 Health and safety in the coatings industry

In recent years there has been a considerable change in the coatings industry. Now the formulator has to satisfy many requirements in addition to that of producing a varnish, paint, adhesive, or ink that satisfies the stringent performance characteristics required by the customer. Today all companies and staff have to be fully awareof the health, safety, and environmental factors involved in producing a productfrom the initial formulation and development, manufacture, marketing, and technicalbackup. Every stage in a coatings development now involves the technologist to answer four basic questions:

- What? What materials are going to be used in this product? Are the substances and intermediates hazardous and what are the risks? .
- Where? Where is this coating going to be manufactured? Is special plant or precautions necessary?
- Who? Who is going to manufacture this coating? Do the personnel involved with manufacture require additional training or equipment?
- How? How is this coating to be used? How is the coating going to be applied and cured? Is the coating for industrial or public use? This can restrict the use of some materials and may require re-formulation.

When reviewing any aspect of safety involving coatings one must first decide whether a hazard exists and what is the risk.

When one considers any aspect of safety the possible hazard that exists with a particular substance or process is considered. A hazard may be biological, chemical, or physical, including fire and explosion, the consequences of which can cause damage, injury, or loss. Having considered the hazards with a substance or process the actual risk needs to be evaluated, e.g. the risk from handling a very toxic chemical delivered in drum quantities is high unless very stringent safety precautions are taken.

The same substance used in a closed circuit system with no manual handling and where personnel do not come into physical contact with the substance is minimal, thus for the hazard of the substance is the same but the risk is low.

RAW MATERIALS AND INTERMEDIATES

The majority of starting materials, intermediates, additives, and solvents used for the manufacture of coatings present some degree of hazard. Information in the form of a safety data sheet (SDS) giving details or the health, safety, fire, and explosion and environmental hazards associated with material has to be presented to the user either with or preferably before delivery. This information is necessary for all materials considered to be dangerous and must be supplied, even for small laboratory samples. Currently the safety data sheet only has to be supplied with materials considered dangerous and for industrial use; however, it is expected that data sheet will eventually be required for materials not considered to present any risk and for public as well as professional users.

The possible risks associated with these substances and products may be subdivided into two main groups: physicochemical properties and health effects.

Properties

Flammability

Many of the solvents, intermediates, pigments, and additives have the potential to cause fire or explosions if stored or handled incorrectly. The possible risk from fire or explosion is greatest from the volatile solvents used in coatings but pigments and fillers may also give rise to dust explosions.

Solvents

The solvents used in coatings and intermediates usually present more possible hazards and risks to the users than other materials used in coating formulations. The major problems occur with flammability of solvents, although some do have acute and chronic toxicity effects which will be considered later.

Many solvents and intermediates are considered to be highly flammable, or combustible. Fires are usually caused by the creation of a flammable atmosphere involving the substance mixing with air under atmospheric conditions in proportions sufficient that it may be ignited by sparks, arcs, or high temperatures. Organic solvents are usually defined by their ease of flammability which is determined by a physical property known as the flash point. This is the minimum temperature at which a material gives off sufficient vapour to form a flammable atmosphere above the surface of the liquid within the apparatus used for the determination.

Many different standard methods can be used for the determination of flash point; usually when a flash point is quoted it is suffixed by an abbreviation indicating the method used for the determination.

Solvents also possess characteristics known as lower and upper flammable limits (LFL and UFL). The LFL is the percentage volume of a material mixed with unit volume of air below which the mixture is too weak to sustain the propagation of flame. The UFL is the percentage volume of a material mixed with unit volume of air above which the mixture is too rich to sustain the propagation of flame.

Another important parameter when solvents and other combustible materials are considered is the autoignition temperature. This is the lowest temperature to which, under ideal conditions, a substance has to be heated to initiate self-sustained combustion independently of any ignition source. Autoignition is also not an absolute property but varies with the method of determination and conditions.

Table .1 — Flammability properties of common solvents

	Flash point (°C)	Boiling point (°C)	LFL point (% v/v)	UFL (% v/v)	Autoignition temperature (°C)
Acetone	–17	56	2.1	13	538
Amyl alcohol	42	128–133	1.2	12.4	325
Butyl alcohol	33	117–118	1.4	11.3	340
Butyl glycol	68	167–173	1.1	10.6	240
Butyl glycol acetate	88	186–194	0.5	3.7	386
Diacetone alcohol	54	150–173	1.8	6.9	600
Diethyl ether	–45	34	1.9	36	160
Ethyl acetate	-5	76–78	2.2	11	425
Ethyl alcohol	12	78	3.3	19	365
Ethyl glycol acetate	54	145–165	1.7	10.1	390
Methoxy propanol	34	117–125	1.9	13.1	286
Methyl ethyl ketone	-7	80	1.8	11.5	514
Ethylene glycol	116	193–205	3	28	400
Toluene	4	111	1.2	8	550
Trichloroethylene	None	87	8	10.5	410
White spirit	36–41	125–198	0.6	8	230
Xylene	25	138–142	1.1	6.6	490

Combustible substances, although not flammable, cause fire by undergoing an exothermic reaction with air when ignited. Typical physical properties for some of the common solvents used in coatings are given in Table .1. Inspection of these values shows that all solvents have very specific flammability characteristics. For example there is no relationship between flash point and autoignition temperature; acetone has a low flash point but a very high autoignition temperature, whereas white spirit has a reasonably high flash point but one of the lowest autoignition temperatures. Even a reasonably stable substance such as ethylene glycol is flammable over the widest range of concentrations in air.

Another often overlooked property of some solvents is their ability to generate static electricity by flow or movement.This generation cannot be prevented but precautions can be taken to ensure that incendive sparks are not generated. Liquids of low conductivity, e.g. toluene and xylene, can become charged when handled. Charges can be induced by flowing through pipes, strainers or filters; spraying, splashing, stirring or mixing, free fall, and in many other ways. The distance a liquid is allowed to free fall or the velocity at which it is pumped or speed of stirring are all factors that can influence the generation of static.

Various precautions are necessary to prevent the generation of static electricity. The most effective method is to ensure that all conducting parts of the plant and equipment are bonded to earth.

Pigments, fillers, and other substances

Many other chemicals in addition to solvents possess flammable characteristics. Some of the chemicals used to manufacture resins and intermediates are equally as flammable as many solvents; even the solid materials have the potential to cause dust explosions during handling. Details of a typical selection of these are given in Table .2.

Many powders are insulating and can easily become charged during handling. Those that are conducting can become charged if they are pneumatically conveyed or form a dust cloud during charging. Some powders are difficult to ignite but ignition becomes easier as the particle size is reduced.The range of concentrations over which powders mixed with air can propagate flame is defined in a similar way to solvents except that with powders they are called lower or upper explosible limits.

With powders, again the essential precaution to ensure that sparks are not generated is for all metal or conducting components associated with the process to be bonded to earth.

Explosive and oxidizing properties

Some of the raw materials used in resins and coatings have explosive or oxidizing properties. Explosive substances or preparations may be in solid, liquid, or paste form. They react exothermically, evolving gases which under specific conditions may detonate, burst into flame or explode when heated. Fortunately very few of the materials used in the coatings industry have explosive characteristics.

Some of the organic peroxides and other substances used to initiate free radical reactions in the manufacture of acrylic resins or to cure unsaturated polyesters or acrylic systems have the potential to be explosive and are classified as explosive substances.

For this reason the majority of these materials are sold mixed with inert solids or liquids, e.g. plasticizers to reduce the risk.Typical of this class of substances are dibenzoyl peroxide and cyclohexanone peroxide.

Table .2 — Flammability properties of common chemicals

Substance	Flash point (°C)	Melting point (°C)	Boiling point (°C)	LFL (% v/v)	UFL (% v/v)	Autoignition temperature (°C)
Acetic acid (glacial)	40	14	118	5.4	16.1	485
Acrylic acid	54	14	141	2.9	0.8	390
Acrylonitrile	0	-82	78	3	17	780
Adipic acid	196	151	338	–	–	420
Benzyl alcohol	101	−15	206	1.7	15	428
n-Butylamine	−12	-50	78	1.7	9.8	312
Castor oil	229	−12	313	–	–	453
Dibutyl phthalate	157	-35	340	–	–	400
Diethylamine	−18	-50	57	1.8	10.1	312
Diethylene glycol	124	-8	244	–	–	225
Formaldehyde	750	−118	−19	7	73	424
Formic acid	69	8.4	101	18	57	520
Glycerol	160	18	290	–	–	396
Morpholine	38	-3	128	1.4	13.1	310
Phenol	79	41	182	1.5	9.0	605
α-Pinene	33	-55	155	0.8	–	255
Styrene	32	−31	146	1.1	6.1	490
Vinyl acetate	-8	−100	73	2.6	13.4	385

Oxidizing Substances

These are substances which react exothermically when in contact with other substances, particularly flammable substances.

Effects

Although the risk of fire or explosion is a potential risk associated with many raw materials, intermediates, and finished coatings, many possess health risks which are often overlooked to some degree, as they may be longer term and not less tangible than fire which is instant and recognized by everyone.The health effects may involve the toxicity of the substance or preparation: it may be classified as being very toxic, toxic, or harmful. Sometimes the material may have additional corrosive, irritating, or sensitizing properties. The possible risks involved during use may be due to inhalation, ingestion, or contact with the skin or the eyes.

Simple descriptions of these health effects are given below.These may help when the terms are encountered in assessing the hazard of any materials or preparations (coatings) encountered.

The terms very toxic, toxic, and harmful are applied to substances or preparations

which may cause death or acute or chronic damage to health by inhalation, or when swallowed, or by skin absorption, Acute effects are those that develop rapidly as opposed to chronic effects which usually the result of long and continuous exposure.

Corrosive is the term applied to substances that destroy living tissues on contact; they also corrode and attack many metals rapidly. Irritant materials are those that, on contact with the skin or mucous membrane, cause inflammation. Sensitization is a property that is becoming increasingly recognized as a potential heath risk. It is the ability of certain substance or preparations when inhaled or when they penetrate the skin to cause a reaction by hypersensitization such that further exposure causes characteristic adverse effects. The quantities of a material and the degree of exposure necessary to produce a sensitization response is very small and the effects, similar to a severe asthma attack, can be life-threatening. Toxic substances or preparations may have carcinogenic, mutagenic, or reproductive effects.

REFERENCES

Patton T C, Paint flow and pigment dispersion: A Rheological Approach to Coating and Ink Technology. 2nd edn, Wiley-Interscience, pp. 365–7 (1979). Gglass J E, J Oil Col Chem Assoc 58 169 (1975).

Wolff H & Zeidler G, Paint Varnish Produ Manager 15 (8) 7 (1936).

Taylor R & Foster H J, J Oil Colour Chem Assoc 54 1030 (1971).

Mckay R B, Prog Org Coatings, 22 211 (1993).

Priel S & Torriano G, Rheology 1 223 (1991).

Weltmann R N, in Rheology: Theory and Applications, Vol III Ch 6, pp.

236–40, Ed Eirich F R (1960).

Oldshue J Y, Fluid Mixing Technology, McGraw-Hill (1983).

INDEX

www.ingramcontent.com/pod-product-compliance
Lightning Source LLC
LaVergne TN
LVHW062110200926
848484LV00007B/1631